Geoforensics

Geoforensics

Alastair Ruffell and Jennifer McKinley

Queen's University, Belfast, UK

with contributions from
Dr Laurance Donnelly
Professor Sergeant Mark Harrison MBE
Antoinette Keaney

John Wiley & Sons, Ltd

Copyright © 2008   John Wiley & Sons Ltd, The Atrium, Southern Gate, Chichester,
West Sussex PO19 8SQ, England
Telephone (+44) 1243 779777

Email (for orders and customer service enquiries): cs-books@wiley.co.uk
Visit our Home Page on www.wileyeurope.com or www.wiley.com

Reprinted October 2009

All Rights Reserved. No part of this publication may be reproduced, stored in a retrieval system or transmitted in any form or by any means, electronic, mechanical, photocopying, recording, scanning or otherwise, except under the terms of the Copyright, Designs and Patents Act 1988 or under the terms of a licence issued by the Copyright Licensing Agency Ltd, 90 Tottenham Court Road, London W1T 4LP, UK, without the permission in writing of the Publisher. Requests to the Publisher should be addressed to the Permissions Department, John Wiley & Sons Ltd, The Atrium, Southern Gate, Chichester, West Sussex PO19 8SQ, England, or emailed to permreq@wiley.co.uk, or faxed to (+44) 1243 770620.

Designations used by companies to distinguish their products are often claimed as trademarks. All brand names and product names used in this book are trade names, service marks, trademarks or registered trademarks of their respective owners. The Publisher is not associated with any product or vendor mentioned in this book.

This publication is designed to provide accurate and authoritative information in regard to the subject matter covered. It is sold on the understanding that the Publisher is not engaged in rendering professional services. If professional advice or other expert assistance is required, the services of a competent professional should be sought.

Other Wiley Editorial Offices

John Wiley & Sons Inc., 111 River Street, Hoboken, NJ 07030, USA

Jossey-Bass, 989 Market Street, San Francisco, CA 94103-1741, USA

Wiley-VCH Verlag GmbH, Boschstr. 12, D-69469 Weinheim, Germany

John Wiley & Sons Australia Ltd, 42 McDougall Street, Milton, Queensland 4064, Australia

John Wiley & Sons (Asia) Pte Ltd, 2 Clementi Loop #02-01, Jin Xing Distripark, Singapore 129809

John Wiley & Sons Canada Ltd, 6045 Freemont Blvd, Mississauga, Ontario, L5R 4J3, Canada

Wiley also publishes its books in a variety of electronic formats. Some content that appears in print may not be available in electronic books.

Library of Congress Cataloging-in-Publication Data

Ruffell, Alastair.
  Geoforensics / Alastair Ruffell and Jennifer McKinley.
      p. cm.
  Includes bibliographical references and index.
  ISBN 978-0-470-05734-6 (cloth : alk. paper) – ISBN 978-0-470-05735-3 (pbk.)
 1. Forensic geology. 2. Environmental forensics. I. McKinley, Jennifer. II. Title.
  QE38.5.R85 2008
  363.25 – dc22
                                                                          2008011004

British Library Cataloguing in Publication Data

A catalogue record for this book is available from the British Library

ISBN   978-0-470-05734-6 (HB)
ISBN   978-0-470-05735-3 (PB)

Typeset in 10/12pt Sabon by Laserwords Private Limited, Chennai, India
Printed and bound in Singapore by Fabulous Printers Pte Ltd

Contents

Acknowledgements ix

Preamble xi

1 Background to the work, organization of the text and history of research 1
   1.1 The scene 4
   1.2 The victim and materials 5
   1.3 The suspect 6
   1.4 The scope of geoforensics 8

2 Physical geography, geomorphology, landform interpretation, archaeology, stratigraphy and hydrodynamics 13
   2.1 Physical geography 13
   2.2 Atmosphere 16
   2.3 Types of landscapes, landscape change and human influences on the landscape (short and long term) 18
   2.4 Soils 33
   2.5 Hydrodynamics of rivers, lakes, estuaries, seas and oceans 47
   2.6 Geography, geomorphology, geological and soil maps, and other resources 50
   2.7 Groundwater 53

3 Geophysics 55
   3.1 Seismic methods: macro to micro 57
   3.2 Gravity/gravimetrics 64
   3.3 Electrical 66
   3.4 Magnetic and electromagnetic 70
   3.5 Ground-penetrating radar (GPR) 77
   3.6 Radiometrics 85
   3.7 Review of why some methods are favoured and others not 88

4 Remote sensing 91
   4.1 Definitions 91
   4.2 Conventional aerial photography: rural and urban examples 92
   4.3 Geoscience use of light photography 93
   4.4 Infrared photography 102
   4.5 Elevation modelling 104
   4.6 Photogrammetry 105

4.7	Synthetic Aperture Radar (SAR) and interferometry	105
4.8	Multispectral and thermal imaging	106
4.9	Hyperspectral imaging	109
4.10	Satellite mapping	109
4.11	Long-distance LiDAR (satellite, aerial)	114
4.12	Laser scanning of scenes and objects	117
4.13	X-ray imagery, X-ray tomography and neutron activation	117
4.14	Field Portable X-ray Fluorescence (FPXRF) spectrometry	122
4.15	Some conjecture on the future of remote sensing applications	124

5 Spatial location and geographic information science — 125

5.1	Geographic location and crime	125
5.2	Spatial data and GIS	137
5.3	Spatial analysis within GIS	140
5.4	Use of Google Earth in criminal investigations	152

6 Scale, sampling and geostatistics — 155

6.1	Scale and spatial resolution	155
6.2	Sampling for geological materials at urban and non-urban crime scenes	157
6.3	Timing of the crime	157
6.4	Sample size	157
6.5	Lateral variation	159
6.6	Use and misuse of statistics in forensic studies	160
6.7	Statistical sampling	161
6.8	Number of samples required for robust statistical analysis	164
6.9	Comparing 'like with like'	165
6.10	Addressing the issue of comparing related material	165
6.11	Spatial and temporal variability in nature	166
6.12	Spatial awareness and use of spatial statistics: application of geostatistics	168
6.13	Geostatistical techniques	172
6.14	GIS and geostatistics	179

7 Conventional geological analysis — 183

7.1	Elementary analysis of rocks	185
7.2	Hand-specimen analysis – case studies from Murray and Tedrow (1991)	187
7.3	Sediment analysis	190
7.4	Fossils and microfossils	220
7.5	A paradigm shift in geoforensics?	237

8 Trace evidence — 241

8.1	What is geological trace evidence?	241
8.2	Scanning Electron Microscope (SEM)	249
8.3	Laser Raman spectroscopy	253
8.4	Inductively-coupled plasma spectroscopy	253
8.5	Isotope analysis	254
8.6	X-ray diffraction and trace evidence	254
8.7	Manufactured or processed materials that geoscience techniques can analyse	259
8.8	Some conjecture on the future of trace evidence	266

9	The search for buried materials		**269**
	9.1	Introduction	269
	9.2	Possible methodologies for non-urban underground searches	274
	9.3	Underwater searches and scene mapping (remote sensing, geophysics)	276
	9.4	Gas monitoring, organic remains and the decomposition of bodies	288
	9.5	Weird and wonderful burial locations	290
10	**Circuit complete**		**293**
Appendix 1.	**Search methods**		**301**
Appendix 2.	**Soil sampling**		**305**
	A.1	Sampling protocol suggestions	305
Glossary			**313**
References			**317**
Index			**330**

Acknowledgements

We are very indebted to Laurance Donnelly and Mark Harrison for their invaluable advice, training, and for showing us the error of our ways. Likewise all those who have helped with advice and criticism of this work: Peter Bull, Mike Goodchild, Chris Lloyd, Duncan Pirrie. This work would not have been completed without patient help in accessing data and images. Catriona Coyle (Chapter 3 images); Ana-Maria Gomez and the Equitas Foundation (Chapter 3 images), Josef Minár (his own figures in Chapter 2), Elisa Bergslien (Chapter 3 images), Chris Leech of Geomatrix Earth Science (Chapter 4 images), Conor Graham (Chapter 3 images), Graeme Swindles (Chapter 8 images and data), Bernd Kulessa and Paul McCarthy (Chapter 4 images).

Over the past 12 years we have had the fortune to work with some excellent people who provided help, advice and tenacity. These include: Steve Bell, Trevor Cooke, Graeme Kissock, John Gilmore, Sam Harkness, Raymond Murray (PSNI), John Meneely, Roy McComb, Will Kerr, Mark Irvine, Steve McKendry, Steve McIlroy. Martin Grimes, Ed and Keela – thanks! Some of these people can't be named, or we were never informed of who they were. We thank you nonetheless.

The help and advice of Bernd Kulessa, Chris Hunt, Colm Donnelly, Eileen Murphy, Jane Holmes, Jennifer Miller, John Hunter, John Meneely (again), Keith McCoubrey, Margaret Cox, Marianne Stam, Mark Dornan, Mark Yates, Neil Ogle, Pat Wiltshire, Paul Cheetham, Paul Moore, Richard Lacey, Jon Shears, Roland Wessling, Sean Wright, Wolfram Meier-Augenstein and Helen Kemp was of massive benefit. We thank those who have been kind with their help over the years and in writing this book: Paul Carey, Martin (Carlos) Gallego, David Jamison, Paddy Gaffikin, Pat McBride, Mark Russell, Wayne Isphording, Jim Riding, Emily Hodgkinson, Barry Rawlins, Matt Power, Maureen Bottrell, Clare Meharg, Eric McGreevey, Bill Schneck, Dave Favis-Mortlock, Dave Barclay, Nicola Farmer, Barbara Kesson, Bernard Anderson, Brian Whalley, Ian Hanson, Dave Nobes, Gary Olhoeft, Ian Meighan, Jamie Pringle, Claire McGrenaghan, Tommy Crawford, Lorraine Barry and Ronan McHugh. Maura Pringle and Gill Alexander are both thanked for all their help with graphics and photos.

The help in providing so many opportunities can't be fully acknowledged, for which we thank nonetheless Ann Blacker, Fr Denis Faul, Jean Faunier, Jim Gemmell, Richard Gray, Ruth Griffin, Chris Hunter, John Logan, Walter McCorkhill, Ruth Morgan, Jonathan O'Hanlon, Peter Parsons, Mark Preston, Val Semple, Jim Speers, Frankie Taylor/Pat Steele (Chapter 8 case study), Peter Bull, Raymond Murray (Missoula) and Nuala Johnson.

Many of those who have provided legal advice and opportunities also can't be named or don't wish to be. Mr Creeny and Mr Mateer, Tony Whittle and the staff of the OSCE are thanked.

Thanks to all the students who attended GGY349 'Geoforensics' in 2007: you were excellent critics! Tony Hallam and Richard Worden know nothing of this book, yet were instrumental at the start of it all. Special mention must be made of one of our finest undergraduates, Ruth Annette, who provided much of the excellent historical information. Fiona Woods and Rachael Ballard – thanks for your patience. The initial proposal for this book was reviewed by 10 very honest and helpful people; the finished chapters were read and criticized by Mark Harrison, Chris Lloyd, Laurance Donnelly, Peter Bull, Jennifer Miller and Mike Goodchild, all of whom we thank. Brian Lavery and Sam (Sammy) Jamison are especially acknowledged.

Preamble

Many experienced forensic geoscientists have their own 'an investigator walked into my office one day with a bag of sand and asked where it was from' type story. In our case, three events coincided to start us on the course that ends with this book. A man walked into the Geology department office one day with a loyalist pipe bomb, constructed from the pup-joint of an oilrig drill-stem. This strange man was enthusiastic about science, hence discussions turned to geophysics and later the whole episode was forgotten. Sometime later an archaeologist walked in and asked if geophysics could assist in a humanitarian search for one of the IRA disappeared, as a man with a pipe bomb (above) knew we had such kit but could not be involved. Why? Because although times were changing in Northern Ireland, involving the police and army would have been counter-productive and no criminal charges were to be brought; the operation was humanitarian, for the sake of the family, with community support.

It sounded like a good use for the equipment (the department had already been working on assessing the vacant plots of a nearby Presbyterian church), so the job was agreed before the caveats were heard. The job was to be done as soon as possible (it was February), at daybreak (it was February) to avoid media and too much community interest. Come the foggy day, quite a few observers, church officials and police officers turned out to see the survey. The second-hand kit we had back then made an annoying beep when it failed to work – which is all that happened the first day. Undaunted, returning the next day, with noticeably less onlookers, we once again had the annoying beep. The kit nearly ended up in the sea on the way back to the department! On the third day, the priest was busy inside his dry church, leaving only a faithful, if un-nerved police officer (you are thanked in the acknowledgments!) and the gravedigger as assistants. Of course the device worked perfectly (nothing was discovered), the graveyard was left alone and the missing person turned up six months later in dunes on a County Louth beach, saving a lot of money, work and community tension.

Finally, someone really did walk in with a bag of 'sand' one day not long after. It had been seized at a petrol storage depot and was suspected to be kieselguhr, or diatomaceous earth, as all good micropalaeontologists know. Kieselguhr can be used as a filter, and will remove the red or green dye placed in diesel petrol to avoid paying tax. If used, and the clear(ish) petrol sold for domestic use ('fuel laundering'), a major tax offence has been committed. The owners claimed the material to be fuller's earth (or bentonite – the same as kitty litter and mud face-packs) to be used in the soaking up of oil spills. It sounded like an easy one – surely diatomaceous earth would be like diatomite −100 % silicon oxide and fuller's earth would be 100 % clay (a type called smectite)? Diatomaceous earth commonly contains a high clay proportion, usually smectite. Fuller's earth often contains a high proportion of silica, from fossil sponges, diatoms and other organisms. The simple forensic test suddenly got more complicated (like all the cases in this text) because the kieselguhr had up to 50 % clay and the alibi clay (fullers earth) had about 50 % silica. The moral(s) of the story: criminal casework is inevitably more complicated than at first sight and purchase your petrol from a reputable vendor!

In March 2003 we were both involved in assisting the search for a presumed murdered person in northwest Ireland. When all their intelligence had been gathered, the police carried out an extensive search of a known offender's house. This was at a time when police scientific coordinators were very keen to have all their 'ologists' present, so we were invited along, to do what exactly we were not sure! We took some samples anyways, and as the day progressed, the Victim Recovery Dog (also thanked in the acknowledgments) located human remains in a makeshift grave, covered with slabs of rock and concrete. These were removed and we examined the soil above and below the victim: it was patently from the surrounding excavations. Comparing the soil to anything from the suspect would only prove he had been to his own house. Then, with night approaching, a friendly scientist from the forensic laboratory asked if maybe we ought to look at the slabs. Of course they turned out to be very distinctive in manufacture and could be linked to slabs in the offender's garden. This crucial item of evidence assisted the conviction of a highly dangerous serial offender, one of only a few people in the United Kingdom who will never be regarded for early prisoner release. In this case we used standard geological analyses in a criminal investigation: which is what geoforensics is all about.

1
Background to the work, organization of the text and history of research

This book has at its core *the application of selected geoscience techniques to criminal (domestic, international, terrorist, humanitarian, environmental, fraudulent) investigations of what happened, where and when it occurred and how and why it took place*. The book's opening ought to have a more precise definition (than that above), but it does not because the applications of Earth science methods to different problems (of a criminal, humanitarian, disaster-related nature) are so wide-ranging that the definition would probably end up being as long as the book itself. The text is not a book on criminalistics, geography, geophysics, or microfossils. As a result, specialists in any one of these fields are going to be disappointed or angry that the discipline they work in is not covered comprehensively enough. Many of the chapter topics fall into this bracket: in more than one case we have issued the challenge to such cynics – please write a review article, or a short or long book on your discipline. Only then will they see the challenge we have (perhaps foolishly, but let's ask the non-specialist 'users' of this book) taken on. If we had 10 or so review articles on the topics we have discussed then this book would not exist: we could merely set up a web site with a link to each topic review and within a few minutes the interested party would have a selection or comprehensive collection of 'geoforensic' articles for very little cost. Unfortunately, not one review of the specific application of one geoscience discipline to criminalistics exists at the time of writing. If any book or review article exists entitled: 'Forensic Geophysics'; 'Forensic Remote Sensing'; 'Geological Trace Evidence'; 'Microfossils in Criminalistics', etc., then our literature searches are very poor or the article is published in too obscure a journal! Conversely, as we state above, specialists in each area are likely to be disappointed by the cursory manner we discuss their disciplines, in which case we offer a gift with our challenge: write the review you would like to see published and a large number of citations are guaranteed: Ruffell and McKinley (2004) is still one of the most highly cited articles of *Earth Science Reviews*.

Geoforensics Alastair Ruffell and Jennifer McKinley
© 2008 John Wiley & Sons, Ltd

In this text we have tried to be as comprehensive as possible while maintaining the overall structure of the work. Various wider-ranging and very specific reviews do exist, and these are vital resources, cited at appropriate locations in the text. Some of the geological topics covered here (in Chapters 7 and 8) are covered by Murray in his fantastic *Evidence from the Earth* (2004) and in Pye (2007). Various aspects in this book, excepting details on Remote Sensing (Chapter 4) and Geographic Information Systems (Chapter 5), can be found in scientific paper form in the edited volumes of Pye and Croft (*Forensic Geoscience: Principles, Techniques and Applications* (2004)) and Mildenhall, Wiltshire and Bryant (2006: on palynology).

The sequence of chapters in this book is arbitrary and many can be read alone, as cross-references will occur to previous and subsequent chapters. Within chapters, major divisions of text are indicated by numbered headings in **bold** text, with case studies and examples denoted by section headings in smaller, unnumbered bold text. Minor subdivisions occur in italic font. Key points are emphasized by use of *italics within the text*, and instructions or warnings by UPPER CASE.

This book presents different types of case study. These can be summarized as: (i) descriptions of investigations into criminal activity, locations and materials; (ii) published articles on techniques and their possible application to cold or historic cases; and (iii) a synopsis of a relevant published paper, book chapter or report that is relevant to the chapter section. In the latter case, we hope not to have offended those not included – this is unintentional, merely a product of how the works 'fit' with the way our chapter is developing. Published case studies have the author's names in the section title: those from the author's personal files have no author names. *Case studies carry a warning! They are specific and thus cannot be used as a template for another investigation. Information from other cases will help, but each case must have its own, evolving, investigation.*

Three books have influenced the style we have developed. Haralambos *et al.* (1990) have individual sections summarizing and providing a critique of what they consider to be key sociological works. Murray (2004) uses case studies to illustrate the use of each technique: we do the same. Both works make for easy, memorable reading, especially useful to students. Canter (2003) is a blend of these two texts, with few, illustrative figures and well-constructed, focused sections of text. We use all three types of format in this work. Although our style is derivative of this rather elective collection of texts, our content is a blend of our own case studies and experiences, published papers and four important books that the reader may consider reading in order to obtain a fuller picture: Hans Gross's[1] (1891 *System der Kriminalistik* (*Criminal Investigation*) or any of the later editions to 1907, or

---

[1] Gross's book has a complicated and long history of publication, re-edition, and translation and thus confused publication dates. The first and second pages of Jackson's (1962) fifth edition entitle the work 'Criminal Investigation. A Practical Handbook for Magistrates, Police Officers and Lawyers', Adapted by John Adam and J. Collyer Adam, from the *System der Kriminalistik* of Dr Hans Gross. First Edition, John Adam and J. Collyer Adam (1906), Second Edition by J. Collyer Adam (1924), Third Edition by Norman Kendal (1934), Fourth Edition by R.M. Howe (1950), Fifth Edition by R.L. Jackson (1962). Jackson (1962) is published by Sweet & Maxwell (London), yet the second page states 'Printed in Great Britain by Eastern Press, London & Reading'. If this seems confusing, Chisum and Turvey (2007) state that the book was published in 1894 (other authors have 1893), with reprint dates of 1906, 1924, 1934, 1949 and 1962, some of which must be the same editions as Jackson cites, yet the 1906 publication is cited as being published. A book could probably be written on the history of Gross's book itself!

Jackson's 1962 translation), Murray and Tedrow (1975, as well as later editions, 1986, 1991, 1992), Pye and Croft (2004), and Murray (2004). Sugita and Suzuki (2003) provide a comprehensive review of forensic geology.

Criminal investigations and enquiries into disasters are required because one or more facts or pieces of evidence are unknown: it is the job of the forensic scientist to produce theories about this missing information, and the job of the investigator to assimilate these theories into a story of what happened and why. Missing information may comprise the scene (the original location of the crime is not established), the victim (covertly buried or hidden by disaster, or missing above ground), materials (vehicles, clothes, weapons, drugs, contraband) or the suspect (yet to be apprehended, or in custody). The geoscientific methods described in this text can assist in providing information to the investigation on all four of these aspects.

This book aims to show how various geoscience techniques may be used in investigations, regardless of any background. Geoscientists are trained in their discipline: it is for us to apply this training to assist investigations. These enquiries will be generated, monitored and reported differently in various parts of the world, with, for example, North America; Russia and Europe; the UK, Australia and New Zealand; Japan; India; and many countries of the Middle East having different legal and social requirements. Geoscientists must take advice on the constraints imposed on their activities, analyses and reporting from the investigators in each country. Behind this, the internationally recognized scientific method must proceed. These may come to conflict, but ultimately, we would be remiss in not stating early on in a text such as this, that geoscientists apply their method to an investigative problem first, and adhere to the method of reporting and court appearance in the country thereafter.

Two messages run through this text. First, we face an increasingly litigious society, where accurate, contemporaneous recording and reporting of anything that may end up in court, enquiry or tribunal is essential. What better way to increase awareness of this than through the application of well-known disciplines to forensic science? Second, forensic pertains to the law, and in all the chapters that follow, data or materials are collected and analysed. At this point geoforensics diverges from the regular pursuit of science, in that original data is stored and sealed (digital media, photographs) to be processed and analysed later, or samples are sealed, labelled and their origin described. Each examination and analysis of data or materials thereafter, the same procedure must be conducted, with a record of what has been done. Most crucially, as we shall see later in the book, some of the most robust analyses of data and material are carried out by different, comparative means. This requires a multidisciplinary approach, often involving different analysts. Thus the chain of custody (who gave what sample to whom and when/where) becomes as important as what analyses and interpretations are made. There are now specialist groups and expert panels at scientific meetings (*https://rock.geosociety.org/meetings/2007/pf-bios.htm*) who work with police officers and other law enforcers (e.g., environmental agencies) and investigators (e.g., structural and civil engineers) to advise on such matters.

1.1 The scene

Identifying a possible scene of crime or other activity from a sample soil, sediment or rock remains one of the great challenges in Geoforensics (see Chapter 7). Yet bizarrely, this application was one of the first applications of microscope-based comparison of sand from suspect materials back to its likely source. Below, we outline how Conan-Doyle, Gross and Popp can each lay some claim to developing methods of the classic geoforensic technique of comparing soil on a suspect's clothing or footwear to a possible source that excludes the suspect from, or compares him or her to, a scene of crime. The comparison of sediment in a sample of unknown origin (the *provenance* of the *questioned* sample) was undertaken some 40 years before Conan-Doyle wrote his famous Sherlock Holmes stories. The 1856 (Volume 11) issue of *Scientific American* has the following section called 'Curious Use of the Microscope'.

> Recently, on one of the Prussian railroads, a barrel which should have contained silver coin, was found, on arrival at its destination, to have been emptied of its precious contents, and refilled with sand. On Professor Ehrenberg, of Berlin, being consulted on the subject, he sent for samples of sand from all the stations along the different lines of railway that the specie had passed, and by means of his microscope, identified the station from which the interpolated sand must have been taken. The station once fixed upon, it was not difficult to hit upon the culprit in the small number of employees on duty there.

Christian Gottfried Ehrenberg was a famous zoologist and geologist (1795–1896) from Leipzig in Germany, a correspondent of Humboldt and Darwin and an expert on diatoms, although whether Ehrenberg ever used these micro-organisms in other geoforensic studies has yet to be established. What can be concluded from the above account, is the re-positioning of Ehrenberg, through the high academic standing of *Scientific American*, as a predecessor of Conan-Doyle, Gross and Popp in the application of geology to forensic casework.

Hans Gross (1847–1915)

Hans Gross published his *System der Kriminalistik* (*Criminal Investigation*) in 1891, in which geographical and geomorphological maps were used to show possible covert locations for activities (forests, brothels, hidden areas) or storage/dumping (ponds, wells, streams, forests). Gross recommended that investigators new to the job or area should visit and familiarize themselves with such places: he did not give an explicit rationale as to how a search should be conducted, but this was obviously his meaning. Perhaps he was intentionally vague, as each case is different, so that deploying standard methods might thwart innovation or open-mindedness, or perhaps the method of his search, emanating from last-seen locations, suspect dwellings, and then focusing on the sorts of places mentioned above, was so ingrained in Gross's psyche that he did not see the need to explain the obvious.

Gross, writing in the late 1800s said 'if we compare a recent [late-1890s] scientific work, with an analogous book written some decades ago, we shall notice a great difference between them arising almost wholly from the fact that the work of today is more exact than that of yesterday' (Gross, 1891). It is testament to GIS (Geographic Information Systems, some say Geographic Information Science, depending on context) and laboratory technology that Gross's words are providential, and yet damning of all the works cited in this text (including all the authors), that Gross's ethos, recording, warnings and advice are as good, if not better than ours of the present-day. One probably leads to another: over-reliance on the machine's ability to record information makes the investigator sloppy.

Following Gross, few works specifically on the geography of the scene have been published. However, identification of the unknown scene is implicit in many of the historic cases recounted by Murray (2004), who shows how questioned materials have led to identification of a definite or range of probable scenes. These are somewhat exceptional cases (as we shall see): the underlying tenet being the 'holy grail' of forensic soil and geology studies – the case where one sample (typically, 'a lump of mud', but equally, mud-stained clothes and vehicles) has such diagnostic properties that a location, or range of locations may be provided. Rawlins *et al.* (2006) test the possibility of this predictive power (see Chapter 7). Another means by which possible scenes are identified may be remote sensing and geophysics: these are intimately associated with covertly hidden objects, especially victims, and will be dealt with below. Should the hidden materials be subsequently moved, so these methods assist in providing search locations for the scene alone. Remotely sensed aerial or satellite data, disturbed ground and vegetation may provide information on the movement of vehicles or people that have long-since gone from a scene.

1.2 The victim and materials

If absent from an investigation, information concerning a missing victim or missing (possibly hidden) materials is either achieved at the macro-scale of remotely-sensed data (photography, geophysics, search dog indications), or at the micro-scale of providing possible contact locations between trace evidence and the suspect. Very little has been published on the range of macro-scale search methods (see Chapter 9), although the cumulative works discussed in Chapters 2 and 4 show how changes in landform, vegetation and geophysical response have identified search locations or shown them to be not worth examining. Murray's (2004) account of the case of USDEA (United States Drug Enforcement Agency) officer Enrique Camarena is a good example, where the soil on his clothes, following exhumation (at the behest of the US government, who suspected a cover-up), did not compare to the given burial location, suggesting a previous burial location and thus an earlier scene of crime. Block (1958) describes the work of Oscar Heinrich ('The Wizard of Berkeley') who identified sand on a shovel thought to be associated with possible kidnappers, to be beach sand. This led to coastal searches and the recovery of the body of Father Patrick Heslin, who had been killed but a ransom still demanded.

1.3 The suspect

Comparing the suspect to the scene (or his or her alibi locations) is the oldest and best-established of the forensic geology and soil science techniques. It extends back to the stories of Arthur Conan Doyle and the fictitious work that his character, Sherlock Holmes, did using soil and stones, an early example being the 1887 publication of 'A Study in Scarlet', followed by 'The Five Orange Pips' (1891) and 'Through the Magic Door' (1907), which all included Holmes relying on evidence provided by soil and rock (Conan-Doyle, 1988). Following Doyle's work, but partly overlapping, *Hans Gross (1847–1915)* published his *System der Kriminalistik (Criminal Investigation)* in 1891, which included discussions of forensic medicine, toxicology, serology, and ballistics, as well as forensic geology. The latter included the use of microscope petrography in the study of materials, such as soil recovered from shoes, to link suspects to scenes of crime or routes, which Gross combined with geography and geomorphology. To what extent Gross was influenced by Conan-Doyle's writing, or vice versa, is perhaps unknown. In 1912 Gross opened one of the first specialist forensic laboratories in Europe at Graz in Austria. *Georg Popp (1867–1928)* used geological evidence in a criminal case for the first time, the now-famous case of the murder of Eva Disch and how a handkerchief (dirty with coal, snuff and hornblende mineral grains) was used to associate a suspect (one Karl Laubach) to a killing by strangulation. Popp is in a unique position as a founder of forensic biology and microbiology as well as forensic geology: his role is sometimes underplayed in that Conan-Doyle was so famous, and Gross's book has been reprinted so often. Yet Popp published some important works on geology and botany (1910, 1939). Of course, other early workers used such evidence, but Popp used it conjunctively. The Heidelberg Mayors Murder of 1921 started when the Mayor of Herford (Germany) and his friend failed to return from a walk, and a diligent (some might say nosy) landlady spotted papers belonging to the mayor in the room of a lodger (one Leonard Siefert). More of the missing mayor's personal belongings were found in Siefert's possession. When students found the shot and bludgeoned bodies of the mayors nearby, Siefert denied involvement, claiming to have had the mayor's belongings planted on him while on a train. Popp noted resin, snail slime, hazelnut shells, mosses, leaves from different beech tree varieties and other organic debris both adjacent to the crime scene (in a probable ambush location) as well as adhering to Siefert's clothing. Popp successfully denied Siefert's alibi reasons and he was convicted.

August Bruning was trained by Popp for two years (1910–12) and in June 1913 proved his credentials by solving a robbery case using what we would now term a conjunctive approach, examining tool marks and material adhering to the tool as well as the marks and host material at the scene (Bischoff, 1966).

Edmond Locard (1877–1966)

Edmond Locard is best known for his generic Exchange Principle, which states that:

> whenever two objects come into contact, there is always a transfer of material. The methods of detection may not be sensitive enough to demonstrate this, or

the decay rate may be so rapid that all evidence of transfer has vanished after a given time. Nonetheless, the transfer has taken place.

(Locard, 1929)

Pye (2007) makes an interesting analysis of this apparent quote from Locard, explaining that the literal translation from Locard (in French) places a slightly different emphasis on the Exchange Principle. For a discussion of the interpretation of Locard's explanation of the Exchange Principle, the reader is referred to Pye (2007), who translates directly from French and considers Locard's words to be more about the actual transfer of materials, rather than the more open-ended concept of all contacts leaving trace evidence.

Whatever the literal translation or derivation thereof, Locard's ideas are as relevant to soil, mineral dust and micro-organisms, as they are to hairs, fibres and biological material (Corre, 1968). Locard's words were truly providential, as we can see in the present-day with historical DNA evidence being used in criminal cases that took place decades ago. In Chapter 8, the nature of earth trace evidence assumes greater and greater importance, also demonstrating how Locard was ahead of his time. Published accounts of using geological materials to compare a suspect to a scene reach a hiatus following Gross's work, although it is known (Murray & Tedrow, 1975) that with the establishment of the FBI in the 1930s and various European crime laboratories (before and after), geoforensic work continued (Chisum & Turvey, 2000). Locard published his most comprehensive examples of the Exchange Principle in 1930, not long after many American academics had begun using all manner of trace materials in establishing suspect to scene comparisons (see below). The lack of published material in this period is a product of wartime secrecy (as we shall see in later chapters, geology, geomorphology and soil studies have been used extensively by military tacticians); the importance of the work to criminalistics; and lack of an appropriate vehicle for publication. By the 1970s, Murray and Tedrow's (1975) *Forensic Geology: Earth Sciences and Criminal Investigation* partly redressed this hiatus, although few papers were published on geoforensics around this time. Murray and Tedrow (1975) described the rationale for their work, as well as providing scientific background and case studies. In recent years, the importance of geography, geology and geophysics to military operations and intelligence has also been publicized: the CIA (Central Intelligence Agency) web site (*www.cia.gov<http://www.cia.gov>*) has extensive pages on the key role that geography (human and physical) plays in its operations. Even the casual reader will find it interesting that such a dry-sounding subject as geography can be key to a nation's security.

The Berkeley scientists (1914–1940)

August Vollmer typifies the 'Forrest Gump' ethos of the American Way: a mailman who through hard work rose to become the Chief of Police in Berkeley, California during World War I. Like Conan-Doyle and Popp or Gross, we cannot ever be sure of what each personality knew of the other's work around this time. Nonetheless, according to Thorwald (1967), Vollmer applied scientific methods to criminal investigations for the first time in North America, not long after his European counterparts. Vollmer was astute in collaborating with a professor of law (Alexander Kidd) and a biochemist (Karl Schmidt, Kirk's teacher), both at Berkeley, making an investigative, legal and analytical team. One pities any criminal operating in California in the time around World War I! The team was joined by one Oscar Heinrich, a 1908 chemistry graduate of Berkeley who lived in Tacoma and Denver before returning to California and eclipsing his senior colleagues through the 1920s and 30s with some seminal cases. Heinrich was the master of scene and trace evidence analysis, although it is the latter that he pioneered, along with his then state-of-the-art microscopes, often combining the two (how does trace material on the suspect or victim relate to the crime, alibi or control scene?). Block (1958, p. 42) recounts a theoretical conversation between 'people who pressed him to explain his mode of operation' and Heinrich himself:

> 'How do you attack a crime problem?' they enquired. 'How do you begin? What questions do you ask yourself?' Heinrich liked to answer. Always he spoke in his slow, calculating way, measuring words carefully. Often he drew a pencil from his pocket, pointing it for emphasis as he talked. 'Understand this first,' he usually said. 'Crime analysis is an orderly procedure. It's precise and it follows always the same questions that I ask myself. Let's consider what they are: Precisely *what* happened? Precisely *when* did it happen? Precisely *where* did it happen? *Why* did it happen? *Who* did it? The average investigator seems to give immediate attention to the why and who but takes what happened for granted'.

1.4 The scope of geoforensics

Type into a web search engine 'definition geoforensics' or a similar phrase and, if your experience is similar to ours, about 20 to 30 'hits' will be displayed, with roughly a half devoted to engineering-related problems and the other 50 % to psychology, offender profiling and geographic profiling. This might be interpreted as Earth scientists (geologists, mineralogists, micropalaeontologists, geophysicists, archaeologists) letting their important role in investigations be dominated by other disciplines. It could equally mean that environmental engineers and psychologists have advertised and practised their legal work for longer and more effectively than Earth scientists. As we shall see in later chapters, psychology, criminal geographic profiling, environmental crime and legal investigations of engineering problems have a very strong role to play in how the 'geo' disciplines can be applied to legal and criminal matters. These disciplines and applications have to be integrated with others in the 'geo' family, so that fit for purpose methodologies and philosophies[2]

---

[2] The conjunctive use of trained personnel and equipment is embodied in what Earth scientists do. The geologist (and geophysicist) is trained to look for materials in the ground (oil, gold, gems) and has skills suited to the search. The archaeologist is trained in excavation and interpretation of the past, ideal for the recovery of buried remains and objects. The palaeontologist and mineralogist use appropriate equipment to discover the makeup of materials: their skills are in excluding samples from locations, and thence comparing suspect to scene. All three disciplines should work together: the analyst needs to know how the sample was found and how it was recovered. The implication is that one cannot train as a 'geoforensic' scientist: better to deploy skills to the problem than develop skills for the problem.

are applied to investigations, appropriate multi-disciplinary methods are used and no one group dominates the application of their science to investigative casework for criminal, environmental or humanitarian purposes. The first pages of Murray and Tedrow's (1991) important text on Forensic Geology includes a number of sub-disciplines (soil science, sedimentology, microbiology and micropalaeontology, stratigraphy) that provide roughly half of the scope of what we are embracing as geoforensics. These applications are mainly concerned with the historically-important, well-set in legal precedent concepts of comparing a suspect to a scene, locating an original death location of a moved body, verifying alibi locations and tracing the movement of goods. Most use physical materials such as soil, sand, industrial products in a comparative, exclusionary or predictive (where did the material come from, where has the suspect/victim been?) mode.

An alternative way of expressing what geoforensics is all about is to avoid the narrow definition and relate how suitable materials and analytical methods can be examined by the 'geo' specialist to provide additional information for investigators. Examples include geological industrial products, or innovative uses of remote sensing and geophysics in areas not traditionally subject to such scrutiny. The paper by Graham *et al.* (2004) in Pye and Croft's *Forensic Geoscience* on the examination of impact marks on spacecraft surfaces is a classic example of such innovative thinking. Who would have thought that macro and microscopic impact marks made in space could be usefully studied by methods used routinely in geology and physical geography?

Geology as a subject in North America tends to include what many Europeans would term physical geography. Although all the sub-disciplines of geography (physical, human, geomorphology, social, cartography, GIS) have been written about from the point of view of criminal, humanitarian or civil investigations, the geography word is rare in books or titles of scientific criminalistic, forensic, environmental articles, except (as we state above) geographic profiling. Thus one aim of this book is to summarize and integrate established forensic geology with the other 'geo' words that ought not to be separate any longer. Why? Because geophysics cannot be understood without soil science, nor remote sensing without geomorphology, or sampling without geostatistics. As Croft and Pye (2004) demonstrate with regard to traditional soil and sediment evidence, using a number of analytical methods to compare samples ensures good science, and provides convincing testimony to the courts. This book has an integration of many 'geo' disciplines at its heart. Some are covered lightly, if at all, as we can never be comprehensive. Similarly we know that forensic biology, toxicology, chemistry, psychology and the work of search and rescue (dogs and personnel), drug and explosive/munitions monitors, health and safety workers can all be integrated with our 'geo' disciplines to provide better understanding of the 'what, where, when and how' questions. A common misunderstanding is that geoforensics is about using soil and rock dust adhered to a suspect's shoes to link them to a scene of crime: the new, 21st century application of the various 'geo' disciplines to human activity includes many other materials than soil or rock, uses all our analytical techniques to establish the guilt or innocence of suspects, and is applied to aspects of criminal, terrorist, war-crime, genocide, environmental and legal investigation. Jago's (2002) definition, that 'Geoforensics is all about helping the police', is thus narrow and incorrect. Geoforensics is the application of Earth science to problems of criminal, humanitarian, war crime and environmental nature, especially with regard to assisting the pursuit of truth and justice, be it for prosecution or defence. Following on from this is a major reason for writing this book, because since the authors' first experience of applying geology to criminal investigations over 10 years ago, a recurrent snobbishness from some practitioners permeates the geoforensic arena. This is exemplified by a 'what I do is more important' attitude, expressed as defensiveness concerning casework. At a recent forensic science meeting the following conversation was overheard: Person A: 'I assisted in tracking down a serial killer – you don't know what real forensic work involves'; Person B: 'So what? Pollution in the case I just completed cost the lives of over 2,000 people'. There are two ways of looking at this conversation (and underlying attitude). First is that any suspected harmful act, be it genocide, intentional pollution, corporate negligence, murder, requires a fair and just investigation based on rigorous science. Second is that of course money pays for expert time and analyses and less finance will be available for less serious criminal investigations, mirrored by media attention. While accepting the ethos of the first and reality of the second points of view, arguing about the status of one's work belittles the greater scientific good expounded in this text: the sharing of experiences, methods and ideas between practitioners can only be of benefit. Hans Gross recounts cases concerning 'The Larceny of Tomatoes' and the 'Larceny of Fowls'. We may well scoff, but for those whose livelihoods depended on tomatoes or fowls, full and proper investigation was essential, and in addition provided Gross with yet more experience of the geoforensic crime scene. Personal experience demonstrates how investigators into illegal waste tips in the eastern Mediterranean did not concern themselves with continuity of evidence; those searching for mass graves did not know about some technology used to find illegal waste tips; homicide investigators had not heard of methods of trace evidence assessment developed by drug squad/narcotics investigators. The cynic will respond: 'maybe they did not need to know about these other methods/investigations: necessity is the mother of invention'. Again, experience shows that such close-mindedness will be exploited by the legal system to develop a weakness; best explained using a case study (below). The inclusion of environmental crime is not only a reflection of popular concern for the environment, but that activity is now causing such extensive health (death, disease), financial and global climate change problems that INTERPOL now include such matters in their discussions (Suggs & Yarborough, 2003).

Case study: Geologists and engineers

High-value, ornamental stone was stolen from a stonemason's yard, where only some off-cuts remained. The stolen material was recovered from a competitor's yard, who maintained that his material was purchased legally over many years (so no receipts existed). The materials were compared using a range of geological analyses such as thin section petrography linked to image analysis and multivariate chemical analysis. The defence employed an engineer as an expert witness, who used different methods entirely such as material tensile strength, manufacturing method and availability of similar materials. The jury understood the expert witness engineer far better than the geologist, delivering a 'not guilty' verdict. We hope that this story enrages many geoscientists, who respond 'our methods are far more exacting than the engineers' and 'we are trained to compare and exclude materials'. Why should they be angry? Should they not have learnt from this – to use available experts/methods where available, include them on their applicability to the case, exclude them for good scientific reasons ('fit for purpose') and use them effectively in court? An excellent example of the all-embracing attitude is included on the web site of Dr Robert Hayes, the owner and president of *www.geoforensics.com*, whose range of work embraces many of those cases explored in this text. They include using geological, chemical, geomorphological and engineering analyses and experts in each field[3] for a range of applications. These include: identifying where chemical pollutants were released into a river; testing the grip provided by a roadway following a fatal road accident; comparing a hit-and-run driver to a scene; comparing a suspected rapist to a scene; re-locating a body at the murder scene following removal; comparing riverside construction activities to a fish kill; locating the cause of a collapsed roadway (following a fatal crash), among many others. These serve to demonstrate how wrong Jago (2002) is in writing 'geoforensics is all about helping the police'. Geoforensics is all about the appropriate use of geoscience in assisting the course of domestic, international (genocide, war), terrorism-related and environmental justice.

---

[3] Never write a report, investigate a problem or go to court using methods of which you are not confident, and if such methods are included, comment on only those aspects of which you have experience.

2
Physical geography, geomorphology, landform interpretation, archaeology, stratigraphy and hydrodynamics

> The Investigator ought to study as accurately as possible the local topography ... There are localities which the Investigator must examine in the light of future events – hotels, public-houses, clubs and brothels, because of brawls that may take place in them, mortuaries because of post-mortems that may be carried out there, ponds and wells in villages on account of accidental drowning, forests because of poaching and illicit felling.
>
> Hans Gross (1891)

2.1 Physical geography

Physical geography examines the natural processes occurring at the Earth's surface that provide the physical setting for human activities: in this definition we see the strong relationship that physical geography has to criminalistics, for without human activity at a location, a crime cannot have taken place. The concept of *Earth systems*, or inter-related spheres of the Earth, is central to how we view the surface of the Earth: the *atmosphere* extends from space to the land surface, the *geosphere* from the centre of the Earth to its surface, the *hydrosphere* encompasses the oceans, seas, rivers, lakes and groundwater and the *biosphere* is the organisms that inhabit the three physical spheres. Physical geography takes account of the processes occurring in and between the spheres and their effect on the landscape. *Climatology* describes and explains changes over time and in space to atmospheric heat, wind and moisture. Some people regard weather as the product of the short-term climate, most especially at the Earth's surface, and thus the importance of both weather and climate to the study of criminalistics becomes clearer: indeed, the central importance of a location's climate and weather conditions at the time of an incident is shown in the idle commentary on the weather that seems to be an obsession in many countries!

Precipitation, (rain, dewfall, snowfall, etc.) wind conditions and temperature are all critical facts that must be established in the early stages of any investigation as a priority: moisture dripping from clothes on entry into buildings, footprints in snow, the likely time suspects or injured parties will tolerate being outdoors and whether material could have been blown from original locations are all obvious examples of why weather, and especially the unusual weather events within a long-term climatic regime, are of importance. Weather also plays a part in the social geographic aspects of crime, an interesting aside dealt with below. Practitioners of environmental forensics, especially investigating activity that is thought to have harmed the environment (e.g., alteration of water courses, illegal extraction) also have to disentangle the effects of natural weather effects (e.g., floods) from human causes (e.g., illegal diversion of water). This requires as complete a knowledge of weather and climate as possible: how foolish would the expert prosecution witness appear in trying to establish blame for a flood, only to find that the cause of the flood is found to be an extreme weather event in the fluvial hinterlands, after which river water was channelled past an area of housing. Is the weather event the result of climate change? In which case is the climate change part of a long-term cycle or the result of greenhouse gas emissions? What effect did the engineered river channel have on the flood? Did it alleviate flooding in one area, only to cause inundation in another? Or was an area bound to flood anyway?

Historically, physical geography and geology have contributed more to the investigation of non-urban crimes: this is changing as the micro-geographies of parklands, gardens, wasteland, and studies of urban geochemistry and particulates (for things such as health studies) increase. Traditionally, the farther from the rural or 'natural' landscape, the less direct relevance physical geography has had to the investigation. Increasingly, however, we find that geoscience techniques can be translated to the urban environment. To be more direct, criminal activity occurring in a remote location will leave abundant imprints on the landscape, from vehicle and footwear tracks, to vegetation breakage, digging and dislodging of loose material. Urban human activity does not have these same unique qualities in altering the landscape: contacts have still been made, traces of material left and picked up, but these are far harder to isolate from everyday urban activity. As we shall see in Chapter 7, the principles of contact still exist and can be found using sophisticated methods or laborious searches. Thus the principles of physical geography, and their application to criminalistics, are best defined and practised in non-urban environments; we translate what we can of these methods and principles to use in other locations.

Geomorphology, the science of Earth surface processes and landforms, is central to the investigation of criminal activity, accidents or natural disaster, occurring outdoors and especially in the non-urban environment. The geomorphologist is effectively a landscape detective, who asks 'forensic' questions such as: 'what happened here?'; 'when did this or these events happen?'; 'why did they happen?'; and 'what was the effect on the landscape?'. The geomorphologist is of course asking these questions from a pure point of view: should an application to criminalistics be forthcoming, so be it. Nonetheless the forensic scientist asked to reconstruct events at a scene of crime asks the same questions. The answers may not all be related to the investigation, but this is good: all too often, natural phenomena create patterns in the landscape that can be misinterpreted by the naive as related to suspicious activity, just as legal activity can, following the completion of say the digging of test-pits for building works, look like covert activity. Landscape analysis and interpretation is a critical first step in many different investigations, from the sampling of soils to the locating of trial pits or geophysical surveys, and is thus considered at greater length later in this chapter (Section 2.3).

The spheres of the Earth interact most directly in soil, historically the most important physical material in the development of our geoforensics. Soil is a solid medium capable of supporting life (biosphere), and is the product of weathering rock (geosphere) by water and air (hydrosphere and atmosphere) (Figure 2.1). Soil is generated from rock, by air, water and organisms and often reflects all four in a unique way. The geography of soils is thus also of major importance and likewise given separate description in this chapter (Section 2.4).

The oceans and seas cover some two thirds of the Earth, and although not populated in any significant way, the coastal zones of the Earth do contain over half the world's population. This fact, contrasted with the remote nature of the remaining coastal regions and the possibility for covert activity on shorelines and at sea, make study of coastal and marine geography an important part of the application of

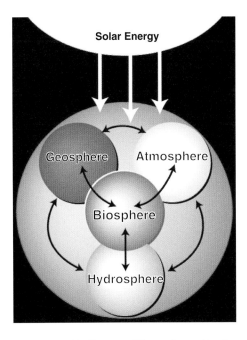

Figure 2.1 The spheres of the Earth. These will be referred to throughout the text, and come back to play an important role in our discussions of the future of geoforensics in Chapters 7 and 8

geographic science to criminal investigations. For this reason, the activity of rivers, lakes and oceans is also discussed separately below (Section 2.5).

On the margins of Geoforensics is biogeography, the distribution of organisms in time and space. The use of organisms in criminal cases, especially diatoms (diagnostic for drowning) and pollen (trace evidence) dovetails exceptionally well with geology and geography, but is now so well developed that it merits separate description as a set of 'bioforensic' techniques. Geography is critical in influencing the distribution of organisms, but the ecological reasons for the development of habitats and diversification of plants, fungi and animals is so dominated by biology that it is considered as relevant additional material in the following chapters. As we recommend in our final chapter (Chapter 10), a text on Bioforensics is overdue.

2.2 Atmosphere

Case study: John McPhee on the importance of meteorology and mineralogy – Japanese incendiary balloons

McPhee (1996) gives a synopsis of Mikesh's (1990) book describing the incredible situation faced by the United States military and population in 1944. Following the bombing of Pearl Harbour, the Americans had launched a massive air raid (using B25s launched from the aircraft carrier Hornet) on five major Japanese cities, embarrassing the authorities and causing widespread panic in the population. The Japanese sought retaliation, and desired their own version of the German V-1 intercontinental ballistic missile. Major General Sueyoshi Kusaba thought that the jet stream (which was known about but not yet named as such) could carry a balloon to America in three days (Figure 2.2).

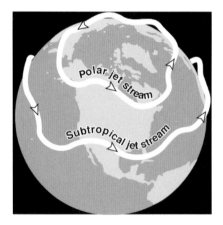

Figure 2.2 The polar and subtropical jet streams of the northern hemisphere. Major Kusaba of the Japanese military command intended using both air-flow masses to carry incendiary balloons from a Japanese beach launch-site to the United States

His team also calculated a means of using electronic switches linked to an altimeter to control the expansion and release of hydrogen in the balloon to control ascent and descent. The final switch was linked to a discharge of gunpowder that lit a fuse and ignited the envelope, creating a massive incendiary. Ballast comprised sand and the balloons were constructed by schoolgirls using mulberry wood pulp-paper. Prior to release, wind conditions were closely monitored, because the balloons had to rise through low-level weather to then enter the jet stream. Thus the importance of meteorology was appreciated but perhaps not studied well enough: a major US air raid on Japanese-occupied islands in late 1944 forced Major Kusaba to agree to a balloon raid. Through late 1944 and into January 1945 balloons were sighted, shot down and seen exploding all along the North American west coast and inland as far as Michigan. Two questions were raised: how were the balloons travelling so far, and from where were they being released? Some suggested that they came from Japanese-occupied Pacific islands, others that they were being launched from Japanese submarines, by shore parties that had landed on American beaches, from German prisoner of war camps, or from Japanese-American citizen internment camps (to which many Japanese-American citizens had been forcibly relocated). However, the Japanese research into meteorology, plus their use of beach sand-filled ballast bags led to the release spot. As soon as the Americans realized how high the balloons flew, the US military saw the potential for an origin in Japan. The specific location was made using the fossil and mineral makeup of the sand, which is recounted in Chapter 7.

Case study: Kaplan's (1960) critique of the crime – weather relationship

This article, although over 40 years old, is remarkable in its functional approach to reviewing previous work on weather and crime. Kaplan's main conclusion is that the weather to crime relationship is a difficult one to establish, but analysis of previous works in which over 40,000 criminal incidents were compared to weather showed that the climate (long-term weather) to crime and season to crime relationships could be supported, with a hierarchy of good to poor comparison.

At the better-established end of the spectrum, Kaplan notes the good correlation between assault/murder and summer months. Specifically, low humidity, low pressure, low winds and high temperatures appeared as relevant factors in the homicide rate. Subsequent workers have supported this view, with modifications, such as social disturbance and riots also occurring in the warmer months, but commonly associated with high humidity. The concept of reserve energy seems plausible, with daylight hours and a reduced need to use calories in keeping warm contributing to acts of violence that result in minimal personal gain. Equally supported was the phenomenon of violent storms and social dysfunction, looting and opportunistic crime. Break-ins and larceny showed more complex patterns: in the summer, open-windows are more common and the opportunity to loiter is higher; in the winter, darkness allows covert activity and the need for valuables (drugs, drink, food) becomes greater. Kaplan mentions the link between the presence of ozone and suicide/homicide rates as being more tenuous. Although sociologists have continued considering the complex relationship between crime and weather, Earth scientists have been less active in applying their meteorological knowledge and observations to criminalistics. This may change in the near future as evidence is emerging of a relationship between optimum sampling times and episodes of high and low pressure. L. Donnelly (personal communication, September 2007) and M. Harrison (personal communication, October 2007) both provide examples of gas sampling in upland bogs, where the shift from high to low pressure creates the optimum time for methane emission and thus gas sampling for cadaver detection by probes or dogs.

2.3 Types of landscapes, landscape change and human influences on the landscape (short and long term)

What controls landforms?

Landforms, or geomorphology, can be considered as the shape of the surface of the Earth, distributed across a space. Thus the three-dimensional representation of these two (shape, location) variables is required as the first consideration of landform. This has traditionally been the contour map, and is increasingly being represented by computer digital terrain or digital elevation models. In the first instance, some idea of scale is required. The scale at which a landscape is portrayed in map or digital form is limited by the size of the area to be investigated. Should a map be required of an individual patch of ground, say 1 metre by 1 metre, this can be mapped at 1:1 scale, or the size of a large piece of paper. This is rare, and for individual sites, maps at 1:10 or 1:100 are common. Landscapes that are being investigated also need to be represented on paper or a computer screen, and are commonly mapped at 1:1,000 to 1:10,000 (the common scale for ordnance survey maps). The movement of vehicles on roads, boats and ships or aircraft cover much greater distances, and require maps that cover whole countries (1:1,000,000 and above) through to the whole Earth.

The primary control on landform is underlying geology. The processes of erosion in a gently sloping area of say $1\,km^2$ will not change significantly given regional climate and weather patterns, insubstantial changes in vegetation and limited human influence. However, should this theoretical area be underlain by rocks of different resistance to erosion, the harder rock type will form uplands, whereas the softer material will be preferentially eroded. This general rule does not just apply to different rock types, but also to weaknesses within otherwise homogenous rocks. A good example is a fault or fracture, which introduces a weakness in the rock that will be exploited by the processes of erosion. Hence it is better to suggest that weak rocks will occupy low ground and hard rocks uplands. This concept can operate at a range of scales, from a few metres to tens or hundreds of kilometres. Both scales have implications for geoforensics. For example, in burying covert materials (weapons, drugs, stolen goods, murder victims) a criminal is automatically restricted to the soft material in low ground, be these isolated areas of low ground (ponds, peat bogs, dry valleys, wind deflation hollows) or more extensive areas that are underlain by moveable material (floodplains, fields, coastal dune fields). The same rule applies to the transfer of soil, rock dust and other material: contact between vehicles and people on hard ground creates limited opportunity for transfer, whereas movement on soft ground (at whatever scale) has the opposite effect. Most important in this analysis is to combine what the landform determines as available material for the scientist and what the accused is suspected of doing. It would be very wrong to base a search of an area on the above criteria of examining low ground with moveable, transferable materials: should the suspects be involved in covert activity that required isolation, then such low ground will also likely be the site of communication routes, agriculture and be viewable from elevated ground. Instead, our suspects are likely to approach inaccessible ground, possibly elevated, first, and then look for places to hide, bury materials or sink objects. A good example is the covert practice of weapons training, common to domestic criminals and terrorists alike. Such activity requires an isolated location with limited sound travel: upland valleys and low hollows in mountains are good examples.

Rock structure also influences the location of streams, the position of catchments and their divides (the watershed), the location of ponds and bogs or mires, as well as coastal landforms such as headlands, spits, bays and estuaries. Geology provides the ingredients for developing landform, and climate and weather provide the process by which the landforms develop. Good examples are the different processes occurring in arid as opposed to humid climate regimes, based on the same rock-type. In an arid location where vegetation is sparse and periodic water flow (flash floods) highly effective in removing debris, sharply-defined landforms with cliffs and scarps separated by plateaus are common. In a humid region, the action of chemical weathering is dominant, leading to the creation of deeply weathered rock and soil, with rounded landforms ('sugarloaf'). Such arid and humid landforms may persist long after the climate that shaped them has ceased: such relict landscapes are common.

Simple landforms

Common geological structures produce the most widely seen landforms, be they modified by climate or vegetation. Some typical regional landforms are summarized in Figure 2.3. Within each of these, erosion through weathering, slope failure, wind and water drainage networks play a strong role in the meso-scale development of a landform. One example is the fluvial (or river) dominated landscape, where the range of confluences (dendrite, trellis), river bends and meanders, waterfalls and sinkholes control both the historic and contemporary shape of the host valley, as well as the intervening uplands.

Landforms begin their development with some common ingredients. Variations in the underlying rock type and structure create initial weak–strong divisions in the landscape, upon which physical and chemical weathering acts to disintegrate rock. Physical weathering includes the action of frost, salt crystal growth and unloading as rock masses are uplifted. Chemical weathering includes hydrolysis, oxidation (see Glossary) or acid/alkali reactions. Physical and chemical weathering rarely act totally alone, although in arid and glacial environments physical breakdown products dominate and vice versa in humid climate regions. The net result of this disintegration is the creation of broken rock debris, or regolith, which when colonized by plants and animals becomes soil. Large-scale downslope movements of

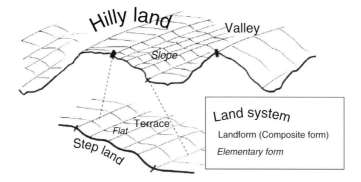

Figure 2.3 Typical regional landforms showing the elementary segments used to represent landform hierarchy. Scale is intentionally left off, as such features may be metres to kilometres high and across. Data originally from J. Minár, and after Minár and Evans (2007)

soil and soft sediment are known as mass wasting: slow soil creep and earthflows as well as more rapid landslides are both problems and opportunities for the forensic geomorphologist as they obscure the landscape but also preserve evidence. Arctic/glacial and arid/desert environments are generally less populated than those more humid temperate and tropical locations. Thus although the very cold and very dry regions of the Earth pose particular challenges to those conducting searches or collecting evidence, we shall concentrate here on those areas where water plays a dominant role in the formation of landscape types.

Important elements of the hydrological cycle include the action of precipitation, groundwater, fluvial action and water storage in lakes and oceans. Precipitation is a major consideration when we are reconstructing events at a scene: the transfer of evidence (soil, pollen, manufactured products) is often dictated by moisture. Groundwater not only causes caves and sinkholes to form (of obvious interest to those conducting searches), but also is a valuable and vulnerable material in itself, open to illegal abstraction and pollution, in which the environmental forensic geologist may be involved (see Chapter 1). The study of caves and cave-forming processes (karst science) became of interest following the September 11 terrorist attacks, as Pinsker (2002) summarizes:

> Geologic interpretation of photographs and videotapes can also shed light on the location in which a photograph or a recording was made. A notable example of this kind of forensic geology occurred shortly after the September 11, 2001 terrorist attacks on the World Trade Centre in New York City and the Pentagon in Washington, D.C. American geologist John (Jack) Schroder (Pinsker, 2002) who had worked in Afghanistan was able to identify rocks in the background of a videotaped message from the terrorist leader Osama bin Laden, and therefore the region of the country in which the message was taped. The *New York Times* reported these caves as being in granite, when in fact bedded limestone can clearly be seen, and is more consistent with cave formation. The use of geological knowledge to infer location was widely publicized, however, and subsequent

messages were recorded against a cloth background in order to prevent the location of the taping from being discerned.

Fluvial activity is of major interest to geoforensics (Figure 2.4). Rivers and streams are common hiding places for materials and dumping grounds for murder victims and waste. Like groundwater, flowing water is also susceptible to theft in the form of watercourse alteration and covert pumping. Also, as we shall see in our case study (see p. 26), streams and rivers divide the landscape into manageable units for the purposes of search and sampling. Finally, ponds, lakes and the seas and oceans not only store water, but also can be used in covert storage and hiding. The activity of these water bodies is given separate treatment in Section 2.5 at the end of this chapter.

Complex geological structure and landforms

Above, we examined the interplay between simple geological structure and landform development. Just as the underlying structure is complicated, so too are the landforms above. This need not deter the investigator however: what happens is that the simple rules become applicable to smaller and smaller areas, as structure breaks up the landscape into different units, each with its own rock and landform character. These complex areas require more time investment in analysis. The same rules of hard

Figure 2.4 Geomorphology and the search. This upland landscape has been dissected by a meandering stream, allowing us to use these physiographic features as domains, mapped by the composite landform mapping approach. The uplands are highly visible, and have rock at shallow depths, the slopes are unstable, but with loose material, the stream is likewise unstable, but with small, loose point-bars that are easily excavated. Those involved in the burial of objects have commonly used the riverbanks as areas to excavate: the point-bars being too wet and the uplands not having thick enough soft material. Field of view at base of photo is approximately 250 metres

rock–soft rock apply, but when faults or folds are introduced, the strong and weak rock is broken up into increasingly chaotic areas. Nonetheless, these can be mapped, using common features to separate areas in order to create a landscape classification. Typically complex landscapes are those dominated by ongoing tectonic activity and/or underlain by complex geology, and volcanic landscapes.

Landform (or geomorphological) mapping

Geomorphological mapping is one of the fundamental research methods of geomorphology (Cooke and Doornkamp, 1990; Minár, 1992; Waters, 1958). Landforms can be summarized with regard to three types of relief units on the basis of their increasing complexity. The *elementary units* represent the smallest and simplest units, which are indivisible at a defined level, determined by their geometric simplicity (e.g., linear and curved slope or horizontal plain, see Figure 2.5). Many traditional geomorphic forms are made up from several such elementary units, termed *composite forms*. Composite forms usually represent basic single geographic forms that we are all familiar with (e.g., rills and gullies, valleys, basins and mountains of various orders). Characteristic patterns, created by form associations, provide a third level of complexity, and are termed land systems, equivalent to the relief form associations, terrain systems, landform patterns or types of relief, of other authors.

Modern geomorphological mapping was defined by Crofts (1974): 'Land classification and evaluation should be made on the basis of geomorphological mapping

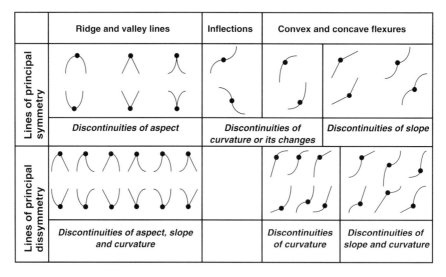

Figure 2.5 Cartoon cross-sections across types of structural lines. Data originally courtesy J. Minár, and modified after Minár (1992) and Minár and Evans (2007) and their interpretation as lines of discontinuity. Note how the 'points' in such a simple classification are easily defined at sharp inflections, typical of jagged topography or incised valleys. No scale indication is intentional because these features can be centimetres to tens of metres

and it has to be the aim of geomorphological mapping.' Identification of elementary landform units is the first stage in the study of spatial aspects of interaction among landforms, soil, vegetation, local climate or hydrological regime. Although remote sensing, GIS and GPS technologies have aided and advanced geomorphological mapping, its theoretical basis has remained unchanged for some decades (e.g., Cooke & Doornkamp, 1990). This relies on the concept of dividing up the landscape using the boundaries between landforms, most especially changes in relief. Perhaps the easiest example applicable to geoforensics (in terms of carrying out searches or discovering illicit activity) is to consider how fluvial processes (streams, rivers) break up the landscape into mappable areas. Precipitation, on contact with the ground may either infiltrate or form overland flow. On slopes, the latter become self-organized into streams that join at confluences to form rivers and so form a drainage network (or watershed), with boundaries both at the edge (outer divide) and within (drainage divides). Within streams, flow is most rapid in the centre, where the moving water encounters least resistance: thus floating or submerged objects move most quickly in this region. The underlying topography modifies stream and river flow, where waterfalls, plunge pools and rapids may form. Unconfined rivers tend to spread into broad braided networks, whereas confined rivers meander across floodplains. The boundaries to these small, upland and large, downslope drainage basins form mappable areas, or composite forms. The same division of the landscape, based on topography but with other features added on, can be applied to glacial, desert and even coastal areas.

The continuous field model of landform (represented by isolines, or contours) can be used very effectively to assist understanding of topography, and divide up the landscape, in many areas of the Earth. Cox (1978) concisely termed this approach as the *continuous* hypothesis (slope profiles are continuous curvatures without definite breaks) in contrast to the *atomistic* hypothesis, where landscapes are essentially a mosaic of discrete units. Segmentation can then occur at natural divides such as the top, depression and saddle points, and ridge and valley lines. Landforms and their discontinuities evolve through time and will thus contain relict topography and deposits as well as those actively forming.

Traditional geomorphological mapping was mainly developed on a morphological basis and became over-complicated when too many characteristics of the landscape were included in the process of segmentation (soil, surface material, drainage). Similarly, many papers through the 1960s and 1970s spent time discussing and defining segment boundaries, when simple morphology is usually sufficient, onto which other information can be added later. The identification of a unit's boundary is a primary goal and the character of the interior does not influence the determination of its limits. We can term this the *graph-based approach*, the most useful way of beginning to understand the landscape and exceptionally useful as a first stage in the search process (see Chapter 9).

Morphological mapping is based on landform segmentation, which uses the simple assumption that the ground surface comprises planes bounded by morphological discontinuities. How these planes and boundaries are organized is a wider, geomorphological question that is still currently being debated. A morphological unit is either a facet or a segment, where the facet is a plane, horizontal, inclined, or vertical

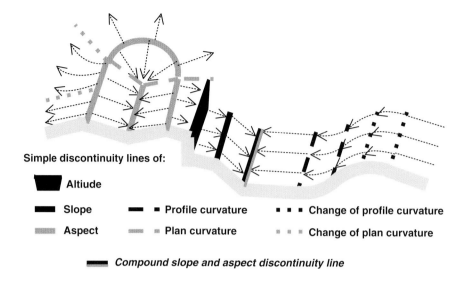

Figure 2.6 Cross-section (from left to right) of a theoretical hill (with hill-top valley), cliff and valley with variable slope, together with simple discontinuity lines. Adapted from Minár and Evans (2007), with permission prior to publication from J. Minár (2006). No scale is intentional

surface area and the segment is a smoothly curved, concave (negative) or convex (positive) upwards surface area. Facets and segments abut at discontinuities: breaks of slope, changes of slope and inflections (lines of maximum slope). Figure 2.6 shows how such discontinuities may be used to divide the landscape, as a first pass at gaining an understanding of each segment. Such division is not a goal in itself however: following this, any number of additional analyses, dependent on the investigation in hand, will be undertaken. In geomorphology, these additional analyses may be of fluvial activity, slope stability or perhaps soil types (Cox, 1990); in a criminal investigation, these may be sampling for pollutants or tracers, geophysics for buried objects, or soil sampling for comparison/exclusion to suspect/victim materials. Most crucially, the segment approach must then be deconstructed and the whole landscape, or links between segments, be established. The splitting up of the landscape, for the investigator, is a means of control. Dropped into the middle of a landscape with no dwellings or obvious landmarks, the human who is used to urban locations can feel overwhelmed. The segment approach allows some measure of control to be established, if only in one's mind, segments being a little like streets or railway lines, and thus easily handled by the human mind. For instance, broad areas (metres to decimetres) of soft ground will be ideal locations for geophysics: much smaller areas (a metre or less) will prove problematic for geophysics (Chapter 3), but are suitable for investigation by probing (see page 36 for a description of the probe), or by excavation.

Morphological units are fully defined geometrically, but their genetic and dynamic interpretation requires the addition of further, qualitative and quantitative characteristics, including position, slope, profile curvature, actual processes, significant microforms and the specific characters of rock, soil and vegetation (Figure 2.6). A sensible approach is to try and define elementary surfaces that can be divided by lines representing local extremes of altitude (ridge lines and valley lines), local extremes of the slope (inflection lines involving not only lines of maximum slope but also lines of minimum slope) and local extremes of profile curvature (convex and concave flexures involving breaks and changes of slope). The basic structural lines are classified in more detail into many types and kinds on the basis of the shape (linear, convex and concave) of bounded surfaces, which in actual fact represent various kinds of discontinuities of morphometric parameters. Key locations in the definition of a land surface include high and low points as well as points of contact and ending of structural lines (ridges, valleys). This approach works best in incised terrain, where sharp topography is present. In more mature landscapes, where rounded landforms are present, landscape division using changes in slope curvature is more appropriate. Maps of geometrical forms are usually created by overlap of maps of zero isolines of curvatures (Figure 2.7).

When the same method is applied to a wider set of morphometric parameters (altitude, slope, aspect, curvatures, etc.) landform segments defined by a set of values of morphometric parameters are produced. The early approaches to landform segmentation used a crisp classification of landform. *Continuous (fuzzy) classification* appeared during the 1990s as a means of reflecting the complex variation in landscape characters in space and time. The resumption of interest in the

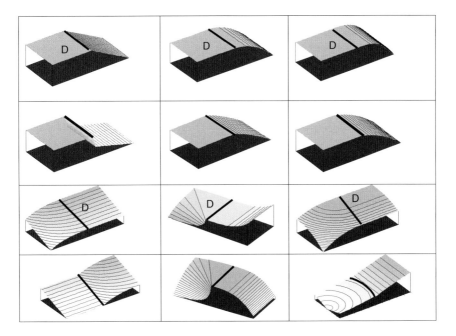

Figure 2.7 Examples of compounded boundaries of elementary forms. D-form-determined boundary. Data originally from J. Minár, and after Minár and Evans (2007). No scale is intentional; these features can range in size

problems of landform segmentation is connected with the recent development of GIS techniques, with many authors stressing the methodological aspect (automation) and the theoretical background of segmentation. Simpler conceptions that are more readily automated were often preferred with new publications (e.g., scientific papers) often representing a new look at old conceptions from the point of view of their feasibility in the GIS environment. Some algorithms combine earlier mentioned principles, but the general theoretical unification of the various conceptions was missing until the publication of Minár and Evans (2007, and included works).

Working example from Minár and Mimian (2002)

An example of the application of elementary form conception is presented on Figure 2.8. The section of complex geomorphological map of Devinska Kobyla Mountain (Minár & Mimian, 2002) is interpreted from the point of view of elementary form conception. Field research and visual analysis of digital elevation models were used for the delimitation of elementary forms.

Case study: Temperate landform classification – search methodology (no body murder enquiry); the background geomorphological approach to sampling (Harrison, 2006)

Background. Harrison (2006) describes how, in late September 1995, a teenage girl attending a club at a seaside town in the west of Ireland accepted a midnight lift home (inland) by a mature man who was known informally to her, although she did not know of his previous criminal records, including abduction and rape. Neither can be named for legal reasons. The man's estate car was noticed by an off-duty member of the police some three hours later, parked in a lay-by to a farm track on a quiet country road. The officer, two colleagues and a dog team returned to this location some 30 minutes later, where the dog indicated at a point ('vehicle turning circle', see Figure 2.9 below) along the farm track where fields passed into scrubland. Concerned at their proximity to the Northern Irish–Irish Republic border, which caused problems of jurisdiction and the presence of terrorists, the police abandoned their search until daylight. A follow-up search was made, but heavy rain and tractor and cattle movement limited the use of tyre marks or footprints. The 15-year-old girl was reported missing some days later, because she was known to occasionally stay with friends or even abscond to her separated father's house in Dublin.

Enquiries led to the mature man being the last person to see the girl, but conflicting reports of her being seen elsewhere in Ireland and England were inconsistent with her disappearance. Only when the mature man was arrested four years later for the abduction and rape of another 15-year-old girl was a possible connection made, and a search made using his flat at the time of the incident as the focal point. As the search widened from this point, so the area of his last known position on the night in question was considered.

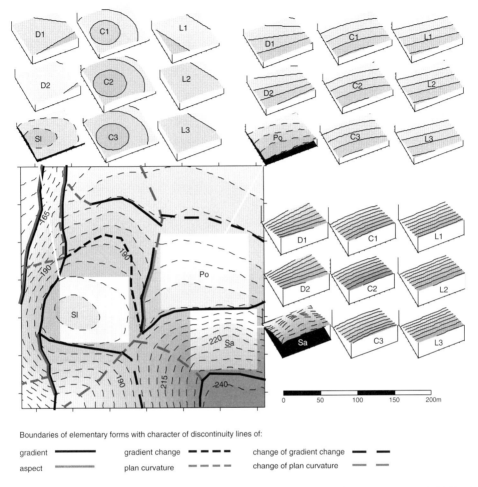

Figure 2.8 How to view topography: an example after Minár and Mimian (2002) of Devinska Koblya Mountain (Slovak Republic)

Search location. The search location is shown in Figure 2.9 and comprises a range of vegetation and soil types. These were initially used to classify the landscape, but proved unsuitable as the basis by which a search may be focused.

Trees comprise planted pine adjacent to the main road and silver birch along the track and around the ridge that formed the core of the search area (Figure 2.10). The main road forms a border to the east. Scrubby grassland with metre-high stands of gorse or reeds, with some bedrock exposure, occurs throughout the area. Low-lying ground is characterized by reeds growing on bog, some of which showed evidence of being drained in the past: an extensive bog that cannot be walked upon even in summer borders the area to the north. A major river occurs to the south, with housing some 500 metres distant, with open, arable farmland to the west.

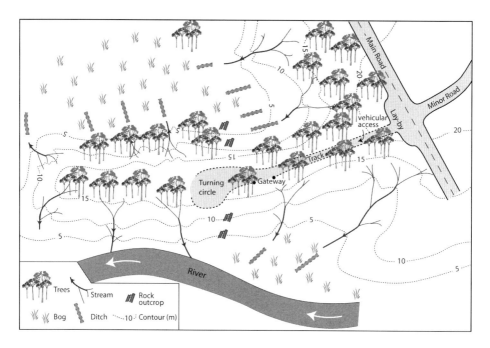

Figure 2.9 Sketch map of the search location used in the temperate landform classification case study. Scale is intentionally vague, but is approximately 500 metres across

Figure 2.10 Photograph of typical vegetation in the search location, comprising scrub silver birch, gorse, reeds. Human scale is 1.8 metres tall

Search limitations and domains. Some fundamental controls on the search were established. No surface remains of clothing, persons or personal items were found, indicating that if the victim was in the location, she may have been buried. Four years had passed since the area was last searched: the location has high rainfall (1400 mm/year) with extensive cattle movement, masking the expression of small (decimetre-scale) surface features. The search area was geographically-limited by road and dwellings to the east, river and house-views to the south, impassable bog to the north and to some distant (from the vehicle turning point) open farmland to the west. Within the search area, individual locations were limited again by views, inaccessibility (thick trees, bog, river) or rock (insufficient depth of soil for burial). A *landform mapping approach* was used to create a priority of search areas, or *domains* (Figure 2.11).

Domains were defined using a mixture of some or all the criteria outlined above: we describe them in reverse order, because the lower numbers have the highest priority; in other words we exclude the least likely domains first. Domain 10 comprised the northern bog, in which a dug hole was quickly waterlogged. Domain 9 was the opposite – having bedrock exposed or within 20–30 centimetres of ground level. Nonetheless, pockets of deeper, softer material could occur. The bedrock comprised hard, fractured sandstones with limited hollows and pockets. The domains collectively grouped as Domain 8 were all on grazed land that could be viewed from houses. Under cover of darkness this ground could have been excavated, but the resultant scar on the landscape would have been visible for

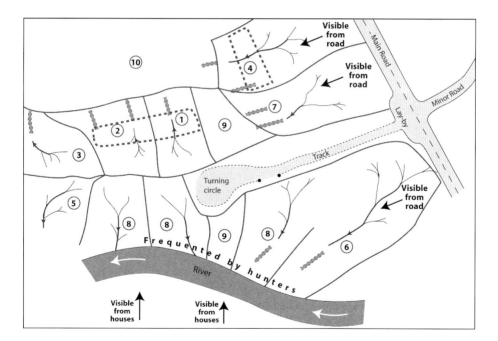

Figure 2.11 A landform mapping approach used to segment the area into domains, based largely on small watersheds. Scale as in Figure 2.9

sometime thereafter. The potential burial locations in Domain 7 were limited by bedrock and bog to the west and by views from the road, most especially where police vehicles parked on the night in question. Domain 6 had the same limitation as 7 and 8, being visible from the road and the houses. A NE–SW oriented spur protected the centre of the location adjacent to the stream, in which waterlogging would occur when dug but perhaps in dry weather, unlike the bog, burial would have been possible. Nonetheless, Domain 6 was also actively eroded by cattle movement and was an unlikely target area.

Domain 5 had all the correct ingredients for a burial location in being covert and underlain by soft ground. The distance that the victim would have been forced to walk, or dragged through rough terrain, was considerable compared to other locations, and as the MO *(modus operandi)* of the suspect showed him to be a lone-operator, this reduced Domain 5 as a likely body deposition site. Domain 4 had the correct mix of cover and proximity to vehicular access. The possible area within Domain 4 where a covert burial could occur was limited by views from the road and the waterlogged bog to the west. A hatched box within Domain 4 (Figure 2.11) covers the area hidden from the road that is free of bog vegetation and therefore is a search location within a geomorphological domain. Domain 3 was similar to 5 in its location and topography, and what made this domain a possible burial site was access. Whereas Domain 5 was inaccessible because of the thick trees and scrub vegetation (Figure 2.10), Domain 3 was accessible across rough grassland on the central plateau. Thus distance to a vehicle was about the same as Domains 4 and 5, but Domain 4 would necessitate movement over rocks, ditches and bog, whereas Domain 3 was a relatively easy walk westwards over grass. Domains 1 and 2 were much the same, with their northern limits, adjacent to the bog being below the 2 metre elevation of the plateau, and with no dwellings to the north. As in Domain 4, a hatched box indicates where the search limits within these areas could be, with bog to the north and visibility to the south. Domains 1, 2 and 4 all had the advantage of pre-existing ditches, which at their upslope end could have been dry enough to be re-excavated. Domain 1 had priority in being proximal to the vehicle access point.

This combined use of geomorphological mapping, integrated with other search criteria is an example at one scale of how the human mind finds it very helpful to divide the landscape up in order to target resources. Similar methodology could be used at smaller or larger scales, depending on what area has to be searched and on the complexity of the terrain.

Conclusions. Landform classification at this scale used small (metre scale) water catchments as the limits to each domain: slope type (as above) could equally be used. Water catchments were used in this case in order to, first, understand the movement of water, which informed, second, the sampling of waters. A less-important consequence of this is that the search area is split into manageable tracts of ground, each with individual characteristics that enabled a hierarchy of domains to be considered for searching and sampling, saving effort in each instance.

Landscape interpretation as an introduction to reconstructing events at a scene

The first stage in attempting to reconstruct events at a scene of potential crime begins with an easy geological principle, before we use more sophisticated methods. This is the principle of cross-cutting events, and although normally applied to geological successions in vertical plan (where faults cut strata in a rock face, so that the faults are relatively younger than the strata), cross-cutting in plan view is as useful and requires mapping. A simple example is where footprints and tyre marks are observed. Did the pedestrians walk to the location, followed by the vehicle, or was the vehicle there first and the pedestrians walked out? If the footprint overlaps or is indented onto the vehicle tread, the pedestrian was there last (Figure 2.12).

When excavated, we can observe the same principle (in plan view) of buried features. The Roman mosaic at Fishbourne (Sussex, England) is cut by plough-marks. The plough marks are patently younger than the mosaic, and were buried by soil (they are in fact Medieval). Covert graves are sometimes re-excavated in order to further obscure evidence. Cutting of infilled ground and incision of clothing, bones or flesh by digging are again cross-cutting events that give a relative age to activity. Finally, natural events may aid the dating of activity. From aerial photographs taken in southern California, law enforcement agencies wished to establish when a dirt track was last used, possibly by drug traffickers. The last-observed tyre marks are

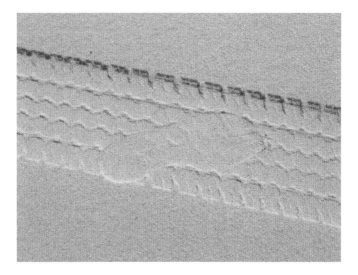

Figure 2.12 Example of cross-cutting features that allow us to begin to create a simple chronology of events. Here, the footprint was left after the vehicle-tread mark. If the footprint-maker claimed to have been at a scene of crime prior to the vehicle movement, then his or her story may be suspect

clearly offset by a major fault. The age of an associated earthquake was known from witnesses and monitoring stations, establishing the maximum age of the tyre impressions.

Environmental analysis, especially of plants and their spores/pollen provides a complementary, bioforensic approach to this use of geoforensic principles: Wiltshire (2006) uses the concept of the 'forensic landscape' based on the various pollen (fresh, reworked, local, transported) found at scenes and on suspects or victims.

Scales of stratigraphy, from covert burials (weapons, inhumations) to micro-stratigraphy of vehicle- and shoe-tread

A refined, if complex, method of establishing a possible succession of events that may improve on the mapping of movements comes from the study of stratigraphy. The principle of the superposition of strata, although developed by the 17th century scientist Niclaus Steno for rock successions, has found application in recent natural successions and archaeological deposits (Figure 2.13).

Therefore, in many texts on Earth Science, stratigraphy is described as a sub-discipline of geology, when the principles are so flexible, they can be applied to soils, sediment layers and microscopic layers (cave-fills, stalactites, mollusc shells). Infilled excavations come most readily to mind (Hanson, 2004), but the layering of material in natural depressions, or the natural infilling of natural or covert hollows etc., also allow for preservation of evidence and for a more complete story of events to be

Figure 2.13 Typical stratigraphic layering found in deep soils. In this excavation in upland basalts (Northern Ireland) during the search for buried weapons, clear layering with no internal undulations or secondarily-disturbed ground was noted by the archaeologists employed. Thus the excavation was stopped at an early stage: initial indications of buried metal objects turned out to be highly magnetized pockets of basalt. Camera case is 17 centimetres long

told. We can split the study of stratigraphy associated with criminal investigations into three types:

- macro-scale below ground (especially of soils);

- macro-scale above-ground, such as removed and tipped or mounded soils, but also mounds of sediment, builder's products, rock slabs and other geological objects built-up over a covertly-hidden object;

- microstratigraphy caught in tyre and shoe treads or microscopically as layers on clothing.

Where human remains are found, the nature of the sediment and soil layers above and below the body/skeleton is critical in establishing a relative age of the inhumation. Thus if the bone(s) or remains have been removed then some idea of the layer from which they came should be established, otherwise the integrity of the grave should be preserved and during destruction (during archaeological excavation) each sequence of the dig and the exposed layers/toolmarks/artefacts must be photographed. This requires appropriate cynicism because memory is rarely accurate and cannot be recorded, sealed and preserved as evidence. Many types of activity create layers (farming, building, natural deposition), each of which contains characteristic materials associated with origin. Thus it is essential that the excavating archaeologist be aware of the surrounding landscape: obsession with the contents of the grave may preclude observation of exit and entry points to the site, materials found near the site that are recovered from the grave, and other features of the landscape that may explain why the burial is in a particular spot (hidden from view, waterlogging, loose ground). Thus the excavating archaeologist may find success in working with a soil scientist or geomorphologist, because each have their own interpretive skills in understanding the 'what, where, when and how' of a scene, be it one of criminal activity or not. Often soil layers are easily distinguished through variations in colour, texture, grainsize and associated materials (roots, fossils). More cryptic layering requires specialist devices (ultraviolet photography, geophysics, soil microscopy – see Goldberg and MacPhail, 2006).

Understanding this layering provides a record of the succession of events at the location being investigated: it is the job of the archaeologist to excavate these materials (in reverse order) and thus reconstruct events before, during and following burial. This depositional tape-recorder is inaccurate, prone to switching off, and difficult to date, but ultimately is like trace evidence, the only (if mute) witness to events at a particular site.

Before we consider these examples, they all commonly involve soil, whose nature we need to understand.

2.4 Soils

Soil is the mix of biological, weathered rock material, air and water that commonly forms the uppermost layer of the geosphere, on and in which much human activity occurs. The mix creates different soil types reflecting the geology and climate of the Earth that dictate agricultural practice. Over time, roots and burrowing organisms penetrate the ground surface and break it up: biological activity introduces organic matter, both improving soil quality. Physical weathering does much the same, increasing the surface area of individual soil grains, while chemical weathering oxidizes and hydrates inert minerals, releasing nutrients. Both weathering processes are influenced strongly by climate and act upon the original ingredient, the rock. Thus volcanic rocks, rich in many minerals, form rich and fertile soils when weathered over time. Limestones, dominated by only three elements, unless enriched by some other process such as glaciation, fluvial activity or application of fertilizer, will in a similar climate and over similar timescales, form calcium-rich but possibly poor soils. Thus the biological component of soils may be living plants, animals, fungi and a range of micro-organisms, or may be dead representatives of the above. In addition, there may be significant dormant biological content (seeds, roots). Plant, fungal and micro-organisms dominate the dead, partly decomposed element of soils that is termed *humus*. Humus is generated at and in the near-surface of soil, but gets carried downward by water percolation and organisms. Thus soil can change over quite short periods of time in response to climate and human activity.

The study of soil is known as pedology (see Glossary): it includes the description and mapping of soil types as well as the development of theories of soil evolution. Soil colour is an important characteristic, often reflecting organic content and mineral weathering state. Thus dark soils of steppe and bog lands reflect high humus content, white soils may have high mineral salt content, such as are found in evaporitic climates, and red or yellow soils contain abundant oxidized iron, typical of tropical weathering. Soil colour used to be a common method of comparison/exclusion, but experiments on varying moisture content and absolute measurement have rendered this method, unless automated, very unreliable (Croft & Pye, 2004a).

Soil grain size is a feature of sediments and sedimentary rocks that we will encounter again. Maximum particle diameter is used to define material above 2 mm as gravel, 2 mm to 0.05 mm as sand, 0.05 mm to 0.002 mm as silt, 0.002 mm to 0.00001 mm as clay and all material below this grain size as colloids. The latter are very important in soils as they have a high surface area to mass ratio and are often negatively charged, holding for plant use the important nutrient (bases) elements such as calcium, magnesium, potassium and sodium. These bases may be replaced by hydrogen or aluminium ions, making the soil more acidic and often thus less fertile. Farmers may introduce lime (calcium oxide) or crushed limestone (calcium carbonate) to increase pH and thus improve soil quality.

Soil colour, grain size, moisture and chemistry vary with depth, producing horizons. These frequently have a more organic content (the O horizons), close to surface, and more mineral-rich layers (horizons A, E, B, C) below. Exceptions occur of course, where high water percolation rates followed by deposition, say on a river floodplain, produce deeper, organic layers, overlain by mineral layers: such soils would not be considered mature, as the upper layers will be vegetated, the lower organic layers mixed or decayed and a new soil profile will develop.

Soil temperature is very important, influencing the biological activity of indigenous organisms, as well as foreign objects and chemical reactions. Biological activity stops below freezing point (0 °C, 32 °F) with seed germination occurring above 5 °C (41 °F) to 24 °C (75 °F) in warmer climates. Temperature thus strongly affects the rate of organic matter decay in soil, and can be radically different on the soil surface, within a few centimetres depth and, because of shade from trees and topography, can vary over a few metres distance. When considering the possible longevity of buried, degradable materials in the ground, these important micro-variations need to be considered and the advice and even experimental proxy work of a pedologist employed. Global soil classifications are advanced and can seem complex to those unfamiliar with the terminology.

More important for the forensic scientist is to be aware of how to describe soils, their constituents, relationship to landscape and engineering properties. Traditionally, these have been split into the physical and biological components, for example, Bommarito *et al.* (2007) state that in forensic work the physical components are concentrated upon yet the biological components are often ignored. These authors advocate the use of chromatography in studying this fraction of soil, whereas Dawson *et al.* (2004) demonstrate how characterization (by chromatography) of the decay-resistant long-chain molecules found in plant waxes called *n*-alkanes can differentiate soils by virtue of the present-day and past plant cover. This method shows great potential for those suspect samples (less so scenes, because the plants can usually be determined here) with high organic contents and minimal pollen or micro-biological information on the plant cover. In this instance, the analysis of *n*-alkanes (as well as other plant biomarkers) provides a good example of an independent variable in soil that is separate to its geological/mineralogical makeup. Similar work, using soil isotopes (Croft, 2002), or in searching for isotopes derived from human remains, has become established in the literature, to the extent that popular books now mention them (Cummins, 2007). The organic and inorganic components of soil together determine colour: a much-debated feature of soil that is used and mis-used widely in forensic work. Croft and Pye (2004a) give an excellent synopsis of some key features of soil colour: 'the blacker the soil, often the greater the organic content; the darker the soil, the greater the moisture content; spotting or mottling indicates 'gleying' or reducing conditions; red and red-orange colours indicate... the presence of iron oxides'. Croft and Pye (2004a) go on to consider the use of automated colour measurement devices such as their spectrophotometer on dried, stored and sieved samples in reducing observer error. The technique is largely non-destructive and from personal experience has great value as an exclusionary tool, yet can be fallible when samples cannot be differentiated by colour/texture alone.

Case studies in soil density: Imaizumi, Owlsey and Ruffell on the use of the probe

The metal probe has been used for many years in searching soft ground. Keating (1991) recounts the case of Samuel Dougal, accused of murdering his common-law wife (Camille Holland) in 1899 at Moat Farm, Essex (England). The disappearance of Camille drew local law-enforcement authorities to Dougal's home, a farmstead with a mud-filled moat surround (*http://www.essex.police.uk/offbeat/o_mu_50.php*). In a classic example of search methodology over 100 years ago, the police and assistants began their search in the house, extending into the garden, where heavy clay impeded digging. While the garden and surrounding lands were searched, the clay-filled moat was also examined: one searcher is quoted as saying 'so much mud!' exemplifying a recurrent problem faced in searches to the present-day. The use of metal probes hastened the search, such that when a hard object was struck in the garden, this was excavated, revealing first Camille's tiny size 2 shoe, followed by the burial Dougal had made.

The probe then is a long-established tool for the subsurface searcher. The method involves the user inserting the rod into the ground and assessing the ease and 'feel' of the ground. In this section, we look at three key works examining attempts at estimating soil density using a probe, or metal rod.

Imaizumi (1974) describes the case of a female university student, who went missing in July 1973 while having an affair with a Tokyo professor. The professor had access to a colleague's villa in a satellite city (Hachioji), which was established as the last likely rendezvous location between him and the student. The surrounding area comprised scrub vegetation (up to 2 metres high), a building site and fields, a fingertip search of which recovered a woman's shoe similar to a type worn by the missing student. This focused the search to a number of fields: still a daunting task for the reduced number of searchers. The search lasted over five months, and culminated in the team using a slim gouge auger as a probe that doubled as a soil-sampler. This device was used to locate an area of soft ground that was the gravesite of the missing girl.

Owsley (1995) reviews the methods developed by Bass (see Bass & Birkby, 1978), himself and others in the location of burials. His work advocates a multi-proxy approach, using botany, landform, geophysics and soil science: in the first instance the investigator looks and records unusual patterns of plant growth, depressions in the ground and changes in soil distribution. These are the locations to which scent (or cadaver) dogs and geophysical investigation teams may be directed. Owsley recommends the auger-type probe as a cheap, rapid method that provides information on soil type. He warns against both the somewhat aggressive approach described by Imaizumi (above), as possibly damaging evidence or the cadaver, as well as allowing inexperienced users to probe ground. He describes search locations where dogs and geophysics proved difficult, time-consuming or ineffective. These were: a buried murder victim in a rubbish-filled backyard, a buried murder victim in one of a number of locations in gullies and steep-sided, heavily forested land, and the underground search of the Branch Davidian Compound, following the Waco Siege (Texas). Operator use of the probe has two problems: first, the invasive nature of the device and, second, the subjective nature of derived information. The invasive nature, as both Imaizumi and Owsley point out, provides information on soil type but may damage buried materials. Ruffell (2005) also mentions how the probe can be used with cadaver dogs in releasing decomposition gases (Figure 2.14).

Figure 2.14 Two types of probe in action. In this case (no body suspected murder) a target (white marker) was identified using ground-penetrating radar. The area was then probed (consideration of ethical issues, above) to a shallow depth (20 centimetres), to get a feel for the softness of the ground, but also for the release of decomposition gases to assist the cadaver dog in his or her scenting

The ethical decision of whether to use a probe or not rests with the investigating officer, who has to balance the need for a result versus the attitude of victim's relatives and the public concerning search methods. One possible solution is that all the non-invasive techniques available are used first, at which point it is clearly documented and communicated (if necessary) why the probe was deployed. The dispassionate, scientific stance, that damage by a probe, sometime after victim death, is nothing compared to murder, must be tempered with the personal attitude of a loved-one resting in peace and being treated with as much respect as possible.[1] Unlike physical evidence, subjectivity in searches is rarely called to question in a court of law, because a covert location is either found or not. What may be questioned is the extent and methods of the original excavator. Ruffell (2005) gives an example of the contentious location of a fibre-optic cable in a landscaped garden. Probe pressure measurements were made, reducing subjectivity and providing quantitative data for the courts, negating defence witness suggestions of 'operator error' or 'getting a feel for the ground'.

---

[1] Those searching for human remains are under constant scrutiny in this regard. Anyone who has visited a post-mortem will know that such matters are of less concern to pathologists, their damage to human remains being borne of a far older, mature science and thus deemed necessary.

Case studies in macro-scale excavation evidence: the works of Hanson, Bass and Birkby, and Boyd

The best synopsis of these types of the application of stratigraphic principles to forensic study is that given by Hanson (2003, 2004), who utilized excavation evidence to show how a detailed analysis of stratigraphy allowed improved understanding of events immediately following inhumation. Hanson follows Bass and Birkby's (1978) early work on this subject in considering the criminal burial akin to those archaeological inhumations that are termed rapid. The latter include the inundation by volcanic debris (e.g., Pompeii), the mass burial by sediment (e.g., Boxing Day Asian Tsunami of 2004), burial inside collapsed buildings and inside sunken vessels (e.g., the Mary Rose [Tudor warship] or Kursk submarine: see Chapter 3 on seismic monitoring). In both the archaeological and the criminal rapid burial, evidence pertaining to the death of the victim, or the hiding of the materials (weapons, valuables), is preserved by the burial. The medium in which material is preserved has a strong bearing on its subterranean longevity. The Man in the Ice (Spindler, 1993) or the Aarhus Bogman, like murder victims in peat, show excellent preservation of tissue, albeit that the Man in the Ice was not in soil. Burials in dry sand may result in mummification, or preservation of dry bone, but are a good medium for burrowing scavengers.

This preservational difference also relates to other buried materials, such as suspect and victim footprints, digging marks and weapons. At the macro-scale of burial, Hanson (2004) gives some excellent examples of mass-burials in the former Yugoslavia, where burials that may have been re-exhumed, robbed and possibly further mutilated (accidentally or deliberately) were often covered (by mechanized diggers) by up to 3 metres of imported soil. All these events could be ascertained by study of the stratigraphy and scraping marks of the sites: Hanson notes that these forms of 'trace fossils' – the marks of former activity (the grave 'cut') – are very useful tools in interpreting events in a burial, because they record what happens on surfaces, and in stratigraphy it is the *surface(s) between layer(s)*, rather than the infill, that is important. Hanson argues that arbitrary excavation of the grave by excavated levels, or spits, is best avoided, because the archaeological interest in sieving and analysing the sediment fill de-emphasizes the importance of the boundaries between these layers. Geologists would be in total agreement with this philosophy, as Derek Ager (Ager, 1973) has said 'the gap is more important than the record', the gap being represented by a surface.

Excavation of a grave is best achieved by removal to successive surfaces. If geophysics can predict these, all the better, although few geophysical techniques have the capability of millimetre-scale resolution. The sieving of evidence is another point of some contention between archaeologists and geologists: the former maybe searching for artefacts, the latter for a record of events in the soil/sediment. Post-burial activity can have the opposite effect, be it human in the form of grave robbing and landscaping, or natural, in the form of animal and root burrowing. Hanson (2004) cites a (1981) grave in Guatemala where such burrowers had destroyed all the stratigraphy presumed to be originally present. The import of soil is less common in single inhumations, especially murder victims, although this does happen when ground is hard, depth to bedrock shallow, and thus extra material is required for the efficient hiding of the victim. The wholesale import of soil to mass graves is known from Bosnia. Perpetrators may add carpet, clothing, lime and straw in attempts to obscure the victim and lessen odour emanation: some of these additives may have the reverse effect to that predicted. *Geologists and geophysicists are trained in exploration – the discovery of subsurface features: a body, wrapped or unwrapped, with or without additives, remains a target for discovery. Current or available methods may not be able to detect the body, or it may have degraded completely ('returned to earth'), nonetheless the body is or was there, and any associated anomaly ought to be detectable.* Hanson (2004) sees the buried object (in his case, inhumations, but equally applicable to any buried material) as part of the stratigraphy, especially in cases of mass burial where bodies are intertwined and the sequence of addition impossible to determine, possibly because the deposition of bodies was virtually simultaneous, as in the Boxing Day Asian Tsunami.

When Boyd (1979) wrote his bulletin on 'Buried Body Cases', the importance of recovering geological evidence had not been widely discussed beyond the publication of Murray and Tedrow (1975), who commented on how soil foreign to the final body deposition site found adhering to a victim indicated earlier, geographically separate, contact with the ground. Boyd (1979) lists those who should be present during the search, excavation and recovery of a buried body. They include (in rough order of entrance to the scene): evidence technician, pathologist, archaeologist, anthropologist, odontologist, toxicologist, psychiatrist, entomologist and botanist. The examination of soil, sediment, stratigraphy is well-practised by archaeologists who will be busy with recovery, and thus it is imperative that someone suitably qualified to describe soil, sediment or loose rock work closely with the archaeologist at this point.

Notes must be taken of events that could impact on this analysis: a classic example (we note, not Boyd) is where the surface layers of an excavation were very different to the material hosting the body. Some of these surface layers were dislodged from the side of the excavation and became stamped into the base of the grave, where they were noted, suggesting movement of suspect or victim in the grave, when in fact the pit was dug by mechanical digger prior to body deposition: no one had stood in the grave prior to the pathologist and archaeologist. Thus the surrounding area, as well as being searched by trained personnel, should be described and preferably mapped by a soil or sediment specialist. Boyd notes that the dimensions of the suspect grave prior to excavation are critical: if an adult is suspected to be buried, the host pit must be over 2 metres wide (Boyd suggests 6 feet) and 3 metres long (Boyd indicates 8 feet). The excavated soil holds many clues to events in the grave: Boyd suggests that here the botanist examines both buried and living vegetation for signs of age and damage; the entomologist will need to search for insect life. They should, in addition consult with a stratigrapher (see above), and successive samples taken prior to any police officers sieving the removed material for weapons, shell casings, displaced items from suspect or victim. The sides and base of the original grave must be excavated with extreme care in case tool marks and again displaced items, weapons, etc. are preserved.

Case study in macro-scale above-ground evidence: Mining fraud (Abbott, 2005)

Abbott (2005) provides a simple example of the use of stratigraphy in loose materials, and his description of what happens in many cases of mining fraud cannot be bettered.

> A number of the cases I investigated are known as 'dirt pile' cases. In order to avoid the statutory definition of a security, which includes the phrase 'undivided interest in oil and gas or other mineral right', investors buy a specific very small volume (a few tonnes or cubic tonnes) of ground or a specific pile of 'ore' that is guaranteed to contain a specific amount of gold and other precious metals. If the pile or volume does not contain the minimum guaranteed amount, enough additional material is added to an investor's pile to make up the difference. Because the investor 'owns' his dirt pile, he could come to the site and mine and process it, or he can hire a supposedly independent contractor to do the mining and processing to recover the precious metals. Investors invariably take the latter option. Typically, each dirt pile has a wooden sign with a serial number so that an investor can identify his or her individual pile. In the case of an abandoned mine in Swandyke, Colorado, the piles were composed of the tailings – rock deemed useless by the miners. Because the tailings did not actually contain valuable minerals in quantities large enough to be economic, the promoters in this particular case salted the surface of the piles with fool's gold, or pyrite-rich samples. Another company promoting such ventures produced a marketing tape showing how these investments worked. The video was titled: 'How to Turn Dirt into Gold.'

Abbott leaves the reader to interpret subsequent events at the site, without explaining that unless simple stratigraphic principles are followed, especially the Law of Superposition (one layer above another will be younger), the 'salting' of the pile would have been far harder to detect. This type of fraud is probably as old as exploration for precious materials itself: Rushton (2004, see: *http://www.geolsoc.org.uk/template.cfm?name=HOGG0954895486*) recounts how entrepreneurs drilling for coal deep below southern England in the late 19th century underestimated the depths needed to drill and ran out of money. To obtain further funds, the cuttings from the drilling were 'seeded' with coal, convincing financial backers that success was imminent. Instead, Silurian shales were penetrated, with no Carboniferous coal to be found. As Rushton dryly adds, the borehole core did recover some excellent Silurian fossils (graptolites). An expensive way of obtaining these little fossils that can be found within a few hour's drive of the borehole in Wales!

Case study in above-ground evidence: Search for victims and weapons in the western Middle East

The details of this mass murder are intentionally vague as the status of this investigation is currently unknown. The details of the background are irrelevant as the unfocused nature of the search, the compromise of evidence and the problems of sediment movement in a highly dynamic landscape are being demonstrated. The problems of ethnic-based violence and genocide in some countries of the Middle East are currently well documented (Iraq, Turkey: see *http://www.brainyhistory.com/events/1993/july_5_1993_168481.html* for an example).

Following a raid by a guerrilla band allied to a minority population, over 30 people were killed and one of the local leaders was kidnapped, presumed murdered. Some months later, fearing his own safety, a goatherd reported suspicious activity on the day in question in an adjacent valley (Figure 2.15). Between the report and the abduction, winter rains and a minor earthquake had occurred. The remote terrain and problems of violence, plus the perceived unreliability of the goatherd's testimony, led to the victim's family and community organizing their own search of the valley in question. Their aimless search concentrated on the valley where soft ground and vegetation occurred, with no results.

A later, humanitarian-based search relied on landscape interpretation (the so-called 'scenario-based searching' of Mark Harrison [personal communication, October 2007]) and offender profiling, with very different results. Instead of the victim being buried in valley, his remains were found under a fresh scree slope. This rock avalanche had occurred not long after his murder (his body was left next to a small bush, in a makeshift hollow, probably the start of a grave). The avalanche followed both the rain and earthquake. The murder weapon, an old revolver, was found wrapped in sheepskin in the hollow of a tree on an adjacent slope, a marker for the perpetrators both of their victim's location and that of the weapon, in case retrieval was needed. This formed the focus of the search (the so-called, 'feature-based searching' of Professor Mark Harrison MBE, personal communication, October 2007). It was later found that amongst the search party were probable accessories to

Figure 2.15 The local population-led search of the valley where a community leader was buried. The villagers can be seen in the valley, when the remains were found under the avalanche slope

the crime, if not the actual killers (locals to the area, recommended to the villagers). Their presence was to observe the 'search' and to guide people in the wrong direction. This had the unfortunate side effect of them being present at the likely scene of the murder, denying the use of physical evidence transferred both from the scene to their clothing/footwear as well as the possible deposition of their DNA close to the victim's remains.

Case study in micro-stratigraphy: Lateral variability in soils recorded in micro-scale footwear tread evidence

Georg Popp showed how the different materials found adhering to a suspect murderer's trousers recorded his presence at a scene of crime and a victim's house: the material captured in shoe-tread can record the movement of a suspect. This principle was tested and described by Locard (1929) and Palenik (1982). This case study documents a typical scene of crime (a quarry track in Co. Antrim, Northern Ireland) where the identification, spatial sampling, analysis and correct association of suspect shoes and scene is dependent on millimetre-scale stratigraphic variation, for brevity here termed 'microstratigraphy'. (The issue of stratigraphy has been examined in forensic archaeology (Hanson, 2003, 2004, see above) at the centimetre to metre (depth) scale.)

Homogenized footwear-tread samples would not link the examined 'suspect' boots to any of the footwear tread locations. Separated layers from the boots demonstrated the passage of the boot-wearer from, and possibly to, the designated scene. The basic geological principle of superposition of strata as a record of events is usually considered in soil or sediment layers but the principle is equally applied to the layers of rock dust, mud or soil in footwear tread.The spatial collection of samples forms part of the exclusionary process (see Chapters 5 and 6) in establishing (a) how common the suspect (or questioned) samples are generally and (b) establishing whether a suspect was at a scene or merely in the vicinity. As we shall see, the established analytical methods for the comparison of soil, rock and other materials as defined by Skip Palenik (in Murray & Tedrow, 1992) include colour, grain-size and petrography as well as many more sophisticated techniques. Here we use simple methods to describe the uses of microstratigraphy.

Microstratigraphy background. Paramilitary groups in Northern Ireland have been involved in capturing and killing both members of other groups (the classic republican–loyalist feuds of the 'troubles', 1968–1999) and their own members (punishment shootings, inter- and intra-group feuds). Some killings are made in urban areas and the victims left as a signal to others. Often, however, to avoid detection the murder victim (or the body) is taken to remote farmland or quarries to be left and usually buried. Thus country tracks (metalled and unmetalled) and their associated fields, buildings and quarries are of particular interest in forensic science as the most likely environments where material that may link scene to suspect will be found. In order to provide proxy analytical materials and control materials, a 'typical' quarry lane and body dumping ground was studied 6 kilometres west of Belfast in the County Antrim countryside, here referred to as the scene. The location provides a laboratory for the study of small-scale variation in soils and sediment, where the range of materials on the ground and adhering to footwear, clothing and vehicles could be studied.

Micro-stratigraphy field and footwear sample methods. A pair of moderately-worn (one-year-old) boots, cleaned of visual debris (there will be some remaining of course: 'every contact leaves a trace') with cleated soles were worn by a volunteer, who visited the site once in climatic conditions when there was night-time dewfall, with damp roadway conditions (early May). When touched or pressed, soil was moist but no standing water occurred.

The location was walked from the road access (Figure 2.16) to the scene, and the return route was run at a brisk pace with no concern as to where each footfall lay, replicating witness reports and crime scene reconstruction of a real crime at a similar location. The sample conditions thus comprised a concrete lane, with a variable-thickness soil veneer (10–30 mm deep), in which there were 52 footprints from the suspect. The site was visited one week after the visit, by the authors, during which time there had been two days of light rain, one day of sunshine and the remaining days comprised mixed weather conditions. The volunteer suspect's footprints were clearly visible along the laneway, these provided 6 of the 10 sample groups analysed, the remainder being unmarked areas of likely access (gaps in hedges, areas of low vegetation).

Each footprint was excavated sequentially; different layers were removed with regard to stratigraphic variation, using a dental pick and disposable plastic spatula to the road surface, over the area of the footprint or a mapped area of 20 cm (length) by 10 cm (width). Slope direction (aspect), moisture, colour, local lateral and vertical

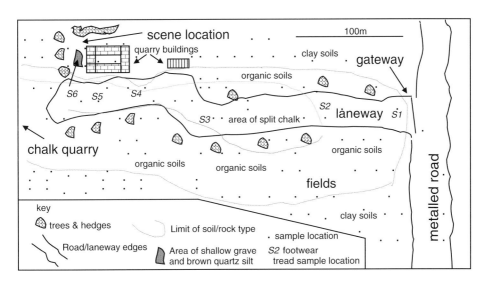

Figure 2.16 Simplified map of the replicated scene of crime (the scene referred to in text), a laneway leading to a chalk quarry in County Antrim, Northern Ireland

variation, and textures were all noted at each visual stratigraphic level. If no colour, grain size or sediment layering was detected, then different layers at 2 mm interval were arbitrarily sampled. This vertical resolution can of course be increased or decreased, depending on time, resources, thickness of soil or sediment, predicted need for subsequent samples, and associated analyses. As this is a case study, designed to test whether movement at a scene of crime can be recorded in footwear, and given that the boot tread at each location averaged 7 mm, 5 samples were taken at each location in order to establish whether gross variation was present. Total weight of the sample varied, depending on rock, soil, organic and other content. If the footwear mark left any unique prints in the material (sediment, rock dust) on the road or surrounding soil, such as cut impressions to the rubber or areas of uneven wear, these were noted as priority footwear sample points. All footwear tracks were sampled using the same methods, regardless of whether the track showed movement into the scene or return from the scene. Samples were stored in cool conditions until examined at a laboratory. On leaving the laneway access to the scene, the boots were worn by the wearer (walking) for another 300 metres to a car and thence into a house, where they dried until 'seizure' was replicated.

The car floormats, house drive and carpets were clean but not given any special treatment in case debris from these locations was detected in the boot tread, because this material would add to the reconstructed history of movement of the 'suspect'. The boots were placed in pristine plastic bags, sealed and taken to the laboratory. The treadwear of both boots was filled with soil to some degree.

Micro-stratigraphy laboratory methods. The resultant samples were split and subjected to visual examination and description. The results from each footwear and laneway sample were similar enough that for brevity they are presented together below. All sampling, preparation and analytical procedures were otherwise identical for all samples. Soil embedded in the footwear tread was picked from the boot using a clean dental pick (Figure 2.17). Sediment was removed sequentially, as each visual layer was encountered. One area was not completely excavated in order to demonstrate visual microstratigraphic variation (Figure 2.17). Once photographed, the boots were completely picked clean of material in tread: any grains lodged in cracks and tread corners were also excavated: in this case study such grains were neither abundant nor distinctive enough for inclusion in the analysis. However, this is a simple case study, designed to raise awareness of small-scale stratigraphic variation. It is possible that single mineral grains in other scenarios could provide links between scenes and suspect.

Each sample was dried for one hour in an oven at 60 °C. The examination sequence used naked eye, binocular microscope (Kyowa, 3–10x objectives and U/V A and B illumination) with regard to colour, texture, grainsize, grain sorting, roundness, mineralogy (where possible), organic content and unusual materials (fibres, glass, paint, asphalt).

Micro-stratigraphy results. The six samples from the lanes and scene showed obvious visual differences that allow division into three groups of material (Figure 2.18). Samples from the gateway comprised dark brown, organic-rich soil with few visible

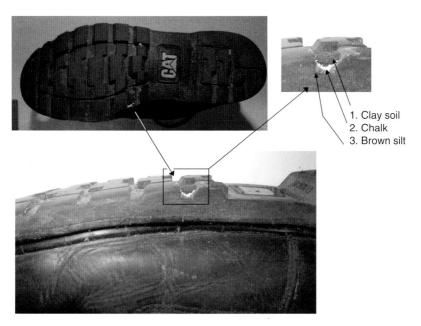

Figure 2.17 The replicated suspect footwear, photographed after most of the tread-infill and covering soil and rock debris had been removed, leaving one area partially excavated to show the simple stratigraphy recorded in the tread. Layer numbers relate to the microscope photographs of Figure 2.18

mineral grains and abundant grass (Figure 2.18). The laneway samples comprised visually-similar soil with variable quantities of crushed chalk (Figure 2.18), defined as a white, friable rock that effervesced with application of dilute HCl observed under binocular microscope. The quantity of chalk increased toward the scene because bends in the lane and uneven ground cause the quarry operators to spill material. The scene itself is dominated by a brown silt (Figure 2.18), with minor fine-grained organic matter. The boot-tread stratigraphy was complex, with an impersistent inner layer of organic matter, brown silt, chalk and abundant outer layer of organic matter. Upon 'seizure' the boots were variably covered with brown organic mud, with some calcite lime visible on boot tread edges. Upon sequential clearance of material, the deep clefts of the boot tread showed a simple stratigraphy of brown silt, chalk and organic matter: this replicated the sequence expected from the lane on the return journey of the wearer (from scene to gateway). Thus correct sequential logging of the microstratigraphy was essential in determining variation. In four clefts of one boot and in three of another, dark organic mud and chalk was found below the brown silt that usually comprised the basal layer of the microstratigraphy.

Micro-stratigraphy case study conclusions. The value of sequential, microstratigraphic-based sampling and analysis is here demonstrated. Consideration of microstratigraphy allows comparison of the mapped distribution of soil and sediment types around the scene (Figure 2.16) with that preserved in the clefts of the

Figure 2.18 Sequential samples taken from the laneway to the replicated scene of crime. 1 comprises the organic mud that dominates samples S0, S1 and S2 (Figure 2.1). 2 comprises the chalk that dominates laneway samples S3, S4 and S5. 3 comprises the quartz silt that is found adjacent to the quarry buildings. Graticule in each picture is in millimetres

suspect's boots (Figure 2.17). In addition, the occasional presence of brown quartz silt and chalk below the more ubiquitous three layers may indicate the adherence and preservation of material *en route* to the scene, as opposed to the record of sediment on return from the scene. The presence of brown silt and chalk below the usual stratigraphy may alternatively indicate post depositional disruption to the normal microstratigraphy due to impact on walking. Other, more sophisticated analyses (palynology, scanning electron microscopy, chemical analysis of small grains, isotopic analysis) could be deployed at finer resolution to give greater accuracy to the results, providing further information concerning the suspect's movements in this case. Footwear samples not taken with regard to microstratigraphy (homogenized) could link to a similarly homogenized series of samples from along the route from the gateway to the scene. However, in a real scene of crime (SoC) investigation, this sampling strategy would be unlikely to occur, with separate footwear tread, tyre mark or general control samples being taken. Homogenized footwear tread samples would comprise a mixture of all three scene mineralogical types, with no comparison between any one area of the scene and the footwear. The concept of variable sediment capture in the accommodation space of each boot cleft, or microstratigraphy, provides a small vertical-scale record of lateral movement through to the SoC. Comparison of the two sequences of materials allows sense to be made of the boot microstratigraphy and the movement of the 'suspect' to be replicated, just as Georg Popp showed over 100 years ago (see Chapter 1).

2.5 Hydrodynamics of rivers, lakes, estuaries, seas and oceans

Streams and rivers

According to Strahler and Strahler (2002, p. 471): 'Most of the world's land surface has been sculptured by running water'. The implications of this statement are that a knowledge of the landforms (see above) and the actions of running water are essential in any understanding we wish to attain of the shape and activity of the Earth's surface. In previous sections we have seen how the action of running water has sculpted some parts of the Earth's surface: we now examine the movement of water itself, something both the search and sample criminalist, as well as the environmental forensic geoscientist, has to embrace at some point or another.

The work of streams is of interest in that although shallow and narrow, streams are capable of having large covert objects (cadavers, packages) thrown into them and of supplying financially-significant volumes of water. Water activity in streams may be summarized as erosion, transportation and deposition. All three are of great interest to the forensic geoscientist because erosion exposes buried objects and relates to changing water flows, transportation means that objects and sediment will be moved from their original site and deposition will cover objects of interest and also accumulate possibly valuable sand, gravel, precious placer (see Glossary) minerals and other objects. Transportation is of particular interest because objects or materials associated with covert criminal activity will be moved by streams: assessing how far, how large (the largest object that can be moved) and over what distance is critical in many criminal investigations where evidence is found in streams and rivers.

The United States Geological Survey has for many years pioneered research into water discharge and the load it can carry. As a consequence, flow in streams and the associated carrying capacity is well-known for natural materials: the addition of unusual objects, associated with criminal activity, is often harder to predict until an experiment is undertaken that caters for the previously-unseen variables. For instance, the erosion of arable farmlands is of major concern, and thus measurements of suspended and saltated sand, silt and clay loads are well-known and can from this be predicted. The introduction of, say, a wooden crate filled with explosives or a dead body, both of which float, then sink, break up, or then float again cannot be predicted from measurements of sediment movement, forcing experiments that allow prediction of the 'what, where and how' associated with these types of material.

River activity can be considered in much the same way, except that the problems and opportunities for the forensic geoscientist are amplified. What do we mean? Well, in terms of problems rivers are larger than streams, allowing the submergence or removal from sight of large objects (large submerged or sunken packages, objects, cadavers). In terms of opportunities, there are often fewer rivers than streams in any given catchment area, making the searching of downland rivers, although technically problematic, easier in terms of ensuring comprehensive coverage. More complicated are those upland rivers whose bedload is so high that they are braided. The multitude of streams within the river channel are continually choked with sediment, causing them to switch course, the former stream filling with sediment. Likewise, the downslope, meandering alluvial river, although simple itself, has numerous associated landforms (back swamps, levees, oxbow lakes) and vegetation that pose challenges to the searcher. The approach to such an environment is to map the area in a similar manner to slopes (see the above discussions of landform mapping), defining domains or areas with similar characteristics. Again we reiterate that although this division of the area is to some degree a psychological device, enabling scientists to believe that they have imposed some control on the landscape, the divisions do contain individual characteristics that are characterized by sediment type, waterlogging or access possibilities. These can then be used to inform representative sampling, exploration methods or suggest where valuable materials may have been illegally extracted.

Lakes, seas and oceans

Non-flowing or slow-flowing bodies of water are all subject to the same water-moving forces of nature. Waves (from wind action on the water surface), tides (from the gravitational pull of the sun and moon) and chemical or temperature-driven circulation or stratification all occur in such water bodies. What distinguishes the dominance of such forces depends on the size (geometry), chemistry and temperature of the water body. Thus it is a common misconception that lakes and small seas have no tides: they are in fact microtidal, sometimes to a negligible extent. The importance of the various dynamic processes operating on water varies with the interest on which each investigation focuses. For instance, in the open ocean, predicting the movement of a floating object (say a wooden box or plastic barrel) will require consideration of wind movements because such an object, with significant above waterline area, acts effectively as a sailboat (Fennessy, 2006). A submerged object, such as a cadaver or mine will be less affected by wind and waves (unless close to surface) but will be moved by tides and deeper ocean currents (Hardisty, 2003). Objects that are in water for long periods change their buoyancy. Most commonly this results in waterlogging and sinking, but in the case of cadavers, can include the generation of decomposition gases that cause flotation. Just as the stratigraphy of buried objects on land has been appreciated, the stratigraphy of lakes, oceans and seas may assume importance: it seems likely that the saponification of corpses causes them to regularly remain suspended at a steep temperature gradient in water bodies known as the thermocline. This changes with seasons and circulation, but nonetheless may hold the key as to why drowning victims, etc. can be discovered some considerable time after their disappearance (M. Harrison, personal communication, 2007). The nature of the problem is therefore of major concern in studies of open water bodies. To demonstrate why lake, sea and ocean dynamics can be useful in forensic studies, we utilize a series of case studies.

Case study: The role of coastal and karst geomorphology in the D-day landings: Utah and Omaha beaches

Intensive studies were made of the D-day landing beaches from the initial occupation of France until April–May 1944 (immediately prior to the combined coastal and aerial assault, Operation Overlord). Although tactical reasons of routes, distance from reinforcements in NE France and Germany, etc. dictated the overall position of the invasion in Normandy, the specific landing beaches were largely selected on the basis of geography, and specifically coastal geomorphology. Geographic elements included location of towns, lines of sight from German artillery and machine-gun posts, rivers that could be blocked by retreating German forces and routes inland. The ideal beachhead landing sites had a checklist of ideal features. These included tidal and subtidal shore gradients that were neither too steep for landing craft, nor so broad and shallow that men and equipment would have to traverse large distances on open ground, where they would be vulnerable to attack by the Luftwaffe.

Even when such beaches were selected, the nature of the sand was critical because any soft or sinking sand had to be avoided or mapped. Likewise the position of rocks and rocky promontories were critical: some areas of rock were suitable for placing bridges and roadways out to jetties, ships and the Mulberry Harbour, yet extensive rocks would hamper the movement of troops, landing craft and tanks in the early stages of the invasion. The position of rock outcrops was thus critical, and mapped in great detail by allied troops and French undercover agents engaged in covert surveillance, as well as aerial photography from modified Spitfire fighter planes. On all such maps, an important note is made:

> Note to coxswain or navigator. Building landmarks, especially near the beach, may be destroyed before any craft land. Terrain features, therefore, are much more reliable for visual navigation from panoramic shoreline sketch.

(This sketch was often a compressed montage of photographs, taken from submarine in a position about 1,000 yards offshore, where the navigator would have to be at their most alert.) Such charts included contours of beach gradient, tide tables and sun/moon conditions and likely current conditions. This demonstrates the importance of a knowledge of coastal landscapes and processes to the planning and success of the D-day landings.

Some veterans of the invasion mentioned the variable accuracy of such maps: some of this is explained by the actions of the violent storm that delayed the operation, which must have displaced quantities of beach sediment, covering rocks in some places and exposing them in others. In basing their beach invasion on coastal geomorphology, and communicating this via a cognitive map, the Allies were actually following some of the elements of the highly successful German method of Blitzkreig, as modified further by the Russian Red Army in 1942. As Rees (2005, p. 265) quotes from an ex-Russian soldier: '"I have to admit that we learnt to fight from the Germans," says Tamara Kalmykova. "Specifically, in coordination of troops, reconnaissance, communications and *cartography*"' (our italics). At the end of Chapter 7 we comment on how geoforensics is assuming international importance in investigation of genocide: here we see the balance to this, that without geographic, topographic and cartographic skills, the pattern of warfare through the German invasions and Russian/Allied counter-invasions of World War II would have been very different. Geography in these contexts was used in assisting war, rather than helping solve crimes and assisting humanitarian ventures.

A second feature of the coastal geomorphology of Utah and Omaha beaches concerns the geology of the cliffs. To the east of Arromanches, interbedded Jurassic mudstones and limestones dominate, forming low, crumbling cliffs. To the west of Arromanches, at the limit of the invasion beach, mudstones pass into thicker, more resistant limestones of the Pointe du Hoc. This area caused a major problem for the Allies because a major German artillery battery was sited on the point, with clear line of fire over Utah Beach. The plan was to reduce as much of this battery as possible by aerial and battleship bombardment, followed by assault by the US rangers. What amazed reconnaissance following bombardment was how the cliff top appeared heavily cratered, and yet with all artillery batteries and bunkers intact. In fact, on closer examination the craters were vegetated and occurred to a depth (over 20 metres) far too great for explosive shells: these were in fact karstic weathering features, or dolines, mixed in with some bona fide bomb craters! The Rangers used ropes and folding ladders to successfully scale the cliffs and reduce the battery.

2.6 Geography, geomorphology, geological and soil maps, and other resources

A surprising aspect to the conduct of geoforensics has emerged in recent years, albeit one used by Heinrich in his work on the Loren (aka d'Asquith) murders (see Chapter 8). This is the use of soil, sediment and geology databases (most commonly, geological survey maps) in a variety of roles. The first role is, as described in many cases throughout this book, to assist in the search for where material came from. For example, mud, sediment or rock appears in a suspicious location and the investigator wishes to find out where the material was picked up and thus continue a geographic chain of enquiry. Another role of such maps is to give context to understanding the landscape. This can help avoid mistakes of interpretation (as in the D-day landings example, above), when those unaware of geological phenomena interpret them incorrectly, or the maps can help understand and plan activity. For example, say a burial is suspected in a region: examination of route, topographic and geological maps will greatly facilitate the search. Once checked for their veracity, and that they do display the accessible routes into and out of an area, route maps provide the focus for how people, materials, etc. got into and out of an area. Topographic maps can be combined, and provide information on whether activities are possible, unlikely or possible, given terrain and distance. For instance, is it likely that a suspect drove to the end of a track (route map) and carried a large quantity of illicit drugs high onto a mountain in a given time period? It is possible, but maybe downgraded compared to the route (on foot) to a nearby river/bog/sand quarry/abandoned house.

Geological maps are commonly provided as solid and drift. Both are of great value is assisting understanding of the landscape (Figure 2.19): areas of soft ground (burials), permeable strata (pollution pathways) and horizons with valuable minerals (mining fraud) may all be identified.

The solid map shows what rocks types occur where, and can indicate the locations of caves, old mine workings and subsident ground, as well as parent geological materials for soils. Drift maps show the extent of (generally loose) overlying materials, and thus possible digging locations, which are likely areas of loose material transfer to suspects. The combined use of solid, drift, soil and topographic spatial geological information is used in understanding hydrology/hydrogeology, and thus the flow of both pollutants and decay products (cadavers, explosives, drugs). Underground fluid flow can have the most unlikely impacts, as we shall see with our case study (below). Other resources may also be useful – archived aerial photography (and increasingly, satellite imagery) will show how the landscape has changed through time. Local knowledge in the form of museum staff and collections, learned societies, etc., can provide critical information that the geoscientist would otherwise

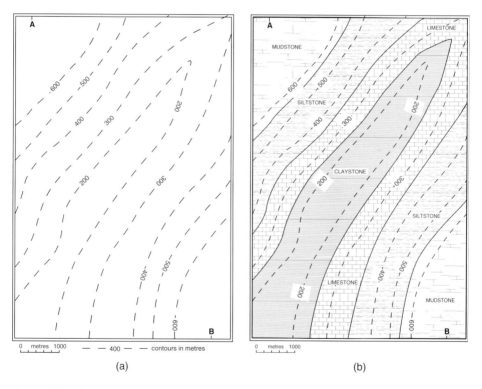

Figure 2.19 (a) A simple topographic map and (b) a geological map of the same area. Note that geological strata in the location are horizontal, or flat-lying (cross-sections in (c)), and show a distinct relationship to topography. This is because hard, erosion-resistant layers tend to form uplands and soft, easily-eroded areas form valleys and lowlands

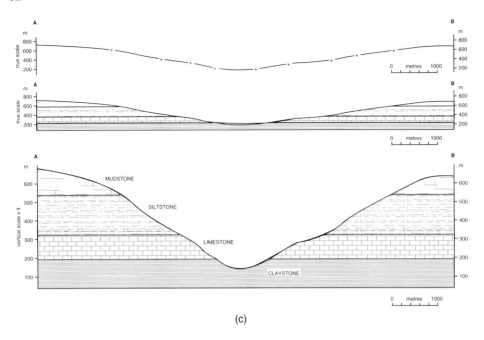

(c)

Figure 2.19 (continued)

take weeks to discover. The combined use of solid, drift and topographic maps in assisting operations of search and sampling reflects the simple application of basic Earth science techniques to 'forensic' problems. This is a central theme throughout the rest of this book.

Case study: The importance of the geological map – the case study used by Lee (2004), after Williams and Aitkenhead (1991)

Lee's (2004) paper concentrates on the philosophy of geoscience, forensic investigations, and the teaching of both. His case study provides a fantastic example of just one use of the geological map in helping to provide the solution to a major event. The background is as Lee states: 'At 0630 on 24th March, 1986, a bungalow at 51 Clarke Avenue, Loscoe … (Derby, UK) … was completely destroyed by an explosion when the central heating switched on automatically. The three occupants, although badly injured, were lucky to escape with their lives.' He goes on: '… after the explosion, gas samples were taken from the collapsed basement and were found to contain methane and carbon dioxide.' In the absence of a gas mains leak, domestic gas source, or evidence of malicious activity, an underground source for the methane was considered. The area does indeed overlie former coal mine workings, with a 30 metre deep landfill below-ground and down-dip of permeable strata: mining activity has relieved pressure on intervening layers, 'stretching' them and increasing their permeability. In addition, from three years prior to the event, a tree died, garden lawns became brown and hot, smells of gas were noted in houses along the geological strike (see Glossary) to Clarke Avenue, and rumbling noises were heard. Lee (2004) uses these facts to ask a number of questions as part of a problem-solving exercise for students, the basics of which were to discount all methane sources bar the coal seams and landfill, with the latter favoured from gas analyses. Combining geological maps (spatial distribution of rock layers, soils and landfill or built-up ground) with cross-sections (showing the vertical arrangement of rock layers, etc.) allows the student in this case study to imagine where coal-seam gases versus landfill gases may flow.

Case study: Geomorphology, the distribution of geology and the British Western Front (1914–18) by Doyle and Bennett (1997)

Doyle and Bennett (1997) show the macro- (kilometre to hundreds of metres) through to meso- (metres) scale importance of geology on military activity in that part of the Western Front known as the Somme-Flanders Lowland. Lowland is a rather non-specific term for this area, it being comprised of either rolling chalk hills (Somme) or plains (Flanders) underlain by clays and sands. The small hills and low, broad valleys of the Somme dictated military activity, with elevated ground as strategic in defensive position or as locations for advance and attack, while the intervening lowlands were major access routes, providing some cover, protection and access to inhabitation (water, food supplies). Thus many marching platoons were susceptible to attack when using only the lowland access roads, yet were visible and without roads on higher ground.

Topography, controlled by geology, was thus a major influence on military tactics at the large scale. In the clay of the Flanders Plain, such topography was non-existent: the location of the enemy was obvious and so topographic protection had to be created by the digging of trenches. The nature of the clay was of key importance in trench construction, and where lacking, access to aggregate (drainage) and woodlands or buildings (wooden props and walkways) was critical. Although the engineering features of the Flanders and Ypres clays were not analysed in the same way as would be approached today, one cannot help but feel that the soldiers who dug, and re-dug their trenches and bunkers learnt more about the engineering properties and centimetre-scale variations in these clays and sands than we will ever know.

2.7 Groundwater

This book is not just about criminalistics in the sense of robbery, rape, murder and genocide. Environmental crime can cause far more deaths or maimings than these acts: a good example is given by Isphording (2006, see *www.geolsoc.org.uk/pdfs/FGtalks&abs_pro.pdf*) who showed how the release of a suite of toxic substances into groundwater near Penscola (Florida) resulted in one of the largest legal settlements in United States history. The illegal abstraction of groundwater is one of the major issues dealt with by environmental forensic geoscientists who usually have a hydrogeology background. Their work may be used to decide whether pumping from one land area denies someone else an adjacent resource, or likewise whether release of toxic materials can be related to degradation of groundwater.

This problem is not new: the importance of groundwater in history is enormous. Military campaigns in the desert are often based around known or predicted groundwater sources. The mythical and religious significance of groundwater has also led to investigations of its true source and nature. Mather (2004, see: *http://www.geolsoc.org.uk/template.cfm?name=HOGG0954895486*) considers the current vogue for bottled groundwater to be the ultimate fraud. There is no attempt by the vendors to cover-up the nature of their product; instead, the description that water is 'millions of years old' or 'filtered through volcanic rocks' is used to imply some special feature of what is very ordinary groundwater. This contrasts with historical claims for the health-giving, curative or mystical and religious powers of springs.

3
Geophysics

> I lingered round them, under that benign sky: watched the moths fluttering among the heath and hare-bells; listened to the soft wind breathing through the grass; and wondered how any one could ever imagine unquiet slumbers for the sleepers in that quiet earth.
>
> Emily Bronte (1818–48) *Wuthering Heights* (1847), last passage.

The 'Geo' in this book's title relates to the Earth, observations of which have traditionally been made by direct observation of surface and shallow materials, observations of the whole Earth and neighbouring planets (geodesy) and by applying theoretical and direct measurements of physics, the realm of geophysics. Geophysical measurements have traditionally been applied to shallow (metres) to deep (kilometres) scales of investigation, with some techniques having more appropriate applications to scale or type of information required. Appropriate geophysical tools depend on both the chemical/physical nature of the target versus its surroundings (e.g., a car buried in a sand-pit), as well as the size or area of the target versus the area to be searched (target to area ratio), and the technique's ability to image. Although many traditional forensic applications have used the investigation of shallow geophysical properties, whole Earth and deep geophysics also have a strong part to play in our fundamental forensic questions – using science to discover what happened, when and how. As is true for physics, geophysics may be split into that pure branch where whole Earth properties and the maths behind them are explored, and applied geophysics, where theory is put into practice: it is the latter branch that concerns us here. Reynolds (1998) makes a useful distinction between the various applications that geophysics has to environmental, engineering, exploration (minerals, oil, gas) geology, hydrogeology (and glaciology) and archaeology. The application to humanitarian or criminal forensics overlaps with nearly all these applications, largely because of the need for the investigator to know as much as is possible about the normal activity of an area, compared to any indications of unusual events. Fenning and Donnelly (2004) provide further detail on the uses of various geophysical methods, including many of those considered here. Geophysical methods are divided into active methods, in which artificial signals are generated and their propagation or attenuation measured, and passive methods, where instruments are used to measure the natural physical properties of the Earth. Both are used in

Geoforensics Alastair Ruffell and Jennifer McKinley
© 2008 John Wiley & Sons, Ltd

forensic studies: an active method would be types of metal detector, where electromagnetic currents are introduced into the ground and the conductivity measured. A passive method would be a gamma-ray detector, in which a tube or crystal reacts when bombarded by radiation, allowing an amplified signal to be generated.

A key aspect to the application of both active and passive methods is consideration of what the investigator is looking for. Good intelligence allows selection of the most appropriate methods: for instance, the search for a buried canister of highly radioactive uranium would be best achieved using a spectral detector that measures the spectra of uranium-generated gamma-rays. Conversely, in an open field where suspicious digging has occurred, but the buried object is undefined, the site should be investigated with as many devices as possible, given time limitations. This 'blunderbuss [a shotgun that fires widely-scattered shot] approach' (Reynolds, 1998) is neither scientific nor cost-effective and should be avoided if possible: investigations of covert activity are by their very nature without significant background intelligence and thus sometimes require not so much the blunderbuss, but a range of methods. The problem is that in forensic, or disaster-related searches, very little time, background knowledge or equipment is available: yet more victims, the general public and search personnel may be at risk. Thus, throwing all that one has available at the problem (blunderbuss) may at least diminish the following criticism: 'you mean you had a metal detector but didn't bother taking it because it may not have worked?'.

A common misconception of geophysics is that the methods provide some kind of photographic areal or depth image of the subsurface. The laws of physics determine signal generation or acquisition, and so the variability and complexity of the Earth make the received data open to interpretation. This too makes the use of many techniques desirable: witness statements in the form of geophysical interpretations based on different methods used at the same locality may contradict, providing barristers much to talk about and making each expert appear unsure of the 'facts'. Reynold's (1998) advice on matching method to target are however very relevant to cases where time is short. Good examples are structural surveys of unstable ground/buildings and the search for victims following disasters. In both cases, time is short, the physical properties of the hidden objects must be predicted or guessed at, and available, appropriate geophysical techniques used. The same may be applied to environmental investigations where no guilt is proven, thus requiring a search warrant and limited time when the maximum information on the site must be acquired. These 'rules' apply to the planning of a survey: this may vary with technique and accessibility, but is again dependent on what is being looked for. As much background information as is available must be acquired, and whether this is informative or not, the flexible approach to data acquisition is deployed. As stated above, the main reason geophysical surveys are used in the first place is through a lack of knowledge of what lies beneath ground surface. Surprises are better expected than avoided, because the technique and survey can then be adjusted when subsurface conditions become better understood.

Geophysical techniques measure actively or passively variations in Earth properties, and so changes from background readings are termed anomalies, which when defined in profile or on a map are targets. Target consistency is measured by the repeatability of the anomaly. Most geophysicists define this property by the concept of *nodes*, or places where intersecting survey lines show comparable anomalies: if these become consistent or make sense geologically or structurally, confidence in the existence of a feature grows. The *conjunctive* method, where different survey methods using the same technique, or different techniques that measure some shared properties are compared, is highly beneficial: critics of this comparative approach claim that each method uses distinct physical attributes, and thus are strictly speaking incomparable. However, many targets are complex mixtures of materials, from reworked soil to introduced metal or human remains, making for combined geophysical signals. Imagine that one method is used for the detection of human remains, and a body is recovered yet remains unidentified. Adjacent to those remains is an identity tag, undiscovered because no metal detector or magnetometer was deployed. Thus the node approach to making sense of target consistency as well as the conjunctive approach provide a target hierarchy, which can be further developed using non-geophysical techniques, as we shall see in Chapter 4.

Targets can remain obscure, even when anomalies are mapped and methods compared. A reason for this may be the problems of other objects obscuring the target. These may be worth investigating themselves, or they may be the product of coherent noise (electrical pylons, radio waves) or incoherent noise (wind, electrical storms, waves). Another reason for target obfuscation is subtle changes in geophysical properties that are only clarified with data processing. This aspect of forensic geophysics is similar to some of the problems of survey method and design (speed) in that on-site results are often required, limiting the time allowed for data processing and interpretation.

3.1 Seismic methods: macro to micro

There are a number of sub-disciplines of the seismic method, based on the shared principles of how elastic strain energy moves through solid media such as the Earth. Geophysicists refer to this movement in terms of wave motion and velocity, variations that can be used for different applications and interpretations. Seismic energy can be considered in terms of both the passive and active methods described above, passive sources of seismic energy including earthquakes, volcanic eruptions or impacts from extra-terrestrial objects. Both are powerful sources of seismic energy whose timing and energy release are not under human control, and thus require seismic detection equipment to be left recording for long periods. A similar source of seismic energy is that generated by nuclear weapons (an active, if 'uncontrolled' [by geophysicists] source of geophysical energy) detonation, especially underground. The instigators of such tests are the only people who know in advance the timing, location and power of the explosion: those monitoring such activity are in the same situation as those studying earthquakes.

Both types of application drove early work on seismic data, with studies of the structure of the Earth being paramount, but important advances were made in fields akin to forensics. In this case it was German scientists in World War I who used the refraction of seismic waves to predict where heavy artillery guns were located

(Keppner, 1991), allowing accurate return fire without direct observation. The need for oil in the 1920s aided work into the use of seismic reflection methods that allow deep geological structure and sometimes direct observation of hydrocarbons to be found. Indeed the seismic method is still the main geophysical technique used in the search for oil and gas today. The collection of seismic data thus always requires a source that generates elastic energy and a detector, most commonly a geophone. Analysis of seismic waves as well as the recording of reflection, diffraction and refraction can provide subsurface attributes on land or water. Two very broad areas of forensic application are considered from hereon. The main application is the search for near-surface (excavation or tunnelling depth) material changes: in this application, the seismic source is usually generated by humans, using a hammer and plate, explosive charge, pulse of air, electrical spark or vibrational device (Figure 3.1). The reflected or refracted seismic waves are detected by an array of geophones (Figure 3.2). The second, less widely-published application is the monitoring of covert, illegal or contentious seismic activity, especially large detonations. Table 3.1 provides a summary of some typical seismic sources, their usual location of use and some limits.

Figure 3.1 The seismic source. Here a hammer and plate source is being used. Photograph courtesy of Chris Leech, Geomatrix Earth Systems Ltd

SEISMIC METHODS: MACRO TO MICRO

Table 3.1 A summary of some typical seismic sources, their usual location of use and some limits

Sledge-hammer	Drop-weight*	Dynamite**	Airgun	Shotgun***	Borehole sparker	Special guns****	Multi-pulse	CHIRP/GeoChirp	Vibroseis
Shallow, land (on metal plate)	Shallow, land	Shallow to deep, land	Various land and water	Shallow, land	Land, shallow to deep	Shallow to deep, land and water	Deep, water	Shallow, water	Deep, land and water
Slow, needs strength and skill	Slow, needs vehicle	Slow, security risk		Source can be poor, security risk	Only in boreholes			Shallow penetration	Large equipment

*can be accelerated;
** other explosives have been used, including detonating cords/caps;
*** other guns have been used, including bolt-action rifles;
**** includes air-, gas- sleeve- and steam-guns

Figure 3.2 The seismic geophone array. Here a land streamer is being used, typically with a 24 geophone array and weight-drop seismic source (the white trailer to the left of the vehicle). Photograph courtesy of Chris Leech, Geomatrix Earth Systems Ltd

Near-surface material changes

Shallow investigations using seismic energy are best made using refraction data, which can be used to measure both the depth to a refracting layer, as well as the velocity within. Seismic velocity can be converted to real depth and also used as a proxy for material strength and thus how easily rock may be excavated or if there is a weaker section within a stronger layer, such as would be found above a covert excavation. There are various ways of calculating the refraction seen when seismic waves intersect various layers with different velocities: the end goal is the same, that is, improved depth, thickness and layer velocity measurements. Reynolds (1997) describes some of his own work on the exploration of landfills that have similarities to many covertly-hidden sites such as mass graves and illegal dumps. They make critical observations for the forensic scientist who needs to investigate similar locations: seismic refraction works better than reflection methods for both the shallow depths and velocity measurements required, and is cheaper. However, the energy sources used are problematic: the ground is soft and vehicles may founder. Explosive sources may cause methane gases to explode and boreholes for shallow sparkers cannot be drilled. Nonetheless, if the sledgehammer and plate method generates enough energy to penetrate the generally slow velocity fill, then good results may be achieved. A modified form of seismic data acquisition on water is the CHIRPS (high resolution seismic) system, which is advantageous over echo-sounding in obtaining information on objects in the water as well as in the top layers of soft sediment. Variations in sediment type can also be mapped, such that if material has been displaced or moved about, the changes in CHIRPS response may be mapped.

Case studies from Reynolds (1997)

Reynolds (1997) provides excellent case studies where seismic refraction was used on landfill sites, with varying success, some requiring the combined use of refraction with other methods (electrical resistivity) to generate acceptable results. Reflection data have been used in contexts similar to forensic studies. The best-known of these is the mapping of a dinosaur skeleton (Witten *et al.*, 1992). In this case, the bones of a diplodocus-like sauropod were found protruding from a cliff in Jurassic strata in New Mexico (1979). The size of the skeleton made excavation likely, but the orientation of the bones suggested rigor mortis had set in prior to burial, making their location beneath about 2.5 metres of sandstone difficult to predict. A specialized form of seismic diffraction experiment was devised called geophysical diffraction tomography, which uses multiple ray paths from the seismic wave to construct an image of a subsurface object. This experiment was performed at the site using four boreholes around the likely location of the skeleton. These produced an image interpreted to be a large bone, confirmed by later excavation.[1] Witten and King (1990) used such technology to image drums, filled with water or sand, buried at around 50 centimetres, in analogous circumstances to covertly-hidden waste or ordnance. This work was published at about the same time as many of the pioneering works using ground-penetrating radar technology to image similar features, with the latter now being used far more commonly than seismic tomography. There has still to be a published account of a controlled site experiment where the seismic method (especially refraction or reflection tomography) is compared to the radar method for different targets, depths and sizes, in which seismic still has advantages over other technology. Hildebrand *et al.* (2002) showed that seismic reflection could effectively locate a dead pig in a wooden coffin at around 2 metres depth. However, they also used ground-penetrating radar at the same site, with equal success, but completed the ground-penetrating radar (GPR) survey in a fraction of the time taken for the seismic investigation. This perhaps is the crucial point when considering 'forensic' applications: the time allowed for the investigation. Just as Morgan and Bull (2007) discuss how the treatment of physical evidence (soil) differs between forensic and 'pure' research, so this applies to geophysical surveys. Many scientists would complain and say that no matter what the extenuating circumstances, the time needed for a full and proper survey is the time needed, no matter what. However, when a burial site is about to be eroded by floodwaters, personal security is at stake, pollutants are moving rapidly toward a town, or public/international pressure is so massive that the survey must be completed, the time allowed for a survey is always limited.

---

[1] A popularized, if impossible, version of this experiment is seen in the opening sequence of the film *Jurassic Park,* where palaeontologist Dr Alan Grant (Sam Neill) uses a shotgun cartridge source to image (almost perfectly) a *Velociraptor* (of course!) skeleton, commenting, 'soon we won't have to dig 'em up'.

Seismology – widescale monitoring

Rogers and Koper (2007, *http://web.umr.edu/~rogersda/umrcourses/ge342/Forensic%20Seismology-revised.pdf*) have an excellent web site dedicated to a study of forensic seismology. From this we can summarize the need for global- to continental-scale seismic monitoring.

1. *Discovering what has caused a seismic event.* This ought to be obvious, but is often quite a problem. Rogers and Koper show the range of events that can cause earthquakes, including human and natural events. Human events include nuclear weapons tests, nuclear reactor accidents, chemical explosions (from terrorist attack, industrial accidents and mining/quarrying blasts), airplane and train crashes, landslides (e.g., mine spoil heaps) and major tunnel collapses. Natural events include earthquakes, volcanic eruptions, meteorite impacts, cavity collapses, rock falls/landslides and tornadoes. Each of these events has a distinct seismic 'signature' or pattern on the seismogram (see Glossary). For instance, underground nuclear weapons detonation causes shocks in all directions, dominated by seismic energy release as P or Primary Compressional waves. Earthquakes, usually associated with faults (linear fractures of the Earth's crust) have a strong directionality to them, dominated by S or Secondary Shear waves. These appear on the resultant seismogram, as seen on Figure 3.3, where seismic traces from a nuclear weapons test are compared to an earthquake. The main reason for the extensive seismic monitoring network around the world at the present-day is (was) the enforcement of the Nuclear Weapons test Ban Treaty, which required verification. These stations are still used for this reason, monitoring both established (USA, Russia, China, France) and new (India, Pakistan, North Korea, Iran) nuclear states. Global seismology has thus long been used by government agencies to monitor nuclear tests undertaken by other countries.

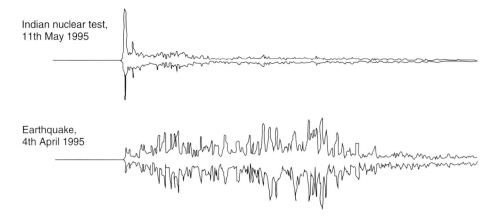

Figure 3.3 Seismograms of a nuclear weapons test and earthquake from India. Re-drawn from original data (*http://www.llnl.gov/str/Walter.html*)

Underground tests have supposedly lesser environmental impact and the transport and installation of devices can be made covertly. However, underground and surface detonations are ideal seismic sources for refraction and diffraction observations around the globe. As the monitoring of earthquake activity uses the same seismic monitoring stations and equipment, and as technology has advanced and recording geophones have been improved, so sensitive and geographically-accurate measurements have been made (Zucca, 2003).

2. *Discovering the location of a seismic event.* This may seem elementary until real events are examined. In remote locations there are few witnesses to give accounts of Earth movement and thus provide a likely epicentre. This was exactly the case in the Gujarat Earthquake (western India; 26th January, 2001), where the remote nature of the area led to numerous problems, including knowing which communities were affected, how many people were dead or harmed, or where the epicentre was. Nonetheless, seismic monitoring stations at least provided some idea of the main areas affected.

3. *Discovering what has occurred during a seismic event.* This is perhaps the most critical aspect to the forensic (what happened, when, how?) use of seismometers. This was first realized when the explosions caused by the 1943 'Dambusters' raids by British bombers on German dams were observed at the seismic station based in Gottingen. The first widely-published test of such equipment was made when a Pan-Am flight exploded over southern Scotland in December 1988, as the result of a terrorist bomb (the Lockerbie Crash). The impact caused an earthquake measured at 1.3 on the Richter Scale (Redmayne & Turbitt, 1990) by British Geological Survey seismometers in Scotland, from where the location of the impact was assessed blind, attaining a 300 metre error. This remarkable result has had important implications for other search operations (Holzer *et al.*, 1996 [Oklahoma City Bomb]; Koper *et al.*, 2001; Koper, 2003 [Nairobi Embassy Bombing]). These include two very important events where a remote observation through seismology was critical in understanding what had occurred. Most relevant to the Lockerbie observations was the Sinking of the Russian Submarine Kursk on 10th August 2000, which was detected by seismic networks in the Baltic area as two events. The first event was smaller than the second event that produced vibrations equivalent to those from a magnitude 4.1 (Richter Scale) earthquake. This information was used to infer that the size of the explosion was equivalent to that which would have been produced by 4000 to 6000 kilograms of TNT. The Kursk data proved that there was no collision or impact. Rogers and Koper also discuss the August 19th (2000) Carlsbad Gas Pipeline explosion (New Mexico). Accident investigators wished to know what had happened in this case, and the seismic traces from nearby monitoring stations showed the pattern of three explosions as one blow-out, followed by two ignitions

4. *Determining the pattern of terrorist activity.* Less relevant to locating the seismic source, but just as important in determining how well-coordinated terrorists are, is the 7th August 1998 bombing of the US embassies in Nairobi, Kenya and Dar

es Salaam, Tanzania, when two truck bombs exploded nearly simultaneously. In both cases the blasts were large enough to cause irreparable damage to both buildings. The Nairobi attack was recorded by a three-component, broadband seismometer, analysis of which allows estimate of the precise origin time of the explosion and the amount of blast energy partitioned into seismic ground motion (Koper, 2003). The Encyclopaedia of Intelligence (2004) shows how similar data can assist in analysis of similar, catastrophic events:

> Seismological data have also been used to help infer the details of 1995 bombing of the Murrah Federal Building in Oklahoma City, the 2001 World Trade Centre attack, and the 2001 Pentagon attack. Analysis of seismograms associated with the collapse of the World Trade Centre towers, for example, suggests that the actual structural collapse occurred over a period of about three seconds. The same principles can be used to obtain evidence of clandestine conventional or nuclear explosions, and in particular to verify that nuclear test ban treaties are not being violated.

3.2 Gravity/gravimetrics

This method depends on the measurement of changes in gravitational pull, or acceleration due to variations in the density of rocks and subsurface materials. The sensitivity of gravimeters is proportional to the size of the object to be surveyed. Thus early experiments measuring the mass of mountains utilized crude pendulum-type devices whose deflection changed as the meter was located closer to the target. Other early gravimeters use a balance or spring mechanism. Recently, the measurement of microgravity has detected voids and rubble-filled chambers only a few metres in diameter. Gravimeters can now be deployed from aircraft, making rapid surveying of large areas possible.

Gravity surveying is tremendously powerful but can be hard to carry out. The problem of masking effects is of major concern: if the target is a small dense boulder, or a drift-covered rockmound underlain or next to a much larger dense body, then subtracting the masking effect requires multiple measurements in different locations or accurate modelling of the density of each object. Variations in rock and sediment density are well-known and alone would provide easy predictive information for gravity surveying. The problem is that the Earth's gravitational pull varies with latitude and centrifugal force, on top of which rocks, soils and sediments are layered or have complex configurations, making the differentiation between the distance from the mass to be measured versus the material below ground (especially in areas of extreme terrain) very difficult. Once applied, these corrections allow the mapping or cross-sectional modelling (depth and mass variation, thus material type prediction) of gravity anomalies. Reynolds (1997) shows how large-scale (kilometres to hundreds of kilometres) exploration of the subsurface has used gravity surveying, yet the smaller-scale applications, especially of microgravity are fewer. Many investigations into covert criminal, wartime or humanitarian burial activities are on a similar small scale to engineering applications. Reynolds states, 'the size of engineering site investigations is normally such that very shallow (<50 m) or small-scale (hundreds of square metres) geological problems are being targeted.' He continues, 'The use of gravity is commonly to determine the extent of disturbed ground where other geophysical methods would fail because of particularly high levels of electrical or acoustic noise'. He cites case studies where back-filled quarries have been mapped and volumes determined (analogous to mass graves, illegal dump sites), underground cavities detected (analogous to covert tunnels or storage) and archaeological features defined (a crypt within a church). The latter has obvious forensic application as a location such as a church interior could provide difficulty for many other geophysical surveying techniques. The difficulty in accessing useful gravity data, plus the cost and sensitivity of the devices, has probably precluded many forensic applications, which is unfortunate as it seems likely that advanced gravimeters and processing could assist in the search for many buried objects.

Gravity/gravimetric opportunities

Most considerations for the use of gravity measurements in 'forensic' work have come from military operations, largely because of the time permitted, the size of the underground voids being investigated, and the need to stay outside/above the perimeter of the area of activity. The latter may be due to the health and safety issues of possible collapse, or because of hostile forces/people/environment. Few domestic or environmental activities are associated with the sort of machinery/workforce required to excavate deep chambers or to engineer/occupy large natural underground caverns. And if this does apply, they are often detectable by other means (e.g., remote sensing, see Chapter 4) or from intelligence. The extremely covert nature of military operations means that intelligence may not be sufficient to pinpoint the location of underground operations, or that very remote methods of detection are needed. Magnuson (2007) suggests that very little technology has yet been deployed in the military location of tunnels and chambers, citing tunnel detection dogs as an effective method. However, Magnuson (2007) is concerned largely with the intermediate scale of human-sized tunnels, rather than the large, covert chambers considered here. An early use of GPR was by the Americans in the Vietnam War to detect Vietcong ('Tunnels of ChiChi') tunnels. Subsequently, successive military forces have also deployed GPR in similar operations. Examples include the search for Hezbollah tunnels under Southern Lebanon (Israeli forces); FARC tunnels under Colombia; various tunnels under the Mexican–US and Canadian–US borders (Department of Homeland Security operations found a one-mile long tunnel near Otay Mesa, California). Magnuson (2007) is perhaps more correct in considering the large voids occupied by weapons production, nuclear engineering and manufacturing plants in Iran, North Korea, the US, China (among others). The non-discovery of weapons of mass destruction manufacturing plants following the invasion of Iraq may be blamed on inadequate geophysical monitoring of the subsurface.

3.3 Electrical

Electrical methods rely on either the active measurement of resistance to induced currents (resistivity) or a subsidiary of this method called induced polarization or self-potential (the latter being a passive technique).

Resistivity

Electrical resistivity is an important property of rocks and soils, especially those containing water. The capability of a material to impede electrical currents over a certain distance may vary in different directions (anisotropy), which can reflect the structure of the subsurface. Currents may be transmitted through rocks and sediments by a number of means, the most important being electronic conduction (rapid movement of electrons) and electrolytic conduction (movement of ions). Dielectric conduction is less relevant here than in Section 3.5 on ground-penetrating radar. The mineralogy of dry rocks, water content of those with pore space and the compaction of rocks and sediments control measured conductivities. These all have great relevance to the searching of ground: an area of altered mineralogy may reflect back-filled excavations where materials have been extracted (legally or illegally), for instance sands containing semi-precious minerals. Water content and compaction in soils and loose sediments may be related: an excavated and back-filled area will contain loose material with higher permeability than the surrounding ground, causing water-logging. This may then compact, or over-compact, with a later opposite effect. Both may be directly observed on the ground, but the depth of such features is hard to establish, where electrical methods may help. Likewise such variations may be subtle, or hidden by a thin, dry, top layer of soil or sand, with the variations below. In this instance, electrical surveying can prove very powerful in delimiting obscure areas of ground disturbance and voids (van Scoor, 2002).

These simple examples belie the mixed fortunes of interpreting variable resistivities, wherein materials in different forms have individual electrical properties, but in simple mixtures, resistivity can be used as an accurate proxy for content. An example is where pure kaolinite clays (low resistivity) are mixed with quartz beach sands (high resistivity) in Cornwall, England. Here, the resistivity of each material is well-known, and the measured values approximate very well to the clay-sand proportion. In over-simplified terms, to measure resistivity, we need to induce a current into the ground using a current electrode, and measure the current recorded at a certain distance away using potential electrode(s) placed in the ground. The type of current, how it alternates and the geometric configurations of the electrodes show that there is far more to this simple statement, but it will suffice for now. Electrode configuration is of major importance to the size of area and likely target. Electrodes may be mounted on survey frames with two current and two potential electrodes with variable layout and spacing (called Wenner, Schlumberger and dipole-dipole – see a standard geophysical textbook for further detail). Other configurations have also been experimented with.

The Wenner and Schlumberger arrays achieve excellent to good vertical resolution in areas underlain by flat layers, with the latter causing curved geometrical results, but can be used for other measurements. The dipole-dipole method is best used for deeper imaging. At first sight then it may seem that the Wenner array is best for forensic surveying, except that it is more labour-intensive than the other two. As we have seen in other sections where access, time and security are limited, this non-scientific aspect of the analytical method can be critical. A second form of electrical surveying deploys a static electrode array that involves the injection of an electrical current into the subsurface using two dedicated electrodes, inserted a small distance (typically a few centimetres) into the ground followed by measurement of the resulting electrical potentials using a further two electrodes, such that acquisition of each individual electrical resistivity data point requires the use of four electrodes. Typically, 24 or 36 electrodes are used, arranged at a constant spacing (1–5 metres) along the line to be surveyed. These data are subsequently processed and analysed using software based on tomographic principles (similar to that used in medical imaging, such as, for example, in MRI scanning). The result is an image showing the distribution of bulk electrical resistivity (the inverse of bulk electrical conductivity) in the subsurface (see the case study on page 74). Each Earth material has a particular bulk electrical resistivity, and test surveys at a known waste sites (Reynolds, 1997) show that the bulk electrical resistivity of household and other mixed waste is typically low compared to that of both soil capping the waste and geological strata located beneath wastes, indicating that many other 'forensic' materials could likewise be located and characterized by resistivity measurements.

Resistivity surveying can generate two types of result, a vertical sounding where depths to different layers can be interpolated (electrical resistivity tomography, or ERT: see the section 'Case study: Resistivity and IP – large (500 metre diameter) toxic waste burial sites' on page 74), or a plan-view of the horizontal changes in resistivity. ERT soundings use electrodes, implanted in the ground, in an array. This can be short (a few metres) or long (a few hundred metres). The principle is the same, regardless of scale: a current is injected at one electrode and detected at one or more further electrodes, providing a measurement of current impedance (resistivity). By using an array, multiple soundings from increasing depths can be overlain, to provide a picture or vertical slice of earth resistivity along the array. The electrode array can then be moved in various ways to generate further depth sections.

The plan-view mapping of resistivity is less contentious, the interpretation often more so! The problem is again one of defining from what depth the variations are coming, although these are (as a rule) shallower than ERT, because the moveable electrode array is frame-mounted and thus has less separation and consequently less depth penetration. Hence this method is very popular in archaeological surveying, where most of the technical advances to the systems and problem ground conditions have been met. It is good practice to map areas and also to carry out some form of depth sounding, in order to evaluate what lateral changes have a strong vertical component (Figure 3.4).

Reynolds (1997) summarizes the main applications for electrical resistivity surveys, and the over-riding importance of the method in hydrology is apparent. This is because the method can be used to map buried water courses and thus predict the location of aquifers and the direction of hidden water movement, but also because in recent years the chemical attributes of groundwater have also been

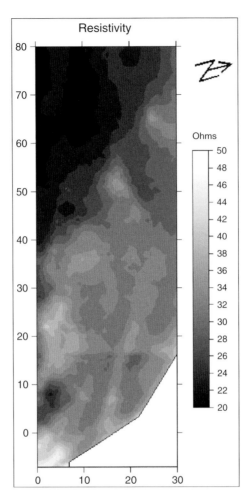

Figure 3.4 A map of resistivity measurements made over a floodplain and adjacent slope (scale in metres). Data courtesy of Paul McCarthy

successfully imaged and mapped. The application of this to studies of pollution is of massive significance, as plumes of contaminated groundwater, moving vertically and laterally, can be monitored with fewer compromising boreholes. Reynolds (1997) cites an additional use for resistivity measurements that does not involve standard arrays: the detection of leaks through membranes. These are most commonly in areas where hazardous liquid or slurry is stored, where a small rupture may, under pressure, cause extensive leakage. The method relies on electrodes being placed either side of the membrane and may find applications in many other situations where remote (i.e., non-invasive) leak detection is required. Other applications often involve the water content of ground, be they mapping of frozen versus semi-frozen ground, the location of landfills, leaking water mains, leaking sewers and ground collapses. Reynolds (1997) also notes that in areas with significant metal (above or below ground) magnetic and radar profiling may not work, but certain resistivity arrays might. This too has obvious forensic applications in problematic locations (Owsley, 1995, see Chapter 2).

Induced polarity

Acquisition of induced polarity (IP) data also involves current injection and electrical potential measurements using two dedicated electrodes. However, in the case of IP, electrical potential measurements are made directly after the current has been switched off (see above: in the resistivity method, electrical potential measurements are made while the current is switched on). In contrast to electrical resistivity (ER) measurements, IP measurements sample the ability of the subsurface to store electrical charge, commonly called the 'chargeability' of the ground. Each earth material has a particular chargeability, and test surveys over known sites show that the chargeability of materials such as toxic waste is sometimes higher than that of the surrounding, 'natural' geological materials, particularly if the waste contains a large amount of metals. For instance, buried cars tend to have a very high chargeability. Induced polarization uses the same electrodes as resistivity surveying, but the principles are different in that it is the decay in voltage between the electrodes, and its subsequent increase on being switched on and off, that is used to predict ground conditions. This decay and rise-time relates to the geological materials in the subsurface, and because of the complexity of both, plus their variation in space, allows a quite complex array of measurements to be made, which can be used to predict variable subsurface conditions. Most applications are in the area of hydrogeology, geothermal energy and metal ore exploration. However, the work of Olhoeft (1985) is of note, because organic contents in the ground were mapped over a few metres lateral and depth distance, equivalent to some of the applications to inhumations that are important in forensic work. The problem of mapping illegal spills and dumps of organic pollutants is also of relevance.

Self-potential

This method uses the electrical properties of the subsurface, but without injecting any current into the ground. It is thus a passive survey method, requiring very little equipment and ideal for remote locations. Self-potential has long been observed in minerals, an example being the piezoelectric effect of applying torsion to a quartz crystal, whereby electricity is generated. As with resistivity applications (above), the dominant control on self-potential is again the presence and movement of water as an electrolyte. Measuring self-potential requires limited invasion of the ground as each measurement requires two electrodes (in pots) to be placed a few centimetres into the ground. A measurement can be made across one location, using just two pots. However, the complexity of any subsurface structure, and the extent to which this can be predicted, requires more measurements to be made. In addition, the electrodes have to be very sensitive (being a passive technique) and are thus influenced by deep Earth currents and the topography of the surface area. The method has been widely used in mineral surveying, hydrogeology (mapping the courses of groundwater) and landfill mapping. The traditional use of the self-potential method has been in mapping large areas and deep targets, as the equipment is basic and light, allowing transport into isolated terrain. This may have direct relevance to forensic work in remote locations where speed and power are important influences on survey design.

3.4 Magnetic and electromagnetic

The Earth's magnetic field is well-known, especially to navigators and explorers. What may seem to be a constant feature of the Earth is in fact highly variable globally, as the magnetic field or flux changes around our planet and its polarization (positive and negative, north and south) occurs. We can see the relationship between this global phenomenon and the individual compass or magnetized needle. Between these two scales there is also variation in the magnetic properties of rocks and sediments, largely because of iron content. This variation is natural, and can be measured by passive methods, or by increasing the magnetic properties by inducing electricity and thus creating an electromagnet. The latent magnetism of Earth materials is known as susceptibility and is most directly related to iron contents in minerals. Minerals are crystalline substances and thus have order to their structure, in which the iron is also ordered, creating a preferred magnetic structure, or fabric. The Earth's magnetic field is generated at its iron-nickel core and has varied in intensity, location and polarity through time. It is the variation between this global field, and that associated with a local magnetic field, generated where materials of different susceptibility and fabric occur, that allows magnetic surveying.

In some ways, magnetic surveying has its equivalence in gravity methods, where the global and local signals have to be differentiated for the anomaly to be found. Simple methods of doing this require the direction and inclination of a compass needle to be measured at different points over the survey area: these variations can be recorded, mapped and anomalies calculated. A more sensitive method of measuring the magnetic field is to generate an electrical current and observe how this is deflected by local variations (the fluxgate magnetometer). This may be affected by temperature, the Earth's daily changes in magnetic activity, and changes in the upper atmosphere, which are problems in daily surveying. Such devices can be shielded and are used extensively in airborne surveys over large areas. These do, however, have a strong forensic application, as buried vehicles, armaments and waste have all been detected using fly-overs, negating the need for ground operations. A second, sensitive device (the proton magnetometer) utilizes the movement of atomic particles as they react to changing magnetic fields (as the device is moved about). These devices are ideal for ground surveys because they are highly mobile: proton magnetometers have been deployed in remote 'fish' from boats and are very powerful aids to the underwater search for weapons and sunken vehicles or boats. The most frequently-used magnetic device in archaeology, and thus traditionally in forensic searches is the gradiometer, in which two magnetometers are mounted on a frame, usually 2 metres apart (Figure 3.5).

The difference in magnetic strength between the two is measured as a gradient and is thus independent from daily changes in the Earth's magnetic field. Rapid,

Figure 3.5 The fluxgate gradiometer – a highly sensitive magnetometer. Image shows the Bartington Grad 601.2 gradiometer being used to map the foundations of a 17th century paper mill near Belfast, Northern Ireland. Note the operator has no metallic objects (spectacles, belt, boots) as these interfere with the high sensitivity of the device. Image courtesy of Paul McCarthy

walk-over surveying is thus possible, making the method very useful to small- to medium-scale forensic investigations (Briener, 1981). Magnetic surveys suffer from noise, creating false anomalies and spikes in the data. These are routinely removed in processing when continuous features are being mapped. Such removal can be disadvantageous in forensic searches, because it may be the spikes that define the search targets. In such noisy terrain where metal is being searched for, the magnetometer offers no advantage over a metal detector. Like many other geophysical devices, magnetic data can be displayed as a vertical sounding, and real depths calculated, or as an anomaly map. Such maps have to be calculated from impingements and shapes of curved original magnetic fields (the Euler Solution), or by measurements in different orientations (derivatives) to supply a target shape model. Magnetometry has numerous powerful applications to both large-scale (regional) and small-scale (local) forensic investigations and searches. The experiments at the Stanford University test site, where a variety of metal and mixed metal (with other) materials were buried at various depths is described by Reynolds (1997, p. 199). These experiments are of major relevance to the forensic practitioner, because the anomalies associated with these materials are shown. Some of these are analogous to the buried metal drums of toxic waste and metal-based landmines described by McDowell (1975).

Electromagnetic methods

Electromagnetic (EM) methods use alternating, orthogonal electrical intensity and magnetizing waves, measured from the same travel direction. Such a field can be generated by a simple electro-magnet, or coil of conductive wire at different frequencies. The most commonly used frequencies are below a few thousand hertz and above, generated by a transmitter coil, which is altered by ground conditions, most especially in conductive media by eddy currents that generate a secondary EM field, detected by the receiver coil. The original EM field, primary and secondary alteration are all detected and can be discerned from one another as a function of the depth, geometry and electrical makeup (resistance, capacitive) of a subsurface mass. Electromagnetic devices are hugely powerful and flexible surveying instruments that have been adapted for use on land, in the air and at sea. Electromagnetic devices can be passive (as in the self-potential methods, above), or active. The latter tend to use EM signals generated locally, from a built-in transmitter, or a very low frequency remote transmission that is being used for communications. A common type of EM system is the ground conductivity meter, which uses a frequency-domain transmitter and receiver coil, separated by a constant distance (a few metres to tens of metres) that is moved along a survey line (Figure 3.6).

Figure 3.6 Electro-magnetic surveying instruments: (a) the EM31 device images the conductivity of the subsurface; (b), the EM31 mounted on a quad bike for rapid, rough terrain surveying; (c) the EM38 is a high-resolution device suitable for smaller survey areas (see Fenning and Donnelly, 2004). Photographs courtesy of Chris Leech, Geomatrix Systems Ltd

The opposite method to this mobile system is to place a large transmitter loop or cable on the ground and move a receiver in transect normal to the long cable. The time-domain EM (TEM) survey is analogous to the induced polarization method of resistivity surveying, with a pulsed EM signal being generated and detection of the primary and secondary field measured. An especially relevant sub-discipline of this TEM method is the very-early TEM system, designed for shallow penetration, high-resolution work such as landfill sites. This system will be very applicable to many forensic studies. Very-low Frequency (VLF) surveying uses background military radio transmissions as source and just deploys a detector. The impingement of the signal on geological features is subtracted from the background: the method has proven very useful in mapping mineralized zones, but is dependent on a third-party source. Many ground-based EM systems are of the 'walk-over' type and thus no contact (surface or intrusive) is made with the ground, speeding up the survey and avoiding contamination of both scenes (criminal) and operator (hazardous environmental). Such systems have limited use in confined areas and poor vertical resolution compared to other methods (see Section 3.5 on GPR).

Some EM systems do have the capacity for deeper surveying, especially in conductive (e.g., contaminated) ground. EM survey equipment has to be specially modified for airborne deployment, which has proven highly successful in mapping metal ore zones, diamond-bearing rocks, groundwater courses, saline intrusions and coal extents. EM surveys of small targets on freshwater have been successful: marine EM surveys are undertaken in the investigation of very large features, up to the scale of the Earth's crust. Geological and environmental applications of EM surveying show some suitable proxy studies for forensic-type studies. Reynolds (1997) shows how ~20 metre-diameter sinkholes were successfully mapped and imminent collapses defined in Northern Territory, Australia. Contaminant plumes in groundwater have also been successfully mapped: using benchmark monuments or differential GPS, return visits can be made to contaminated ground, where the proxy movement of contaminants by EM mapping can be undertaken. The closest archaeological study to a forensic application comprises the work of Frohlich and Lancaster (1986) who deployed the widely-used EM31 ground conductivity meter in Jordan to locate and characterize shafts and tombs. The varying response in conductivities was related (on excavation) to the amount of silt present in the shafts and tombs, resulting in a 100 % success rate for the identification and discovery of the voids. Recent searches for buried victims of murder/homicide have deployed an electromagnetic and resistivity surveying device such as the Geonics EM38, to significant effect. The rapid, walk-over type deployment, ease of use and narrow width of the EM38 make it a highly effective device: future searches for buried victims are likely to use this device, in conjunction with GPR, metal detectors, cadaver/victim dogs and the probe as rapid and efficient search devices/methods.

Case study: Resistivity – Scott and Hunter in peatland and river valleys

Scott and Hunter (2004) discuss how resistivity measurements were used to assist the search for one of the victims of the notorious 'Moors Murders' child killers of northern England in the 1960s, Myra Hindley and Ian Brady. The area investigated comprised three stream valleys with a covering of 0 to 1 metre peat on sandstone bedrock. The aim of the survey was to identify the burial site: instead a more realistic result was obtained and a map of peat thickness, ground-truthed by augering, was produced. This was used to define areas where a cadaver dog could be selectively-deployed, demonstrating a good use of geophysical resources – in assisting the overall search, rather than as a means in itself. A far easier survey was conducted by Scott and Hunter (2004) in a field beside a river where locals suggested four child graves were situated. All the anomalies observed were explained by reference to present-day or archaeological land-use. So insistent were witnesses that a large area was excavated, regardless of geophysical targets. No graves were found, vindicating Scott and Hunter's interpretations.

Case study: Resistivity and IP – large (500 metre diameter) toxic waste burial sites

This survey serves as a proxy for both environmental, mass grave and hidden weapons/armaments/explosives sites. Resistivity surveys (conjunctive with GPR – see below) led to one of the third-ever jail sentences imposed on the perpetrator of a massive toxic illegal waste burial site in Northern Ireland. The project was typical of many environmental forensic cases, where as much data had to be collected as possible in a short time period with hostile landowners. Three types of geophysical data were collected, including electrical resistivity, induced polarization and GPR, conjunctively allowing the extent of the waste to be determined. Modern electrical geophysical instrumentation allows electrical resistivity (ER) data and IP data to be collected simultaneously. The ER and IP images resulting from the survey shows underlying bedrock to be highly resistive (pink colours in Figures 3.7, 3.8 and 3.9). It is readily apparent that the subsurface material located above bedrock is highly heterogeneous, comprising several electrically resistive (pink, red, orange and yellow colours) and electrically conductive (blue colours), with greenish colours indicating the transition between resistive and conductive areas (Figure 3.7).

It is highly unlikely that matural processes deposited such heterogeneous ground. In the absence of a more likely explanation the heterogeneous ground is likely to be fill of human origin. It should be noted that areas of high chargeability (pink colours in figures) correspond well to the electrically conductive zones (blue colours in figures). Comparison of this observation with the results from test survey sites is consistent with the inference that the blue colours in this survey and the pink colours reflect pockets of buried waste material. It is likely that the intermittent zones of higher electrical resistivity (pink through yellow colours) and low chargeability (blue colours) represent 'natural' Earth materials separating these waste pockets.

On the basis of the ER and IP data it is unfortunately not possible to identify the nature of the layers: a full three dimensional (3-D) survey would map the extent of all the layers, but as noted above, given the legal and security constraints, these data are all that are available to make a judgment. Nonetheless, two lines cross (Figure 3.8), giving some semblance of a 3-D concept of the shape of the waste deposit.

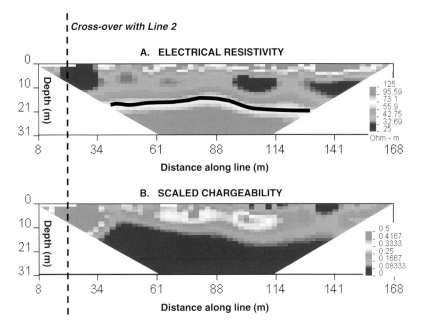

Figure 3.7 Electrical resistivity and induced polarization data from the toxic waste case study site. Low resistivity equates to high conductivity: the waste here was determined from limited excavation to comprise metal, concrete and liquid toxic wastes, the combination of which will be highly conductive. Note the boundary between the conductive (waste?) and resistive (bedrock?) at ~20 metres. Note also limited patches of resistive material at surface – where again trial pits established rock debris and soil placed on top of the waste. The conductive nature of the waste is matched by the ability of the different subsurface materials to store an electrical charge (chargeability) (data courtesy of Bernd Kulessa)

From the ER and IP data alone it is not possible to identify unequivocally the nature of the subsurface material, although it is possible that a considerable amount of waste is present and capped by drier and compacted soil (Figure 3.9), but it is equally as likely that waste is absent altogether and that subsurface materials are entirely natural, such as, for example, a layer of drier soil overlying glacial till or an electrically similar material. The positive comparison between excavated waste pits and ER response suggests that similar responses reflect the extent of waste.

Case study: Abbot (2005) on a murder and a magnetometer

Abbott (2005) gives an excellent case study where the use of a magnetometer was entirely appropriate. 'Two men and their pickup truck were missing from a former mining town … the police had received a tip [off] that the men had been killed and their truck … buried in a mine dump. The investigator wanted to know … where he could rent a magnetometer to find the truck … – a perfectly reasonable, albeit unconventional, use for a magnetometer.' Abbott suggested the investigator

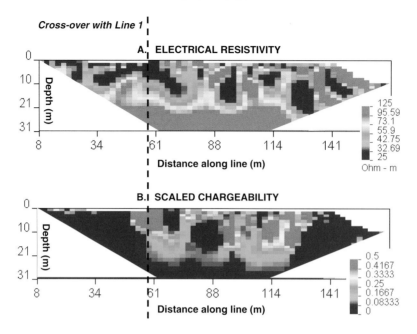

Figure 3.8 Whereas the broad areas of resistive–conductive and chargeable material remain, it is evident from this figure that Line 1 (see Figure 3.7) crossed the main area of the waste deposit, and that here the waste occurs in pockets that are less easily mapped (data courtesy of Bernd Kulessa)

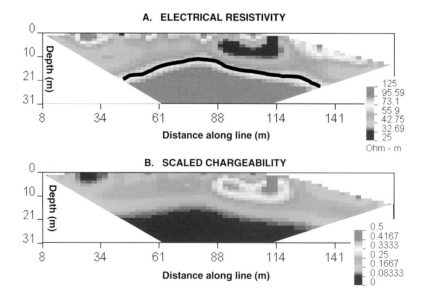

Figure 3.9 This line is parallel to Line 1 (see Figure 3.4) and confirms the overall shape of the deposit, albeit that the main chargeable zone is deeper in this location. This may reflect the main area of buried waste, or where more chargeable material has ended up in the pit (data courtesy of Bernd Kulessa)

persuade a geophysics class at the nearby Colorado School of Mines that this would make an interesting laboratory exercise, involving a small, metal target in unmetalliferous ground. His ruse worked, but the students did not find the truck, as it was not in the mine dump. Three years later, a prison inmate told the authorities the truth: the truck was down a nearby mineshaft, not dump. It was recovered and the investigation became a murder enquiry.

3.5 Ground-penetrating radar (GPR)

GPR uses the transmission and reflection of radio waves (typically 25–1000 MHz) in much the same way as reflection seismic profiling uses seismic waves. Thus a GPR system requires a source antenna and receiving antenna (built to measure the same frequency). The transmitting antenna generates a pulse of radio waves that the receiver detects at a set time interval: the longer the time interval, (potentially) the deeper the waves will have travelled into the ground and back again. Unlike EM, radio waves have far higher rates of attenuation, and thus penetration and reflection depths are low. The receiving antenna has either an electronic or fibre-optic link to a recorder that converts incoming radio waves to digital format and displays these graphically as wavelets. As the transmitter-receiver array is moved, so these wavelets are stacked horizontally to produce a radargram, analogous in appearance (stacked wiggles, smoothed greyscale, colour intensity, 3-D plots) to the seismic reflection profile. Like seismic waves, the speed of radio wave propagation is determined by the makeup of the transmitting medium: in this case the speed of light and dielectric permittivity. Magnetic properties can also influence radar wave speed. Changes in dielectric permittivity can cause radar wave reflection, without which GPR profiling would be impossible. Radarwave attenuation, or signal loss is extreme in conductive media such as seawater, clays (especially hydrous) and some leachate. GPR has good depth penetration (tens to hundreds of metres) in ice (with minor fracturing/interstitial water), hard rocks and quartz sands. Vertical resolution versus depth penetration is of major concern when choosing antenna frequency. Low frequencies (15 – 50 MHz) achieve deep penetration with poor vertical resolution in the received signal, due to the long wavelength. High frequencies (500 – 1000 MHz) show high resolution with weak penetration (centimetres to metres). Low-frequency antennae are large (a few metres long), high frequency antenna are small (tens of centimetres). Again, this can influence the use of the method as deeply-buried targets in enclosed spaces are virtually impossible to survey: see Figure 3.10).

As with all geophysical methods, some intelligence concerning the likely size and makeup of the target is useful: where unknown or questioned, a range of antennae should be used, and in very poorly understood locations, with other geophysical and invasive techniques (the blunderbuss approach). Moisture contents influence radar wave velocity because in homogenous media porosity has a direct relationship to dielectric permittivity. Thus dry sand will allow increased wave propagation: sand with high freshwater content will give improved vertical resolution. A major problem with early antennae was the effect of 'out-of-plane'

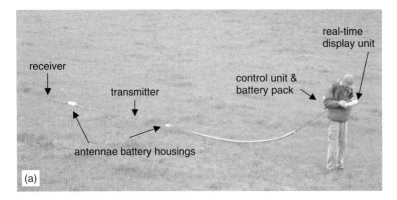

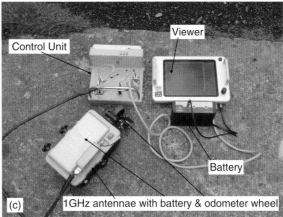

Figure 3.10 Two of the low-frequency (deep-penetration) GPR arrays: (a) shows the Mala Geoscience Rough Terrain Antennae, a 50 MHz system designed for rapid (walking speed) reconnaissance surveying, especially useful for outlining large targets such as mass graves, major burial sites, and the edges of large toxic waste sites; (b) shows the standard 100 MHz twin-antennae array: antennae at the 100–500 MHz range commonly deploy a similar format; (c) shows the shallow-penetration, high resolution 1 GHz antennae and array. Equipment courtesy of Chris Leech, Geomatrix Systems. The monitor (top left) is ∼35 centimetres across

reflections. It is easy to think of the radar wave as a focused beam (the ray-path at right-angles to the wave) when in fact the radar wave as it travels into the subsurface is more like a bubble, hemispherical at first, becoming distorted as it travels at different speeds. Thus lateral to the antennae, on or in the ground surface, may be structures (trees, posts, drains) that cause reflections at ground level. The effect of these surface features can be diminished by altering the orientation of the antennae, or by shielding the above-ground portion of the antennae, such that the radio wave is only allowed to penetrate the ground.

GPR has found its best uses in imaging glaciers, frozen ground, sand deposits (river deposits, non-saline coastal sands), aquifers (porous nature), archaeological features (moats, buried buildings) and concrete/pavements. Modified GPR systems have been successfully deployed in the detection of plastic landmines, often in conjunction with some form of minimum mine detector metal detection device.

Case studies: GPR

GPR has become one of the main geophysical tools for those involved in the search for buried human remains and other non-metallic objects. Prior to 1992, GPR was used by law enforcement search teams and the military, throughout the world but was rarely documented, except in general terms in newspaper articles such as the double-murder on Jersey (Channel Islands, UK) or multiple-murders of Fred and Rosemary West (Sounes, 1995) in Cheltenham (UK). Reynolds (1997) does give information on the use by the US army when locating Viet Cong tunnels in the Vietnam War (see also Magnusson, 2007). One the first discussions on the uses and abuses of GPR in locating gravesites was made by Bevan (1991). Strongman (1992) published a series of case studies from controlled environment burials, using 5 year old bear carcasses as well as actual crime scene profiles. Owsley (1995) evaluated GPR against other geophysical devices and concluded that a physical probe was still a better device for use in the detection of soft ground around inhumations and buried objects. Miller (1996) followed the Strongman (1992) approach, with an evaluation of test sites against actual case studies. The paper by Miller is a very useful starting point, as he provides a good review of work undertaken up to 1996, including using GPR to find Viet Cong tunnels in Vietnam, utility lines, landfill debris, areas contaminated with fluid pollutants and plastic explosives. With regard to burials, Miller recounts how archaeological mapping of Japanese burial mounds, the interior of an Egyptian pyramid, and mud-walled dwellings in South America set the scene for forensic work in the search for graves. Again, historic, or archaeological graves were monitored first, decreasing in age from 16th century, to 200 years ago, to a clandestine burial in 1982.

The application of GPR technology was at an impasse until 1996, as new technology and small, rugged laptop computers and data loggers were developed. A rush of papers in 2000 reflects these advances and their application to actual forensic cases (Nobes, 2000), historic mass burials (Davis *et al.*, 2000) and to the experimental responses of buried corpses (Hammon *et al.*, 2000). The development of other geophysical techniques such as resistivity and magnetometry has caused a re-evaluation of GPR in comparison with other devices (Buck, 2003). Many of the early problems with GPR have now been solved: the development of shielded radar antennae has made usage in forests and urban environments possible. The range of antennae available (25–1000 MHz) now allows investigation of large subsurface features such as walls, foundations, moats, channels and mass-graves down to small (centimetre-scale) objects and features such as plastic landmines, cracks in walls and pavements or even neonate inhumations. Wet, clay-rich and wet, salt-impregnated ground still cause difficulty in obtaining images of the subsurface where the large-scale disturbances of nature need to be separated from those made by humans. In problematic situations, extensive post-acquisition processing, other geophysical methods (Davenport, 2001b; France et al., 1992), penetrative investigative techniques, cadaver dogs and geochemical sampling are required. For initially non-invasive assistance in subsurface searching, GPR exemplifies the problems faced by all users of geophysics, that of controlled experiments. Many published studies use pig cadavers or graveyards for their proxy material. A common problem with both is the difference in geology, soils, vegetation, groundwater and human influence between the test site and the real scene. The dielectric response of a fresh to decomposed pig cadaver is presumed to be the same as human remains, even though actual measurements of this and the addition of materials in human graves (clothes, weapons, lime, disinfectant, embalming fluid) may compromise the results. Practising surveying in graveyards[2] can overcome some of these deficiencies but then create others. Coffins and deep burial are common in legal burials (Ruffell, 2005; Watters & Hunter, 2004) yet very rare in covert murders. Hence some practitioners have used places of rapid legal burial (flu epidemics, pauper graves) sites in preference to regular graveyards (Davis et al., 2000).

Prisoner escape tunnels or sand erosion? GPR investigation of roadway collapse

In May 2004 a roadway in a major border town of Northern Ireland began to subside. The reason for the subsidence was not known, but concerns were expressed that a jail (now disused) was located opposite the roadway subsidence, and that prisoners during either period of major unrest in Ireland and Northern Ireland (1914–1921; 1969–1980) had tunnelled under the road. Three methods of investigation were deployed: a desktop study of bedrock and drift geology; a survey using ground-penetrating radar (GPR); and visual examination of soft ground assisted by probing.

Bedrock geology below the site comprised granodiorite, a hard, often fractured rock that weathers to granitic sand and boulders and kaolinite clays. Drift geology comprised glacial till, overlain by peat and alluvium. The sub-roadway was mainly loose sand, but whether this was imported during construction, or was natural, was hard to determine: a variable-thickness tarmac cap occurred on the road, with concrete paving an concrete gutters on the road-edge. A Mala RAMAC GPR system was deployed in this survey, using 225 MHz shielded and 200 MHz and 100 MHz unshielded antennae. GPR is a standard method used in the assessment of thickness of roadway ballast, sub-roadway damage and subsurface voids and weaknesses. Examination of the playing field opposite the site and water-courses concentrated on one side of the road, where an embankment allowed examination of soil and glacial till. This was assisted by determining the presence of soft ground and voids

---

[2]The experimental surveying of graveyards has a beneficial side effect, of informing church and municipal authorities of any unused ground for future burials.

using a 1.6 metre long probe (see Chapter 2 for advice on the use of the probe). Bedrock was close to surface at the north-eastern end of the survey lines, descending through a series of undulatory steps down to below 6–7 metres below the road at about midway through the survey transect (Figure 3.11). Above bedrock, drift had a number of hyperbolic (dome or mound) anomalies, suggestive of boulders and weathered rock protuberances. Above these, inclined reflections (to the west and south-west) were suggestive of sediment movement in this direction, either as glacial or post-glacial (fluvial) sands, or as construction of the embankment was made. The survey transect could be imagined as soft till, sand, etc. infilling an asymmetric hollow in harder granodiorite. The previously-identified deformed areas of roadway occurred in this soft sediment.

A water-course ran parallel to the road, comprising a land drain adjacent to the embankment between the road and adjacent playing-fields; the most likely catchment. The drain issued from the embankment about halfway along the transect, and was joined by further outflows from the embankment, where a stream ran parallel, to drain away into the sewage system of the town.

Small-scale conditions (anomalies). A number of anomalies were identified in the field, most of which could be related to ducting for electrical cabling, drains, etc., many of which had pre-existing repairs to the roadway. Anomaly A is evident on Figure 3.11(d): the nature of this anomaly required analysis, it being adjacent to the jail and in a location where some subsidence had already been noted. Figures 3.11(d) and (e) show the anomaly clearly on both unprocessed and processed data, the latter having the advantage of also displaying 50-centimetre thickness of roadway ballast above 1 metre of loose material to the top of the natural geology at 1.5 metre.

Possible explanation. Combining the geological, GPR and visual examination showed that the area of deformed road was coincident with where soft material formed the embankment, which infilled topography above granodiorite bedrock. This soft material was sand-grade in nature, as observed in the sediment outflow of the drain. Water, flowing from the north and northwest, had washed this sand away. Badgers and rabbits had dug into this material; the sets had also washed away. Combined, the hydrogeological conditions, land-drain, embankment and animal burrows allowed the soft sandy material to be removed, causing preferential settlement of the roadway some 2 – 3 metres above, and across much of the survey area. No major voids (larger than ~1 metre diameter), indicative of a prisoner escape-tunnel, were found beneath the road. Small voids (a few tens of centimetres in diameter) may of course have occurred at depth and been undetected. Of the shallow anomalies identified and analysed, one was restricted to the northwest (Anomaly A), and others being associated with previously-packed areas. Rather than (metre-diameter) voids from tunnelling being the problem, it was the sandy nature of an infill to hard bedrock topography, being washed away by drainage culverts and next to the topography of the embankment, that may have caused areas of deformation and collapse to the roadway. The lack of a linear aspect to Anomaly A, plus its shallow nature, made it unlikely to be a tunnel. Nonetheless the area was

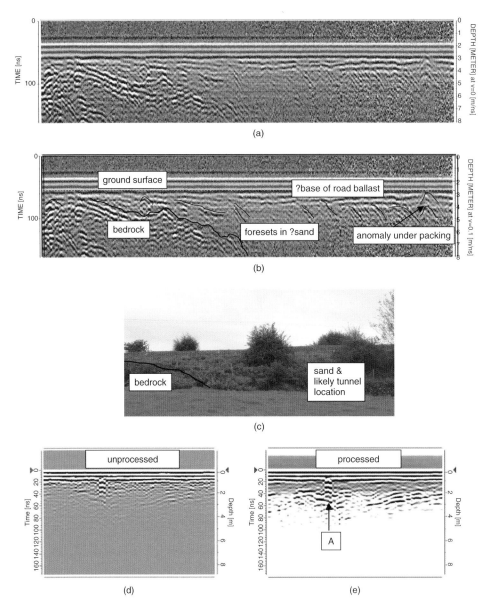

Figure 3.11 (a) Long profile (100 metres long) through the survey area, viewed from the playing field toward the town. (b) Interpretation of (a). (c) Photograph from the playing fields toward the town, with an interpretation of geology observed in the embankment and GPR data from parts (a) and (b). (d) 200 MHz profile (raw data, 25 metres long) over the only unexplained anomaly of the survey (A). (e) Processed data from part (d), also 25 metres long

excavated and a large tree-stump recovered: rotting of the stump had caused the subsidence. The GPR survey saved the excavation of the entire roadway.

Pauper's graves – a proxy for mass burials

Of the many mass diseases or epidemics to have devastated the population of Ireland, the Potato or Great Famine (~1845–1851) stands out as particularly noteworthy, being so recent, having many political repercussions, and for the great suffering of those affected. One consequence of the famine was the extreme pressure put on poor houses and associated hospitals. Most of the afflicted were of rural descent, with no means of support and often with no substantial material goods: many travelled from their homes, looking for means of support, often with children, to find themselves eventually thrown on the mercy of the welfare institutions, including hospitals for the poor, workhouses and debtor's prison. The high mortality rate resulted in what was effectively mass burial. Many victims were either wrapped in a shroud and buried, or placed in crude coffins. Clothes of any value were removed, although this was rare, the unfortunate sufferers in most cases having given up anything of value. Hence there now exist mass graveyards with mixed coffin or no-coffin inhumations of 150 years age. With urban expansion and the building of new hospitals adjacent to former sites, inevitably the land (consecrated and non-consecrated) containing these graves becomes subject to possible use for building, etc. With no headstones and little other visible indication of the age of the burials, less pressure for the preservation of the graveyards exists.

One such situation occurred at two hospitals in the west of Ireland. In the former, the graves were easily defined by excavation (Figures 3.12(a) and (b)). In the latter, all that could be seen was an undulating field (Figure 3.13(a)). At times of low-angle light (dawn and dusk) and when vegetation was of moderate height (spring), collapses associated with likely burials could be observed (Figure 3.13(b)).

The site was surveyed using 200 MHz GPR antennae, together with regional 100 MHz lines (to determine deep geological conditions) and 400 MHz over specific graves (to determine grave morphology). Figures 3.14(a) and (b) show the key features of the site.

All the lines that crossed visible collapses at right-angles north-south, the graves being mostly east-west in the Christian tradition), displayed a distinct variation across the graveyard, with multiple burials on the northern side, a lack of burials through the middle of the field and fewer burials to the south side. Evidence of a former wall, separating the two areas was noted in some locations. Furthermore, local intelligence suggested that many common-ground graveyards had separate areas for Catholic, Protestant and a walled area or area just outside the graveyard for child burials. In the densely-populated, northern side of the graveyard, grave collapses were noted with a consistent metal detector anomaly one third of the length of the burial. Only one or two such anomalies occurred in the southern side. The reasons for this are unclear, but interviews with archaeologists (personal communications, Colm Donnelly, Brian Sloan, Finbar McCormick) suggested that the metal would likely be a crucifix. A number of small collapses were found in the northeastern corner of the graveyard, separated by a wall-like feature. In this,

(a) (b)

Figure 3.12 (a) A pauper grave and intact coffin excavated from a location nearby. This information provided invaluable data on what the form of the target was. Ground conditions were very different, however, with less water at the case study site. Coffin is narrower than modern coffins. Photograph courtesy of Brian Sloan, Centre for Archaeological Fieldwork, Queen's University Belfast. (b) The inside of the coffin, showing top of a child's skull. The good preservation at this site suggested the presence of the cadavers at the case study site may have substantially affected local chemical conditions, causing a different dielectric response to surrounding material, the only problem being the likely dry versus damp enclosing sediment. Photograph courtesy of Brian Sloan, Centre for Archaeological Fieldwork, Queen's University Belfast

(a) (b)

Figure 3.13 (a) View of the graveyard site. Note the gentle slope, lack of headstones and undulations (likely grave collapses); wall is 1.2 metres high. (b) Close-up of one grave collapse, outlined in black. Note the slightly raised vegetation. Blue cross indicates a metal-detector anomaly (approximately 1 metre across)

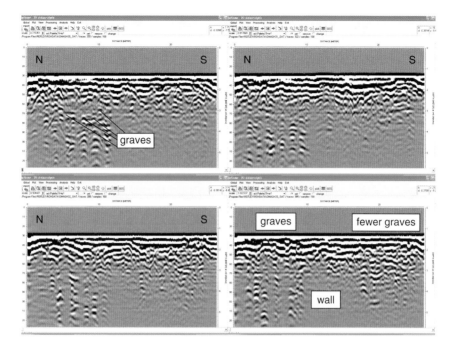

Figure 3.14 Screen dumps (raw data, in the field) of four 200 MHz north–south lines that cross most burials at right-angles. Four lines are selected to show the multiple inhumations on one side of the graveyard compared to the other. A buried wall divides the two areas with some conjecture that this separated Catholic from Protestant burials. Each profile is ~26 metres long

most likely a children's graveyard, no metal anomalies were found, these likely being stillbirth or unbaptized children. Thus some supporting evidence for the Catholic–Protestant–child-burial divides was found, from geophysics alone. East-west regional lines within the graveyard (Figure 3.15(a)) and outside the graveyard (Figure 3.15(b)) showed first how difficult it is to spot the graves when not surveyed at right angles, and second how disturbed the ground is within the graveyard compared to outside its limits.

3.6 Radiometrics

As the nature of radiation was elucidated through the early parts of the 20th century, the uses and hazards needed to be evaluated. This required monitoring devices, which comprised photographic plates and later, the Geiger-Muller Tube, Geiger-counters or other scintillation counters. These devices, used in conjunction with experiments in shielding, allowed physicists to recognize the nature and mobility of alpha and beta particles, as well as identify gamma-radiation, or gamma-rays.

Three main sources of gamma-ray radiation have now been identified: cosmic (from solar and stellar nuclear reactions, as well as background from other sources);

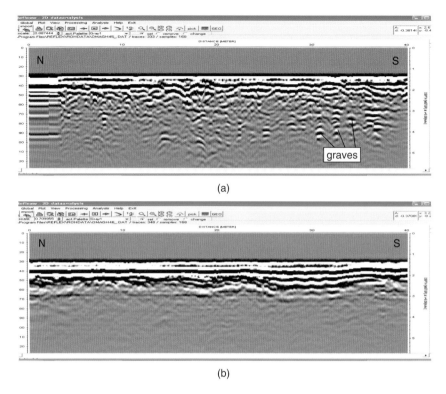

Figure 3.15 (a) East-west 100 MHz GPR line through the graveyard, with lack of definitive anomalies compared to (b) shot outside the graveyard limits. Each profile is 40 metres long

natural radioactive isotopes in rocks, soils and water; and anthropogenic (processed or manufactured isotopes of elements such as uranium, caesium, californium and others for nuclear power, medicine and weapons). Three types of gamma-ray surveying are possible. The first passive method deploys a simple scintillation counter that typically contains a crystal (sodium iodide or lithium), which oscillates when struck by a gamma-ray, or a tube filled with an inert gas that fluoresces. This is detected by sensors or photo-tubes and converted to a digital signal, giving a value for gamma radiation. The second (also passive) method is added to the scintillation device and comprises a spectral gamma-ray detector, which allows the isotopic source of gamma-rays to be estimated by measuring characteristic gamma-ray wavelengths. The third type of gamma-ray measuring device is active, wherein a gamma-ray source of known isotopic composition is placed one side of an object or area to be investigated and a spectral detector deployed opposite. Gamma-rays of known spectral quality enter the area to be investigated and exit with different spectra, dependent on the intervening medium. This method uses very hazardous source materials (often Californium) and requires a database of known materials for a comparison to be made: it is used extensively in the assessment of chemical weapons.

Total count and spectral gamma-ray data are obtained by non-destructive, automated, rapid and inexpensive surveys that rely on spatial variations in Earth materials that contain radiation-emitting isotopes. Gamma-ray surveys have traditionally been used either over large areas (kilometre-scale) for mineral prospecting (Darnley & Ford, 1987) and radiation hazard assessment (Jones et al., 2002) or down boreholes (at centimetre-scale) for rock petrophysics (Kearey et al., 2002). Applications of the gamma-ray technique to smaller-scale (centimetres to metres) spatial archaeological, geotechnical and forensic surveys have been limited to crude spatial measurements of total count, K-, U and Th-based radiation (Moussa, 2001; Ruffell & Wilson, 1998).

In geological studies, gamma-ray transects or mapping have been utilized for many years in the search for mineralized zones, either for nuclear energy or as a proxy for radon hazard (Kearey et al., 2002). These studies employed either total count scintillation or spectral detectors in aircraft or vehicles because the K, U or Th bearing rocks and minerals being looked for gave off high levels of gamma-rays. The application of gamma-ray detectors to surficial sediments has followed two courses of research: airborne, hazard detection studies (Chiozzi et al., 1998; Tyler, 1999) and surface, archaeological and geotechnical studies (Jones et al., 2002; Moussa, 2001; Ruffell & Wilson, 1998). All these applications have been made possible through the manufacture of highly sensitive, weatherproof automated spectral gamma-ray detectors (Figure 3.16). The airborne and surface surveys cited above have utilized total count detection, together with spectral information from specific natural and anthropogenic isotopes.

(a) (b)

Figure 3.16 (a) The Exploranium 320 gamma-ray spectroscope being used in walk-over mode (akin to aerial surveying). (b) The Exploranium 320 gamma-ray spectroscope being used to determine ground levels of radiation. Photographs courtesy of Chris Leech, Geomatrix Earth Systems Ltd

Total count and spectral gamma-ray measurement using a spectral gamma-ray detection device is a frequently-used and standard geophysical technique (Kearey *et al.*, 2002). The method relies on the individual gamma-ray spectra of each isotope being split and detected along a photomultiplier tube in the measurement device (Kearey *et al.*, 2002). Some forensic applications are summarized below:

- *Nuclear materials.* Fly-over and walk-over searches for lost, covertly hidden or illegally-stored nuclear materials can deploy total gamma-ray detectors where the source material is not known, or spectral detectors where multiple sources are suspected, or specialized, single isotope spectral devices for the isolation of one type of material (e.g., Caesium).

- *Mine waste.* Fly-over and sometimes walk-over surveys of radioactive mine-waste can be used where the extent of waste and leachate is under contention. Typical applications include coal mines, whose waste may contain significant thorium. The mining company may state that spoil areas are undefined, being vegetated, or that leachate is not moving near any population centres: spectral surveys will provide an independent map of spoil extent, especially if linked to GPS.

- *Links to health.* Radiation from power stations, reprocessing plants, medical storage facilities and to radon from rocks like granite can all cause problems for environmental health that become subject to legal investigation. In these cases, independent measurements of the spatial extent over time of different isotopes and their decay products are required, using spectral gamma-ray information.

- *Archaeology.* Gamma-ray surveys have proven useful in conventional archaeological prospection, especially of problematic areas such as karstic terrain and coastlines, where airborne or ground-based gamma-ray data is easy to collect and provides a resolution better than other methods.

3.7 Review of why some methods are favoured and others not

Killam (2004) gives a review of the geophysical techniques most likely to be used by the homicide investigator (see Appendix 1). However, the scope of geoforensics is far wider than the search for buried bodies, and thus methods Killam does not include are discussed in this chapter – reflecting a common theme of this text – which is 'fit for purpose'. Therefore, Killam does not include the use of SP or IP data – which can be important in studies of groundwater pollution, mining fraud, toxic waste disposal. This is not a criticism, more a reflection of what we have and will see throughout this text – which again is what method is most appropriate for the problem? In this we know the problem is not always scientific – it may be logistical, legal or environmental. In the case studies given, time (speed of survey), security and legal requirements all play a part in the planning of surveys. Many scientists will balk at such words and demand that scientific protocol be the driving force, which is correct, only, unfortunately, scientists do not run the law courts. Favoured methods depend on the *target* and its *context*. What is being looked for and where is it? Some targets are solid, metal, radioactive or hollow, each of which requires a specific geophysical detection method. Some contexts provide advantages or disadvantages that preclude or enhance data acquisition. GPR will not work on/in salt water; magnetics may become impossible in urban environments. Many methods will work on open, sandy soils but some will be quicker and just as efficient. The advice of a geophysicist who has been provided with as much background data as is available, including a non-survey site visit, may save hours of wasted travel time, or even data acquisition. Geophysics is not magic, and thus, like all search operations, requires full disclosure of all relevant data (from the investigating/search team) to obtain the greatest effectiveness.

The future – multi-sensor platforms and software advances

Fenning and Donnelly (2004) consider the advances in hardware that will initially aid the large area, reconnaissance type search: 'In attempting to survey large site areas more cost effectively and to ensure a uniform survey coverage, geophysicists have developed arrays of geophysical sensors which can be man-carried or vehicle-mounted and walked/towed across site' (Figure 3.17). In addition to this, advances will also be made in advancing individual methods such as GPR (Freeland *et al*., 2003).

They continue, 'Frequently, differential global-positioning systems (DGPS) are integral to such an array in order to provide a location for the geophysical readings

Figure 3.17 The multi-sensor platform, GPS-linked magnetometers (four in total) an EM38, and temperature and humidity sensors. Platform is 3.4 metres in length. Other devices may be added. Photograph courtesy of Chris Leech, Geomatrix Earth Systems Ltd

and to control traverse progress. Such systems involving GPR, magnetic and electromagnetics have been developed for unexploded ordnance (UXO) location and ... archaeological surveys .... The advantages of using mobile multi-sensor systems in forensic studies are clearly apparent where there are large tracts of survey and copious detailed data points required.' Fenning and Donnelly were ahead of their time, with the multi-sensor platform in 2007 fast-becoming a standard means of instrument deployment. Collection of multiple spatial datasets requires a GIS-based format of data storage and display: Watters and Hunter (2004) provide some excellent examples of how the conventional 2D cross-section type GPR profiles across historic graves can be combined to produce 3-D images or 'sliced' at varying depths to give outputs not dissimilar to plan-view resistivity or magnetics surveys.

4
Remote sensing

> Pile the bodies high at Austerlitz and
> Waterloo.
> Shovel them under and let me work –
> I am the grass; I cover all.
>
> Carl Sandburg (1878–1967) 'Grass' (1916)

4.1 Definitions

Remote sensing methods gather data from which we obtain information about the ground surface and subsurface without direct contact. Some geophysical methods are very much remote sensing techniques, and the division between some is arbitrary. Likewise, the simple digitization of paper topographic maps allows their visualization on a computer screen or in a 3-D Visualization Suite, thus deriving remotely-sensed data from direct measurement. In addition, such digitized maps may then be analysed and enhanced by geostatistical techniques that provide a secondary, output that maximizes information use, or as some non-specialists put it, 'makes more realistic-looking maps'. In this chapter, we confine ourselves to remotely-obtained imagery obtained using electromagnetic (X-rays, ultraviolet [UV], visible, thermal [infrared] and microwave) energy in atmospheric conditions. The interaction between electromagnetism and the atmosphere is thus critical to how we obtain and use such information, just like the interaction between geophysical energy sources and the subsurface rocks and soils we examined in Chapter 3. The path that electromagnetic energy takes through the atmosphere can be affected by absorption, in which energy is lost in ozone (O^3), carbon dioxide and water vapour. Scattering is caused by small particles in the atmosphere, is most common in the shorter wavelength radiation types such as X-rays and UV and results in no total loss of energy. Thereafter, electromagnetic energy may continue to travel its path or be reflected. Different energy types show various travel distances, absorption, scattering and reflection characteristics, making the energy use suitable for certain applications. Reflection is of critical importance in remote sensing as the amount and type

---

Geoforensics Alastair Ruffell and Jennifer McKinley
© 2008 John Wiley & Sons, Ltd

(direct, diffuse) of reflected energy provides information on surface topography (roughness, smoothness) whereas the portions of the various electromagnetic spectra that are reflected back following absorption tell us about the physical and biological chemistry of the ground or surface. This latter measurement is called spectral reflectance. Water with variable algal content, buildings made with mixed concrete types and different areas of vegetation (natural or planted) all give unique spectral reflectance. This can be cheap and easy to obtain (for some types of measurement [photography]) or may require specialized sources and detectors.

The importance of remote sensing in land use mapping has resulted in a wealth of information on the spectral characteristics of soil, water and vegetation. In pure form, each of these have very characteristic spectra: the real world however comprises complex mixes of materials, with soil moisture increasing from negligible in arid regions to over 50 % adjacent to flooded areas. Thus confidence in remote sensing information increases with ground-truthed, or reference data. These may be pure observations of what is distributed and where, or they may be analogue measurements of certain electromagnetic reflectance using hand-held or vehicle-mounted devices. The location of such ground reference data is crucial, especially when being used to interpret data with poor horizontal resolution. Ultimately, all reflectance data must be generated from a location, or spot. When transferred to a computer, this becomes a pixel, with an assigned reflectance value. The ground reference data must either relate directly to this location or be an average of locations where ground and remotely-sensed variation can be averaged and/or compared. Reflectance values are stored in a digital format, and can be subject to further manipulation using Geographic Information Science (GIS) (see Chapter 5). Geographic location of both remote-sensed and reference data is critical in GIS, whose primary function is to handle and manipulate layers of spatial information. Without identical positioning, such comparison cannot be made. Prior to global positioning system (GPS) technology, aerial photographs required some form of standard marker, from which ground measurements could be compared. In the following sections, key aspects of the remote sensing system will be addressed. These include the energy source, which dictates the type of measurement device (sensor) and its platform (satellite, aircraft, ground reference survey), which together are affected by atmospheric makeup and thus scattering and absorption. Combined, these produce a reflectance image or measurement. Being spatial and digital, this can be geostatistically described and processed.

4.2 Conventional aerial photography: rural and urban examples

Early aerial photographs (1850s and 1860s) were obtained from stationary and then moving balloons, kites and in 1908 from aeroplanes, with a massive increase through World War I. Spectral data can be obtained from black and white photographs as greyscale intensity: the arrival of colour and infrared film has allowed easier and more precise representation of spectral values by red-green-blue (RGB) and other colour intensities. Other advances have included camera types that have wide apertures, ultra-high resolution film types and digital photography. Continuous imagery has also advanced in recent years, deploying both video-type cameras, from which stills can be extracted, as well as multiband cameras that filter certain light reflectance at source. The most important advance made using conventional photography has been its use in photogrammetry: most typically this is the making of digital maps from aerial photographs, but may include other applications such as measuring building heights or distances.

The traditional means of constructing topographic maps from aerial photographs required two photographs of the same region, taken from different angles, the stereopair. As the height of the aircraft is known, often very accurately, then the same feature will be imaged twice but show a different angle on each photograph. An easy example is a vertical feature, whose height is a simple function of elevation above ground, and thus when the oblique distance, apparent height and height of aircraft are known, so the absolute height can be calculated. More complex are slopes: traditional mapping using photogrammetry used interpolation, either between breaks of slope or between measured markers on the ground. Say the base of the slope and the top are located and the heights measured, then photogrammetry can be used to estimate contours between the two. This crude but effective method of mapping large areas still has its uses where other features need to be related to topography that are not easily accessible (thick vegetation, water courses, dangerous areas). The stereopair and or its equivalent (the dual monitor computer system) have much more to offer than the making of maps. Both provide the optical illusion of giving the photographs depth, or actual topography, allowing meaningful visualization by the viewer before ever having stood on the ground. Remote sensing specialists stress the difference between data and information because they are acutely aware of how an aerial photograph is subject to so many variables that alter the data input and thus the interpreted information output (Grip *et al.*, 2000). Cloud cover, time of day, time of year, height of platform, oblique view, type of camera, type of film and finally, but not least, length and type of experience of interpreter are all influential on the interpretation given. Remote sensors have tried to provide some absolutes for the description of features, including their size, shape, pattern, tone (or colour – hue) texture, and location.

4.3 Geoscience use of light photography

Sources of light photography commonly available to the geoscientist involved in investigations of 'where, how and why' (commonly criminal, terrorist, environmental enquiries) comprise satellite, aerial (aircraft and helicopter) and terrestrial (ground level, mounted or underground). The common-nature of photographs make this appear to be very low-tech and to the uninitiated, uncontentious: surely a picture never lies, what you see is what is there? In fact, photographic interpretation, especially satellite and aerial data, requires a high level of experience and interpretive skill. Military and emergency response intelligence officers are trained extensively before they are used in the tactical and logistical analysis of such data for their purposes. Figure 4.1 demonstrates one use of such photography in predictive analysis

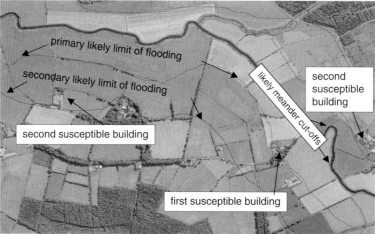

Figure 4.1 Emergency response planning use of digital aerial photography. The upper plate is the replicated raw data, taken at 2000 metres height using standard panchromatic film of a meandering river system (see Chapter 2) and floodplain. The lower plate shows one example of numerous interpretations that may be made of this type of data. Here, flood risk is considered, using reference to topographic maps and stereo-pair images. Two levels of risk are considered, and which buildings are likely to be damaged by each level. Other interpretations and risk levels may of course be applied. Image from Google Earth Permissions, after original Environment Agency images. Scale not provided

of likely flooding hazards; later examples show how retrospective analysis of what happened and where, may be made of the same and similar data.

Why then would anyone think that the analysis and interpretation of vegetation, morphology and geology should be any easier? In fact, it is generally harder, due to the difficulty in dating events, without cross-cutting features (multiple tyre-tracks) or time-series images of the same location. One interpreter's ground collapse is another's bomb-crater (see the D-day Normandy Case Study of Chapter 2); one person's car park is another's mass grave. This demonstrates how much photographic interpretation depends on what is being searched for, or predicted. Some ground rules can be applied such as: what size of feature is being searched out/examined? If small, are the data of good enough resolution? What type of activity is being considered? Can natural processes be confused for suspicious (river channel switching may be natural, or may be the result of illegal water-course diversion)? Is the activity above

Figure 4.2 Two examples of the geomorphological interpretation of aerial photography for both predictive and retrospective purposes. (a) A meandering river system like that shown in Figure 4.1 has been examined for likely locations where a body was dumped into the river and likely locations where the body may be washed ashore. Points of access are determined and then point-bars (locations of deposition) downstream located. Access points may then be secured and tyre-tracks, footwear tread, fibres, blood, etc. be searched for and scent dogs deployed. Once complete, point-bars may be searched or over-flown once again. (b) Areas of possible covert human activity (beyond the obvious) were to be considered. Peculiar features such as the old railway line, different farming practices and infilled lakes were all noted. The diverted water course was the focus of the investigation, although this was not mentioned to the interpreter, who correctly identified the location of illegal activity. Images from Google Earth Permissions, after original intelligence data. Scale not provided

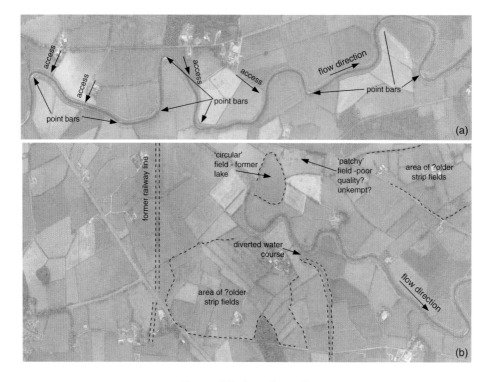

Figure 4.2 (continued)

ground or below? Does it impact vegetation, soil, rock, water? What resources were needed by the suspects (walking, digging, vehicles)? How long ago did the activity take place? What has happened since (human activity, weather conditions, soil or water movement)? Hopefully this arbitrary list illustrates and reinforces the point (above) that what appears a familiar and routine method of analysis contains many variables with which the non-specialist interpreter would struggle (see Figure 4.2). Aerial and satellite photographs are a bit like all observational sciences: the best interpreter is the one who has seen the most.

Case study: Abbott (2005) on mining fraud

Abbott (2005) provides a straightforward example of the power of common-sense aerial photograph interpretation. The case rested on a defendant claiming he had driven a tractor up a dirt track five years previous to the investigation, in order to improve and assess his mining property. Abbott obtained aerial photographs taken before and after the date of the claimed improvements. A visit to the location showed a deep seasonal river channel on the route the defendant claimed he took. This channel ('washout') was too deep for a tractor to traverse and was in existence both before and after the contended date of improvements. To complete the denial of the witness statement, trees were noted on the roadway that was supposed to have been improved. These were visible on the air photos: one cut tree trunk showed 25 annual rings, making the claim impossible to defend.

Case studies in pollution monitoring

In Figure 4.3, three case studies in pollution monitoring are presented. In Case A, illegal importation of fuel dyed red for agricultural use meant the perpetrators needed to remove the dye with kieselguhr (diatomite: see Chapter 7). The import route was along a linear road through a village, and spilt fuel oil was noticed all along the route. An illegal motorbike-racing track was also along the route and became the subject of the investigation. However, use of aerial imagery allowed a thorough search of the area to be made. This search strategy made use of pre-existing boundaries to movement, such as the linear fields shown in the east of the image. These had been searched when a long way (3 kilometres) east of the motorbike track, the fuel processing plant was discovered.

In Case B, shallow groundwater was found to be heavily polluted with hydrocarbons, heavy metals and other toxins at a farm. A visual search of the area established a likely vehicle-turning location and small illegal dump, still not completely infilled. This location was placed under observation, with few results. An examination of aerial photography showed that numerous small streams fed into the groundwater at the affected farm. These were used for a search by independent monitoring officers, who located a far larger, toxic dump, very well-hidden by trees. The incorrect location was probably an initial dump site, abandoned because of compromise and visibility, but used as a decoy, which worked until aerial photographs of the *surrounding* landscape were examined.

In Case C, the soil from a field was removed and sold as valuable topsoil. Farmyard slurry was placed in the metre-deep basin so created, and thin soil placed on top. A third field had similar soil removal and waste deposition, but no soil placed on top. This location was removed from the other two, and close to an innocent farmer's buildings. This clearly implicated the innocent, when the reality was that the perpetrator was intimidating his neighbour. Examination of aerial photography clearly showed the attempts by the guilty party to cover his polluted fields.

Case studies in body deposition sites and animal cruelty

Figure 4.4 shows three examples of where aerial photography may assist in the location of suspect activity. In Case A, police intelligence led them to believe that a missing person had been placed in a lake. The problem was that there were over 50 small lakes within range of the missing person's last known location. Each lake was over-flown for signs of activity, or possibly of the body. No body was found, but in the image A, an anomaly was observed. This was initially considered to be some form of artefact on the lens.

A rain-drop was mentioned: although how a rain-drop can fall onto a lens mounted underneath an aircraft is hard to envisage. Geomorphologists brought in on the case suggested that the anomaly was a plume of sediment, where a large disturbance, consistent with people carrying or using a boat to take a weighted

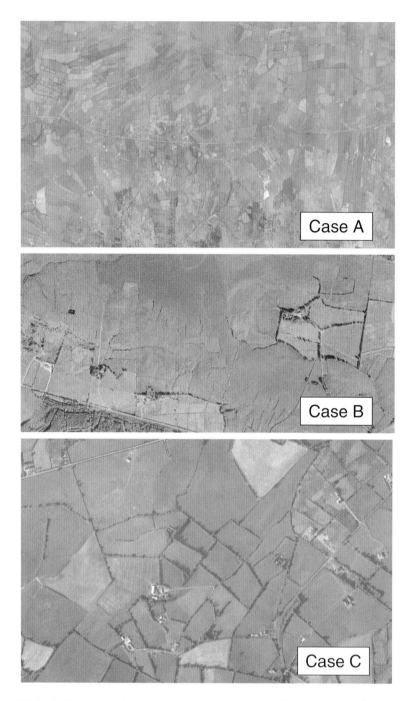

Figure 4.3 Pollution case study imagery and interpretations. See text for full details. Modified, Google Earth Permissions

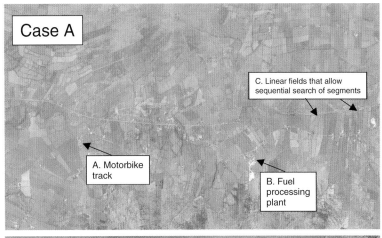
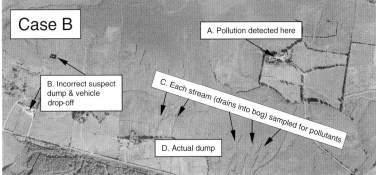
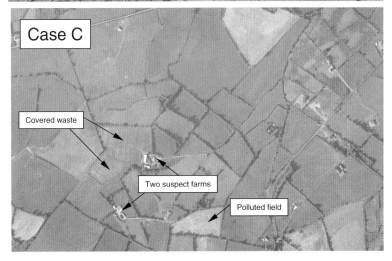

Figure 4.3 (continued)

Figure 4.4 Criminal and other cases where aerial photography assisted in obtaining information on what had occurred and where. See text for full explanation

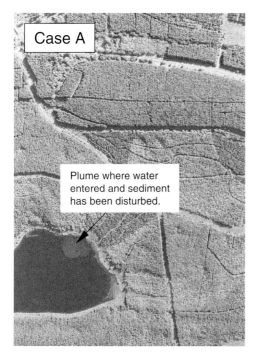

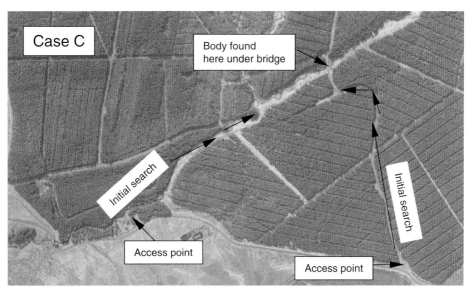

Figure 4.4 (*continued*)

body out into a lake, had occurred. When suspects were apprehended, one admitted to assisting in the murder, and to carrying out two trips into the water, the first to dump the body, the second to dump associated materials (clothing, etc.). When re-examined, two semi-circular plumes of sediment can be observed: the right hand plume being fainter than the left. The associate, when re-questioned, stated that the body (first trip into the water) had been dumped to the east and south of the access lane (fainter, older plume), the clothing, etc. to the west and north (less faint, younger plume).

In Case B, animal welfare officers raided a farm that witnesses stated was used for illegal practices such as badger-baiting and dog-fights. The raid failed to produce any significant evidence, and all areas were searched. An examination of historical aerial photography over the location showed a heavily-used track, leading away from the buildings to a location enclosed by woods. Officers visited this location, where abundant evidence of animal cruelty was found. A later operation, based on this intelligence and location, caught 22 people in the process of setting up a hare-coursing event.

In Case C, a student project was set up where 'suspects' in a murder were apprehended late one night. The mud on all their footwear indicated that they had visited a pine forest and freshwater. Pine forests are commonly quite dry underfoot, so the combination, during a period of dry weather, was unusual. CCTV imagery suggested that the suspects had been to a known forest location. An aerial photograph showed likely access points to the forest: these were searched unsuccessfully. The aerial photography was then used to initiate a search of the area, which was undertaken as part of the student training exercise. One party moved from an access point along the river, the other from a second access point to the river. Both 'search' parties arrived at an intersection – a location that could not be avoided, where the dummy body was found.

4.4 Infrared photography

Visible light is but one part of the electromagnetic spectrum, the red and green portions of which are closest in wavelength to the near-infrared and infrared (IR). Beyond these come thermal infrared and microwave, television and radio waves. These wavelengths overlap, and thus the spectral definition of each part of the electromagnetic spectrum is arbitrary. The near- and mid-IR (closest to light) do not transmit heat, whereas the thermal IR with longer wavelengths does. The reflectance of different portions of the IR spectra relates to the chemical makeup of the ground being imaged: IR is most commonly used in vegetation mapping, where the organic makeup of plant cover cannot be discerned from normal light photography, but is very evident with IR. Near-IR is recorded on regular colour film; it is the extraction of the photographic spectrum that allows the IR part of electromagnetism to be observed. Less complicated, however, is making specific colour IR film, in which green and red are specifically recorded in the dyes. This, a false colour film, reverses the natural colours, so that green objects are recorded as red, etc. This provides a very powerful imaging method when used in conjunction with a yellow filter, as non-IR light does not reach the film.

The development of colour IR film has a distinctly forensic origin, when in World War II the technique was used to differentiate painted camouflage from background vegetation, forcing the use of natural vegetation nets on personnel and armour to the present-day. The sensitivity of IR imaging allows the maturity, or timing of vegetation growth to be ascertained. Figure 4.5 shows just this, with a succession of vegetation contrasts evident that relate to the growth of crops. This tool is immensely powerful in detecting areas of disturbed ground, such as graves, mass graves and other large areas of soil removal/disturbance.

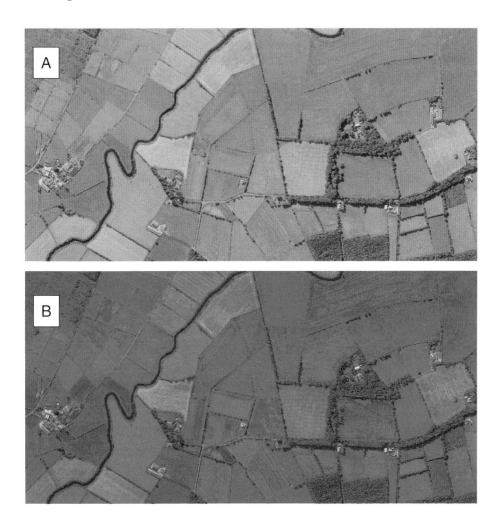

Figure 4.5 Colour composites from digital images acquired by a static balloon-based multiband system. A: normal colour photograph; B: Infrared photograph. Roughly, the ages or maturity of the vegetation is reflected in the redness displayed. The trees (early leaf development) are of deep red hue, with crops of successively lighter red to pink colours. The uses of such images in detecting the relative ages of ground disturbance are immense. Scale not provided

4.5 Elevation modelling

One of the primary uses of remotely-sensed data is to produce maps and computer models of terrain (height, slope angle, rough/smoothness) without recourse to direct measurement by traditional surveying with theodolite (to measure angles) and total stations (to measure distance). The visualization (by map or 3-D model) of topography is all-important in studies of criminalistics (crimes against humanity and warfare), as we saw in Chapter 2. Thus the time expended in traditional surveying is always worthwhile, but in investigations of what happened, where and when, can be self-defeating because once the surveyors are present, part of the job is done – predicting where to investigate. The advantage of remotely-sensed elevation data is that predictions of where and how events occurred can be made and the survey areas mapped prior to actual physical contact, preserving scenes, limiting danger to personnel (in unstable regions) and saving effort.

Satellite data

The most commonly used system that provides location (X, Y coordinates) and height (Z coordinate) is the United States military GPS. A Russian system GLONASS is also well-established and the more recent European Union's Galileo uses both its own and other satellites. Other satellite systems that will become operable are Beidou (China), GAGAN (India) and QZSS (Japan), all of which will deploy some form of connection with the United States and/or Russian systems.

GPS is limited in its areal coverage to the segment of space with sufficient satellites and location of ground control stations to provide the user with accurate X, Y, Z information. There are 24 satellites at some 23,000 kilometres height in stationary orbit continuously transmitting atomic clock times and position data by radio. Satellite position is maintained and corrected by the control ground stations. The end-user of the GPS system has a receiver that may be hand-held or in a vehicle. This is what most people familiar to the system know about, as this device acquires and stores the data. The device does this by taking information from at least three satellites, with a fourth control, whose position is known relative to the receiver. In ideal configuration, such satellites broadcast positional information that can be imagined as intersecting spheres, whose common point provides an X, Y, Z location. The more satellites providing information, the better accuracy is obtained, although the distance between satellite and receiver, and the absence of an atomic clock in each receiver, makes this form of satellite positioning crude, with X and Y commonly 5 to 20 metres in error and Z errors up to 30 metres.

The solution is to have two GPS receivers, one of which is stationary and precisely known, the other of which is used to make the X, Y, Z measurements, which can be corrected relative to the known stationary receiver. This method, known as differential GPS (DGPS) provides far greater horizontal and height accuracy, down to centimetres. This has an additional use in our studies of what happened and how, by allowing continual monitoring of a location, especially for vertical movement such as subsidence, imminent earthquake or volcanic activity. Geospatial imagery, remote mapping technology and the integration of GPS positioning are currently the main geoforensic applications to military and humanitarian monitoring operations (*Encyclopaedia of Espionage, Intelligence and Security*, 2004).

4.6 Photogrammetry

Horizontal measurements from aerial photographs are best achieved when taken orthogonal to the land surface (orthoimagery: commonly with a 15 to −25 metre resolution) and using either measurement of aircraft/satellite/balloon height or some reference distance on the ground. More challenging is the measurement of height, which is achieved by taking a stereo pair of photographs from different positions, but overlapping (Lillesand *et al.*, 2004). The relative displacement of objects allows calculation of their height, so long as orientation of photograph and camera platform are known: in the former this is best achieved by establishment of ground control points, which may be arbitrary objects, or in homogenous terrain, a marker (often called a monument). The hard-copy stereo pair are viewed by stereo viewers or optics, which magnify the same part of each image. Scanned or digital stereo pairs are viewed on a photogrammetric workstation. In each case the output is the same, a digital terrain model. The output of either manual or digital photogrammetry is dependent on image resolution, height of platform, camera type and method of measurement. The highest resolution cameras may achieve as low as 1 centimetre resolution, making sequential (repeated fly-overs) images from this method ideal for the monitoring of subsidence associated with burials or water abstraction, as well as other uses such as earthquake or volcanic stress build-up in the Earth's crust and the height of offshore waves that may cause tsunami.

4.7 Synthetic Aperture Radar (SAR) and interferometry

Radio Detection And Ranging (RADAR) uses radio microwaves of known wavelength to measure the distance between objects. The analogue is the traffic police officer's radar 'gun', which emits a series of radio waves that hit your speeding car successively and are reflected back to a detector on the gun. The compression of the waves, the Doppler Shift, gives a measurement of your speed. With pure radar distance measurement, the Doppler Shift is not needed; all that is calculated is the speed of the microwaves (speed of light) compared to the time of emission versus reflected return in microseconds. Radar devices may be mounted on aircraft or satellites and they can operate in all weather conditions: problems do occur with both the scatter of the wave as well as penetration into water and ice, which may then reflect and give erroneous measurements, unless the ground conditions are known. Resolution of ground features is dependent on the wavelength of the emitted (and reflected) radar wave (controlled by antenna aperture). Long apertures give fine resolution but require large antenna lengths (tens of metres), thus limiting what is possible without modifying the measurement.

This problem has been overcome by returning to the speed-gun idea of measuring the Doppler Shift. Synthetic Aperture Radar (SAR) measures the shift caused not by movement of the target (as with the speed gun) but between the relative movement between the antenna platform and ground target. As the platform flies, the same area is successively 'hit' by radar waves, each from a point when the emitter (platform) was in a new location. Information reflected from the same spot will be offset slightly in time: these reflections (or echoes) can be stacked, to sharpen the resolution. Such information may be used in the same manner as photogrammetry, where adjacent or overlapping images are correlated for height offset and digital data calculated (radargrammetry). A commonly-used alternative method is interferometry, wherein two antenna images are directly compared and the offset between is known and used instead of the image offset. A known baseline height (at the start of the survey) can assist greatly in calculating change, although this can of course be done relative to the successive measurements, with no absolute height above a datum given. Thereafter, the relative height difference between the two antennae may be calculated for every pixel in a digital image, producing high resolution elevation models. The use of digital data allows more sophisticated methods of interpolating between data points in GIS, thus increasing the resolution of the data still further. Differential interferometry is the equivalent of the repeated fly-overs of photogrammetry, wherein a succession (three or more) of images allows height changes (subsidence, tectonic stress, water height, glacier movement) to be recorded. In some instances aerial photography (and photogrammetry) may achieve better resolution than SAR; the quality of the two methods is roughly comparable, so the weather-independent nature of SAR gives this method an advantage. A further advantage is that satellites are not subject to such easy detection from the ground, and thus possible hiding of people and objects or hostile ground-to-air fire. Aircraft are however cheaper and more readily available, and photogrammetric data requires minimal computer processing.

4.8 Multispectral and thermal imaging

Early light cameras and IR film and filtered cameras operated on the basis of capturing all visible light, omitting certain portions (e.g., yellow) and recording others (e.g., green and red). The multispectral scanner also captures many electromagnetic radiation sources, but then selectively senses different parts of the spectrum for different uses. In addition, each part or band of the sensed spectrum can then be compared to another, creating a differential that may be far more informative than either the collective or each single image. A very powerful form of scanner is that which detects the thermal spectrum, although as with all forms of remote sensing, the area scanned dictates what resolution and definition is derived: an aircraft-mounted scanner that is imaging a large area (or footprint) will cover a large area quickly, but with poor resolution, so that thermal objects will overlap and consequently have poor definition. A smaller footprint will pick up discrete targets, but will accumulate a large amount of data, requiring sizeable computer memory and processing capability. Thermal radiation is not a simple matter of heat emission, with variations in storage (radiant and kinetic), material heat release (emission), atmospheric diffusion and movement and diurnal variation. The latter is particularly important, as heat-reflecting objects may be preferentially-imaged in the daytime, and heat-retaining objects at night, depending on the heat of the background substrate.

Thermal image interpretation is subject to high levels of operator experience and interpretation, as well as knowledge of ground conditions. The range of applications to criminalistics, humanitarian and environmental law are enormous, because the thermal image and a range of thermal images captured through time provide information on what people have been doing, often inside buildings or under cover. For example, take the simple matter of heat escape. In illicit drug or explosives manufacture, large quantities of energy and noxious chemicals may be used, requiring air extraction by mechanical means, or windows to be opened. Another major application for thermal imagery is studies of natural phenomena such as volcanic hazards and plumes of warm water associated with problems such as algal blooms in rivers, lakes and seas. In addition, Lillesand *et al.* (2004, p. 358) give an excellent example of the monitoring of warm water discharge from nuclear power stations. Finally, a more direct forensic application is the location of vehicles, most especially from car tyre marks (cars driven at speed, such as those stolen or involved in robbery, give high heat signatures). Lillesand *et al.* (2004: Figure 5.20) show an excellent example of where a helicopter had been previous to take-off in warm sunshine by the 'cool' shadow it left on the tarmac.

Case study: Detection of 'disappeared' victims, Pueblo Bello, Colombia (Equitas team)

Equitas is a non-profit scientific organization dedicated to assisting Colombian families whose relatives have disappeared as a consequence of the internal conflicts in that country. The Equitas role is active (defining, exploring, excavating and recovering remains for repatriation), advisory (providing best-practice methodology for other teams or organizations) and as monitors (e.g., peer-review) of other activity in this field.

Context. For over 30 years the various fighting factions of Colombia have been covertly burying the victims of their activities (Gomez-Lopez & Patino-Umana, 2006). Paramilitary, guerrilla and governmental armed forces have buried victims in a variety of ways and at different scales. These comprise the abducting of individuals, groups or whole sections of communities, thus replicating both the individual 'disappearances' of the Mafia in Italy and the USA or of the IRA in Ireland, as well as the genocide of communities (or parts thereof) in Bosnia, Cambodia and Rwanda. Although some periods in Colombia's recent history have seen an increase in such activity, it has never been as intense as was seen in the latter three countries. This makes Colombia unique in having large numbers of disappeared, sometimes in mass graves, but buried in widely-spread locations over a long period of time. The situation in Iraq shares some similarities, albeit driven by the (past) government and in a very different geographic environment. The methods used to locate burials in mass graves, short-period, and individual disappearance can be applied, with modification, to Colombia. Modifications to known methods include the rapid vegetation re-growth, changing land-use and humid climate, as well as anthropological challenges of ageing any human remains.

Pueblo Bella is in the northwest of Colombia and it was from here in early 1990 that a paramilitary group abducted up to 43 peasants. Witness statements reported that the victims were taken into the adjacent region of Monteria to an area called Las Tangas on the Sinu River. Here they were purportedly tortured and killed before being placed in mass graves. 24 of these victims were recovered later that year by Colombian authorities, whose heavy machinery caused damage to the grave sites (Equitas, 2006). Subsequent searches of the area by government-led groups (2005 and 2006) have not resulted in any further discoveries. The area comprises dry and wet grassland, low scrub, tall mature forest, some cultivated manioc and banana around the meandering Sinu River. A seasonally-humid climate causes flooding of the river with alluvial deposition.

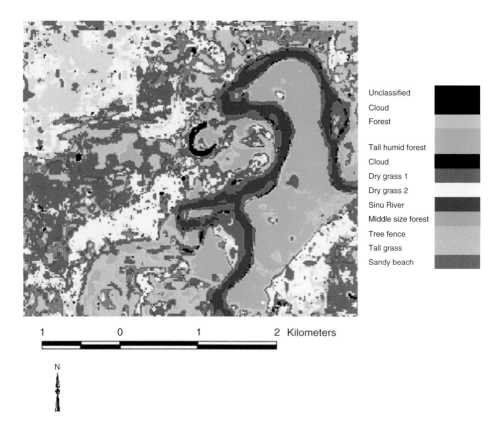

Figure 4.6 The preliminary supervised land classification of 2002 LANDSAT image, showing ground cover. Image by Alejandro Dever of EQUITAS, with permission

Method. The Equitas method follows many of the guidelines developed by previous workers (Davenport, 2001a; Hunter *et al.*, 2005; Killam, 2004), as well as in this work. LANDSAT data were combined with other historic aerial and satellite data of low resolution to provide a land-use and land-cover map (Figure 4.6).

This incorporated witness reports of the area and resulted in priority areas for searching. Higher-resolution QuickbirdTM imagery, filtered for green, blue and infrared bands provided further and refined information on potential burial sites. Two areas coincided with witness reports, access locations, etc., with one area having distinct groups of pixels that could be burials: this area was searched first by digging 6,000 test pits, with no results. The secondary area, with good overall aspects but no individual locations, was investigated by digging 60 test pits in areas with thick vegetation. This secondary area yielded a burial site with three individuals in it. Subsequently it has been assumed that uncontrolled vegetation growth was allowed in this area, obscuring remotely-sensed data and hampering excavation. The success of the work was in defining the prioritized search areas, while interpretation was made difficult by vegetation growth, reinforcing the introductory remarks, that the area is problematic to interpret by reference to other geographic locations (temperate, arid zones). This problem of high rates of vegetation growth resulted in the opposite outcome to that usually observed. The first area had attributes of both scenario-based (witness reports, access roads) and feature-based (specific roads, pixel groupings) search criteria, and should have had a good return on effort invested. Instead, 6,000 test pits produced no results. The second case had only 'scenario-based' (terminology of M. Harrison, personal communication, October 2007) criteria, with no features (high vegetation) and yet was successful.

4.9 Hyperspectral imaging

This advance in imaging from the multispectral methods described above measures upwards of 200 spectral bands from the visible to thermal IR electromagnetic spectrum. These are calibrated against the known spectra of materials of interest (clay, water, rock, plants), such that high-resolution geological, soil and vegetation can occur. Like multispectral imaging, it is the study of natural features that dominates the literature. Multispectral sensors such as the CASI (Compact Airborne Spectrographic Imager) system can also easily identify gas plumes from processing and buried materials, especially when used to sequentially monitor a target over time.

4.10 Satellite mapping

Apart from delivering weapons payloads and putting humans into space, Earth imaging for military and scientific uses has been central to the development of rocket propulsion. The launch of the LANDSAT series of satellites (precursors in 1972, LANDSAT in 1975), with increasing Earth cover, sophisticated cameras and imagers was the impetus for a range of imaging technology to be placed in space. Of particular interest to those monitoring changing ground conditions are the high-resolution generation of satellites, IKONOS, EROS, Quickbird, SPIN and EO (hyperspectral imaging system). These satellites have the resolution to map large burial sites, as well as to provide spectral information on ground conditions, an excellent application of which is the discovery of mass graves, hidden armaments stores and covert waste repositories.

Case study: Silvestri *et al.* (2005) on illegal waste assessment, Veneto Region (near Venice), Italy

The rapid urbanization and associated industrialization/population growth in this area of Italy created massive quantities of industrial, manufacturing and urban waste before recycling and disposal mechanisms were developed. The result is the existence of a range of small to large covert burial sites on a very flat landscape, underlain by lagoonal/floodplain silt, sands and muds with high water-tables. The flat landscape makes oversight and topographic division (see Chapter 2) difficult. Silvestri *et al.* (2005) develop a simple and effective methodology for the classification of anomalies, based on the comparison between thermal imaging and stressed plant communities from light photography using a number of satellite sources. This relies on both the heat generated from such locations as well as the toxins present closer to surface. In this way, deep burials with 'healthier' cover soils, as well as shallow, possibly thinner and cooler but toxic masses, are recognized. Their work sets a standard for similar warm and dry floodplain/reclaimed lagoonal regions: how applicable the work is to other topographies and climate regimes is examined below.

Case study: Spectral thematic mapping – large covert burials

The island of Ireland is divided between the mostly southern Republic of Ireland (26 counties) and 6 counties of Northern Ireland, part of the United Kingdom. Each part has different jurisdiction and enforces European environmental law to different degrees. The border areas (north and south) have always been a focus for lawlessness, problems of jurisdiction and smuggling, partly as a consequence of varying commodity pricing, partly because of the need for extradition (should a crime be committed one side and an escape made to the other) and also through low population concentrations. During the period of the 'troubles' (late 1960s to late 1990s) the area was the focus of the most intense republican terrorist activity, which has left its legacy in difficulties of policing and a tendency to flaunt the law, including environmental jurisdiction. The high cost of waste recycling, especially of toxic (medical, chemical, industrial) materials has given rise to illegal removal of such waste, under the cover of proper facilities for covert burial. The most severe cases involve burial of waste from the Irish Republic (population 3.5 million, with very stringent recycling measures) in Northern Ireland (population 1.5 million), which is hidden in the low-population density, sometimes under-policed border areas where local residents are used to 'turning a blind-eye' to their neighbour's activities. A complex system of removal is used: sealed lorry-trailers are swapped en route to confuse those monitoring traffic; a person is asked to dig a hole or release an abandoned quarry, sand-pit or former excavation (often where land ownership is disputed or unknown) for a favour or small cash reward; a third party then ensures the waste can be covertly brought in and covered. In this manner, such waste disposal practices are analogous to other sophisticated covert burials of genocide victims, dead animals, waste that flaunts international agreements and large amounts of stolen or hidden goods.

The detection of such illicit burial sites has developed using the sequence of methods outlined in this chapter, using publicly-owned data as far as possible, progressing to the purchase of specific satellite or aerial images, or the commissioning of the latter, with dedicated imaging equipment. An obvious question arises, and that is 'why not proceed directly to the most sophisticated method with best resolution output, avoiding the time-consuming background work?' There are three answers. First, publicly-held topographic maps and aerial photography may provide presently-obscured historic information on the location of access routes, former excavations and vegetation and at a scale where an overview of how landforms vary and routes may interconnect. Second, purchasing all the highest-resolution data available may swamp the interpreter with data, who has no geographic focus to his or her work. A large number of targets will be identified by this means, possibly wasting time and money. Third, different successive methods are required as not all waste is the same, and thus a range of remote sensing methods are required to obtain a result. *Domestic* waste decomposes and may ferment, depending on variables such as oxygen and water availability and micro-organism activity, resulting in liquid (leachate) and gas generation. *Commercial* waste is commonly generated by dumping or uncontrolled release of chemicals, hydrocarbons (fuel laundering by-products), oils, acids, dyes and solvents. These are commonly already contaminated (e.g., from rusting containers, by rainwater ingress) and thus of no commercial value incurring a cost to recycle. They frequently have a highly negative impact on vegetation, which can be assessed by remote sensing (see below as well as Brilis *et al.*, 2000a, 2000b). *Medical* waste is often a mix of the materials found in domestic (decaying organic-based materials) and commercial (toxic chemicals) wastes. The likely inclusion of pathogens and this mix of waste types makes medical waste particularly hazardous, but in small volumes that may defy detection by remote sensing. Construction and demolition waste is often non-organic and may be non-toxic in its original state. Minimal thermal heating will occur and slow release of toxic metals such as copper will have limited short-term impact on vegetation. Consequently this is a difficult type of waste to detect: the very large volumes of infill make visual identification possible, although in our experience this can also prove a problem, when one is looking too closely at imagery and not being open-minded to the idea of waste tips being very broad and shallow, narrow and deep, or worst of all, both.

Road or route maps provide information on the possible routes that illegal cargo has travelled: substantial quantities of material can rarely be moved for long distances without firm ground, tracks or roads. Such maps need to be discussed with those who have local knowledge akin to the criminals, such that police vantage posts, check points and common times of operation are known. Topographic maps provide information on land use, especially in the past, and are invaluable as a cross-reference to satellite imagery. These should be compared to soil, drift geology and solid geology maps, in order to establish both 'normal' conditions, as well as what material will have been excavated: databases of sand, rock, clay and other extractive operations are of great benefit in this regard, although they may not be comprehensive. When hyperspectral imagery is obtained, this type of information (actual records of what is on the ground) is critical as a ground-truth to mineralogical determinations. Aerial photography is often patchy in coverage and type: GIS allow the merging of data types such as panchromatic and colour images as well as orthocorrection for accurate measurement. Topographic maps, aerial photographs and satellite images may each be switched in priority of study, depending on how extensive the data are: generally the data type with largest coverage is used first, as a template, onto which patchy data sources are laid. This allows any gaps to be considered for specific aircraft or satellite flyovers. Pre-existing data sources may be sufficient to provide enough intelligence for a ground visit. Good examples are where topographic mapping has shown a disused quarry and yet remote sensing indicates infill, with no obvious reason (e.g., land reclamation).

If suspect areas are identified but not characterized as landfill sites or burials, then specific remotely-sensed imagery can be purchased from a database or by fly-over. Interpretation of these data sources follows that of the pre-existing data, referenced to legal waste dumps, each of which (*in italics*) is now considered with a brief example of why. *Black and white* and or *colour photography* provides information on access routes, changes since topographic surveys or other photographs were produced and, if supported by elevation data, can be used for the creation of digital terrain models (DTMs), such that lines of sight (from roads, dwellings) may be established. *False colour and near infrared* images provide information that is especially pertinent to what cover has been used to hide the burial, especially vegetation types: these two data sources may be used to map either vegetational changes caused by stress, using the NDVI (Normalized Difference Vegetation Index) or variations in soil (using SAVI – Soil Adjusted Vegetation Index). NDVI is assessed by measuring the change in reflectance from visible red to the near infrared, known as the 'red edge'. This is best measured by conversion to greyscale, with healthy vegetation showing white and poor growth in black (Figure 4.7).

Thermal imagery is especially useful where decaying organic matter raises the heat of the burial: porous waste (concrete, plastics, large frameworks such as vehicles) has the opposite effect, allowing cooling water to percolate and lowering the temperature of the site. For 'hot' sites, thermal imagery collected at night or in the winter can provide the greatest contrast, the opposite being true for 'cool' sites. Either way, images are best collected in the early morning of both winter (for mineralogical analysis of soil, SAVI) and summer (to show plant stress, NDVI) counteracting any heating from the sun.

Hyperspectral imagery is especially useful in the identification of burial sites, regardless of heat or very careful burial. This is because the spectral reflectance values of transition metals (such as iron) are very diagnostic for minerals, and thus reworked ground, where naturally buried layers are brought to surface or foreign materials brought in, will be detected. Hyperspectral imagery is also used successfully in conjunction with false colour and near infrared data for vegetation mapping. No illegal (or legal) landfill sites so far recorded by the authors have established a vegetation cover that is identical to the surrounding area. The spatial large warships (aircraft carriers) and in large installations, where direct measurement every few centimetres was impossible. The laser was shone along the length of each turbine in order to determine if any leak occurred, and then if positive, the laser was directed sequentially across the suspect turbine in order to direct those taking direct measurements to the location of the leak. Pollutants and atmospheric organic vapours are now routinely measured with a range of methods including infrared, NMR and Raman spectrum.

The oil and gas exploration method is of special interest, because the Russian scientists claimed they could determine a few parts per million methane (albeit in desert) over a few kilometres. Ruffell (2002) took this claim and reasoned that over shorter distances (metres to tens of metres), with better standards and less wind/other methane sources, smaller quantities of methane should be detectable. This reasoning was mapped onto the US navy method of a grid pattern of laser fire and detection points that would lead to definition of the target. Ruffell (2002) hinted at the implications for the search for buried bodies, certain drugs and toxic waste when he considered the sensitive issue of whether a body was buried in a suspect back-garden, and yet where to issue a warrant and search may cause political instability (Figure 4.8).

Case studies: State of the art for forensic remote sensing

Keith Challis (University of Birmingham) and Julian Henderson (University of Nottingham) (*http://143.117.30.60/gis_web/Frnsic-Rmt_Talksdoc.htm*) have shown eloquently how, in the absence of aerial photography, landscape reconstruction in areas such as the Middle East, depends on satellite remote sensing. Since the late 1990s a new generation of high-resolution civilian satellites have provided access to imagery of sufficient spatial resolution to resolve archaeological landscapes and burial activity. Since 1995 the United States government has made available declassified intelligence satellite imagery, acquired between 1960 and 1975 by the Corona and Lanyard satellites. This imagery, of up to 1.5 metres spatial resolution, extends the high resolution remote sensing record back into a period before the widespread destruction of archaeological landscapes. Challis and Henderson use the Corona imagery in landscape reconstruction in areas where traditional sources of aerial photography are hard to come by and conventional remote sensing is of limited use.

Martin J.F. Fowler from South Wonston, Winchester, has worked extensively on aerial photography and satellite data in archaeology. His work is of great interest to the forensic practitioner. As Dr Fowler observes, 'archaeologists have used aerial photographs acquired by aircraft flying at low and medium altitude to search for the remains of ancient man-made features and landscapes. Over the past 20 years there have been significant advances in the availability and quality of publicly accessible satellite imagery of the Earth's surface and relatively high spatial resolution satellite imagery is now available for most parts of the globe at costs per square kilometre that are becoming comparable with conventional aerial photography. It is therefore not surprising that this period has also seen a growing interest in the exploitation of satellite remote sensed products by the archaeological

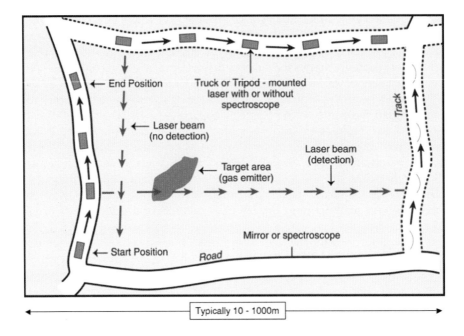

Figure 4.8 In this case, laser fire and detection points could be established that do not require any contact with the suspect ground: in the domestic situation this of course has dangers associated with covert surveillance, and the technique is subject to local and national law. This may not apply to war crimes and humanitarian situations. The same rationale can be made of mass graves, protected by unfriendly militia; toxic waste ground where human contact may cause health and safety problems; and areas of subsidence (karst, quicksand, bog) where personnel/equipment may need floating platforms for direct access. Subsequent to 2002, the known importance of infrared spectra has been confirmed. Coates (2000) makes two important statements: 'The vibrational spectrum of a molecule is considered to be a unique physical property', and 'The spectrum is rich in information'. Coates goes on to show how at the microscopic scale, materials such as hydrocarbons (petrol, oil, gas), amino acids, organic carbonate (e.g., bone) and nitrogen-based compounds (e.g., fertilizer, compare explosives and environmental damage) may be analysed, understood, and compared/excluded by infrared spectroscopy. The problems of open-air, non-laboratory controlled measurement still pertain (as above). Modified after Ruffell (2002)

community' (Fowler, 2004). Remote sensing satellites have advantages to aerial photographs in that many satellite data sources are independent of cloud cover and have both good resolution as well as widespread coverage, allowing the landscape interpreter to see the context of features at a wide range of scales. As noted in the earlier section 'Case study: Spectral thematic mapping – large covert burials (page 110), the range of data ages and types is actually an advantage in terms of observing changes through time but also by creation of a statistically-sounder, multi-proxy hierarchy of targets.

4.12 Laser scanning of scenes and objects

This method works on similar principles to the satellite and airborne systems, with a laser-emitting source and reflected-light detector. Most especially, this land-based method is akin to the side-looking radar developed for all-weather and night-time military operations through the 1940s. These military applications often failed, due to the 'clutter' of trees, topography, etc. that obscured the target: nowadays it is just these topographic features we are interested in! This method uses LiDAR instead of radar – laser light instead of radio waves. The operator chooses an area or object to be scanned, which again uses a pulsed light source, moved sequentially over the target. The time shift between each reflected point gives a highly accurate measurement of distance, which if 'truthed' from one or more points, can be converted to absolute information, presented as X, Y, Z points. The different reflectance of each point can also be measured, giving some indication of changing materials as well as topography. The method is subject to interference from precipitation and intervening objects (trees, posts), and is best suited to the gathering of rapid (a few minutes), dense topographic information, from which digital elevation models can be made and features extracted automatically. The advantage the system has over conventional photography is the acquisition of both topographic information and absolute reflection data (Figure 4.9).

Reflectance is akin to the seismic method in geophysics as it is not the strength of the returned light that is measured, but the change in its wavelength. This provides a measure of LiDAR intensity and can assist greatly in mapping different features on the ground such as vegetation types, rocks and soils, and building surfaces. The method is currently being used in capturing images of scenes of crime, conjunctively with photography, to prevent/monitor contamination and movement. This can also be used sequentially, as long as the laser is left in place for the period of monitoring, or returned to the same location (by marker or DGPS) for comparative data collection. The output also allows visualization from a number of angles, depending on the number of locations from which scans have been taken. Current laser scanning technology does not replace a walk-over and photography of a crime scene, but provides enormous additional resources to the investigator, who can walk about a scene in virtual reality before entering. As with any technology, the need for laser scanning should be balanced against cost, time and effort. The method works in darkness and in enclosed spaces: for fire investigators with limited visibility, or for bomb disposal squads who need to assess where a suspect device is located and how much room is available, the method shows great potential.

4.13 X-ray imagery, X-ray tomography and neutron activation

At the smallest scale of remote sensing we find geophysics and the non-destructive imaging of small objects. It is hard to know where to include such methods – in a chapter on geophysics or on remote sensing. As the main small-scale (tens of centimetres) methods utilize electromagnetic radiation, we include them here to avoid repetition of the basic principles. This is purely for convenience and the reader

(a)

(b)

Figure 4.9 Laser scan of a quarry face and catastrophic rock-fall. Cliff face is ~40 metres high. The data are displayed at the reflectance values, showing that the bulk of the face and fall comprises one rock type (basalt), with some layering visible (lighter reflectance) at the top of the face. From this information the volume of the rock-fall can be quickly and accurately measured, the correct angle of slope of the fall and the faces be measured and a 3-D visualization of the area provided. Thus any secondary hazards can easily be established along with equipment or infrastructure buried by the rock-fall; a plan for extraction may be activated. Quarry face is 40 metres high. (b) Laser scan of the same quarry face, with the laser point-cloud of data points (giving topography and reflectance) overlain onto colour (RGB) values from digital photographs taken sequentially as the laser images the quarry face. This negates the need for the photographs to be rectified (twisted back to assume the camera lens was at 90° to each surface), because the photograph is draped onto each topographic point. Like (a), this image can be manipulated in 3-D to give a 'walk-around' or fly-through view of the area. Images courtesy of Conor Graham

ought to be aware that these forms of remote sensing are almost not remote at all, with direct physical contact between the measurement device and area/object of interest being common.

When medical X-rays are produced, a thin metallic sheet is placed between the emitter and the target, effectively filtering out the lower energy (soft) X-rays. The resultant X-ray is said to be *hard*. Soft X-rays overlap with ultraviolet, hard X-rays overlap the range of 'long'-wavelength (lower energy) gamma rays; however the distinction between the two terms depends on the source of the radiation, not its wavelength. The basic production of X-rays is by accelerating electrons in order to collide with a metal target (usually tungsten, but sometimes molybdenum). Here the electrons suddenly decelerate upon colliding with the metal target and if enough energy is contained within the electron it is able to knock out an electron from the inner shell of the metal atom and as a result electrons from higher energy levels fill up the vacancy and X-ray photons are emitted.

X-rays are detected by photographic plate, and this plate or film is used in hospitals to produce images of the internal organs and bones of a patient. The part of the patient to be examined is placed between the X-ray source and the photographic receptor to produce what is a shadow of the internal structure of that particular part of the body being X-rayed. The X-rays are blocked by dense tissues such as bone and pass through soft tissues. Those areas where the X-rays strike the photographic receptor turn black when it is developed. So where the X-rays pass through 'soft' parts of the body such as organs, muscle, and skin, the plate or film turns black. Contrast compounds containing barium or iodine, which are radio-opaque, can be injected in the artery of a particular organ, or given intravenously. The contrast compounds essentially block the X-rays and hence the circulation of the organ can be more readily seen.

The most important advance in X-ray technology following its discovery in the late 1890s was the development of computed axial tomography (CAT or CT scanning). Of great interest to geoforensics is both the specimen that appears homogenous, and thus frustrates further analysis, and rock, soil or sedimentary material in which objects have been covertly hidden, but for logistical reasons of time, or of preserving evidence, or (as in the PINS method discussed in the 'Portable Isotopic Neutron Spectroscopy (PINS) chemical detection systems' section, below) for reasons of investigator security, destruction of the specimen must not be attempted. Since all rocks, soils and sediments possess variations in grain/mineral size, composition, fabric, cementation, X-rays transmitted through the sample will be absorbed to varying degrees. Often, visually-homogenous materials, such as sediments and sedimentary rocks, reveal an internal structure that has been obscured by weathering, stone-mason cutting, etc. when irradiated.

What these variations mean is often more difficult to find out, requiring measurements of mineralogy, permeability and density. X-ray radiography has a major advantage over many other means of analysis, in that it is non-destructive. Most instruction manuals and textbooks on the X-ray analysis of materials such as rocks consider a 2 centimetre thick slice to be ideal for analysis. The case study (below) shows that military-grade X-ray inspectors can easily penetrate rock of 5 to 10 centimetres thickness. Stereoradiography allows something of the internal shape of the rock or sediment interior to be ascertained, as with CAT scanning, whereas micro-X-ray radiography can be used to examine minute changes in materials, such as precious materials that have been tampered with: examples include faked diamonds or claims to have created diamonds and the embedding of modern organisms in amber (see Chapter 7).

Case study: Fraudulent claims of oil discovery in China

A geologist who specialized in the location and study of oil seeps and natural bitumen returned from fieldwork in western China with outcrop hand-specimens and borehole core of sandstone of the same texture as oil reservoir rocks in adjacent basins. One control specimen had no oil impregnation: his key specimens had patches of oil staining, which he used, along with a mix of real and fabricated maps and notes in attempts to persuade capital investors in his oil exploration venture. Being somewhat cynical of the geologist's claims, one investor sought to test the oil-stained specimens. Ultraviolet light confirmed some surface residue of oil: 'the live oil is held inside the specimen' was the somewhat leading response of the geologist. The outsides of the rocks were cut off and the core 'slabbed' (cut in half, down the middle), leaving the interior exposed. These too showed oil under UV light and were subject to X-ray imaging using a portable X-ray (the Foxray 2000 mobile X-ray source and detector), which confirmed the presence and internal location of 1 centimetre diameter spots of oil stain (Figure 4.10). These were examined, and in some a small hole was noted at the centre. The blocks were then subject to a CAT scan, which elucidated the elongate geometry of the spots, allowing the ends of each oil-stain tube to be examined, some of which had a small (0.5 millimetre) hole. Scanning Electron Microscopy of such holes showed them to be bored after the formation of the rock, and to have cut edges. A raid on the geologist's laboratory (a converted garage) recovered a microdrill with 0.6 millimetre tungsten bit, veterinary syringes with 0.4 millimetre, oil-filled needles and a bottle of commercially-recovered crude oil. The geologist had drilled into the boulders, injected the crude oil, washed and roughened the boulder outer surface, safe in the knowledge that the interior of the boulder would be subject to greatest scrutiny, especially if suggested by himself, while the majority of the microdrilled injection points would be cut off, destroying some of his handiwork.

Portable Isotopic Neutron Spectroscopy (PINS) chemical detection systems (EG&G ORTEC)

Originally developed for non-intrusive classification and segregation of unidentified munitions, the ORTEC PINS system is used as a key tool in a variety of applications where it is necessary to determine the chemical contents inside a container by a non-intrusive method. A nuclear interrogation technique 'prompt gamma neutron activation analysis' determines container constituents without invasive examination. PINS is used in combination with X-ray imaging, which establishes the rough geometry of any infill, after which a non-intrusive chemical assay system such as PINS is required to identify the fill material. The system deploys neutrons from a

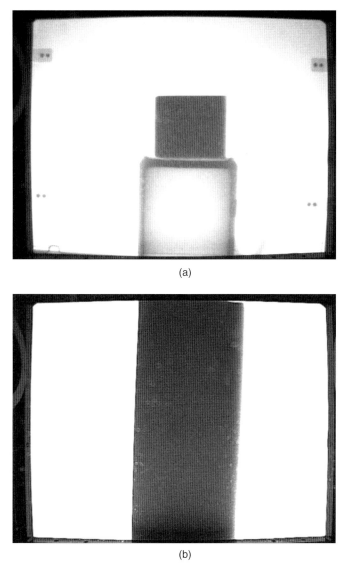

Figure 4.10 (a) X-ray image of the cut block of 'oil impregnated' sandstone. The circular areas of oil can be noted on an upturned cup. Block is 10 centimetres in width. (b) X-ray image of the 'slabbed' core. Similar circular areas of oil are apparent. Core is 25 centimetres long

radioisotopic Californium-252 source to interrogate the contents of an item. The neutrons pass through the wall of the container and collide with the atomic nuclei of the contents. As the neutrons interact with the nuclei of the chemical elements within the item, one or more energetic gamma rays are emitted and detected by a high-purity germanium (HPGe) gamma-ray spectrometer. Each chemical element will emit a characteristic gamma-ray energy/intensity, and the element can be identified from this gamma-ray spectra. The fill chemical or chemical mixture is inferred from the presence and relative abundance of various key chemical elements. These energies and ratios are in themselves characteristic of the contents of the container, whether high explosive, nerve gas, or other. In this way, with a 'library' of known signatures, it is possible to determine the contents of the container.

4.14 Field Portable X-ray Fluorescence (FPXRF) spectrometry

Bergslien *et al.* (2006) discuss some of the uses of the FPXRF device in forensic geoscience (see Figure 4.11). They note that FPXRF

> has become a common technique for many environmental geoscience applications. Its key characteristics are ease-of-use in a variety of environments, low sample preparation requirements, and rapid turn around time, usually in the order of minutes. These characteristics also make it an ideal tool for use at crime scenes where rapid, non-destructive screening of materials may be necessary. A typical FPXRF system uses either an X-ray tube or radioisotope as an excitation source to irradiate samples. The incident X-rays interact with the samples' atomic structure by knocking electrons from their inner shells, leaving vacancies that are filled as outer shell electrons release energy to fall in to new ground states. The energy released will be an X-ray equivalent in energy to the energy difference between the two shells. Since each element has a characteristic arrangement of electrons, the X-rays released by such transitions will be unique to that element, allowing its identification. By comparing the intensities of X-rays from an unknown sample to those of a suitable standard, elemental composition can be quantified.
>
> Bergslien *et al.* (2006)

The main problems with the method are detection limits (small quantities of certain elements), overlap in the generated spectra causing confusion in the assessment of element abundance and the inability of the system to detect elements lighter than phosphorus (silicon, oxygen, aluminium and magnesium, all major elements in rocks and soils). Bergslien *et al.* (2006) discuss the following applications:

- assessment of whether cremated human remains occur at a non-compliant crematorium;

- tracking of toxic metals to their source;

- assessment of arsenic levels from wood preservative in a school playground;

- traditional suspect to scene comparison (conjunctive with other methods).

Perhaps the most significant use for FPXRF has yet to be realized. In Chapter 6 we see how the problem of how to sample a scene of crime in a representative manner has yet to be addressed. Current lines of thought include: consider how the

Figure 4.11 The Portable Field X-ray Fluorescence Spectrometer in action. Photograph courtesy of Elisa Bergslien

questioned sample (from the suspect, victim) compares visually and by grain size at least (Murray, 2004) to materials at the scene; sample access routes and transferable material footprints (Horrocks *et al.*, 1999); sample a grid of a large enough number to test statistically (McKinley & Ruffell, 2007); or use a proxy measurement of soil/sediment/loose material variation and sample each 'type'. The most commonly cited proxy is remotely-sensed data or geophysics because such sampling exercises have been most common in rural locations at large-scales with fly-over, satellite or gridded geophysical possibilities. An urban carpark or basement burial location presents major problems for the deployment of such sampling proxies. Simple proxy measurements for general background variation could include pH, resistivity, gamma-ray activity, all of which may not measure features of (most likely dried, possibly mixed and compromised) soil or sediment from suspect or victim. The use of the FPXRF over the area under question provides data on the surficial chemistry that can be related to the chemistry of a sample: in Figure 4.11 we see just such a scenario, where rapid assessment of onsite variation has to be made in order for sufficient samples to be taken. The converse is also true: say a questioned sample shows anomalously high or low elemental values of something like potassium (K) and lead (Pb) : the existence of such material can be searched for using a device such as a FPXRF, avoiding the laborious collection of hundreds of samples, most of which are excluded. Anomalous (high or low) concentrations of key elements may act like the unique particles or identifiers discussed in Chapter 8 (trace evidence), which provide a non-uniform distribution compared to background and thus some measure of exclusion–comparison between suspect, victim and scene. For sampling, the FPXRF has really shown forensic geologists the future: sample representativeness can only become more of an issue in assessing how robust a criminal case is; any use in court or investigation of the word 'bias' will generate unbalanced opinions.

When a portable mineral identification machine and field-based assessment of organic particles can be made like the FPXRF, these problems will start to have been solved.

4.15 Some conjecture on the future of remote sensing applications

Scale/resolution/accuracy

Google Earth is an example of how future advances in scale and resolution will be made with regard to research-level remotely-sensed data. Scale issues include the fact that the whole earth has some form of image that can be successively zoomed in on, becoming patchier in resolution depending on use and interest. Eventually larger and larger areas are displayed at increased resolution, when the whole dataset is again improved. The ultimate resolution is unlikely to be limited by technology but more by legal privacy issues. Access to and the resolution of other remotely-sensed data (UV, IR, spectral) will likely mimic Google Earth, again to the same limits. The future thereafter lies in the hands of the data owners, not the scientists who gather and produce such data. GPS links between satellite, aerial and ground-based data points will allow GIS integration: the ground-based measurement of spectra for both its own sake and also 'truthing' of aerial data is already well-established. The next stage in things such as spectral image classification is to actually sample the ground: advances in technology like FPXRF (for solid geology and soils), gamma-ray emission (for radioactivity) and volatile measurement using infrared spectra will make physical sampling a third (not second) tier of remote and then ground-based remote sensing.

Interpretation

The increase in automated classifications either from single sources or through multiple sources in a GIS has advantages and disadvantages for the assessment of remotely-sensed imagery. Advantages include spatially-comprehensive and mathematically-consistent machine examination of all the obtained information. This of course replaces neither the complexity of the human eye nor interpreter experience. The latter has advantages and disadvantages: the best interpreter is the one who has successfully interpreted images, but then how does a person whose positive experience and success with one interpretation adapt to be open-minded and see the unexpected? Machine-dependent classification systems can be taught, thus enabling both identification of one layer of targets from instructions provided by the human operator (e.g., look for all pixels/groups of pixels with feature x), as well as a set of targets determined to be anomalous by the machine. In this way criticisms of over-automation can gradually be limited.

5
Spatial location and geographic information science

> In an important case the circumstantial evidence had been brought together and conclusions thereby suggested drawn, results which might have been of decisive importance in clearing up the case. At the last moment it came in to the head of some outsider to ask if the distance between two points was really two thousand paces. That was one of the grounds of the argument so artistically built up: in fact two witnesses had declared the distance to be two thousand paces. It was decided to visit the ground, and when the distance was found to be only four hundred and fifty paces, the new conclusions rendered necessary contradicted the former ones.
>
> Hans Gross (1891)

5.1 Geographic location and crime

Geographic location is a key element to many crimes, as is implicit in the naming of famous murders, mass killings and thefts. Indeed, the predominant naming of serial killers[1] falls into two groups: those with a personalized name (Acid Bath Murderer, Black Panther, Jack the Ripper, Metal Fang, Night Stalker, Son of Sam) or with a geographic reference (A6 Murders, M25 murders, Axeman of New Orleans, Beltway Sniper, Boston Strangler, Green River Killer, Moors Murders, Yorkshire Ripper)[2]. Although we tend to think of the bulk of such terrible killings as a modern phenomenon, Jack the Ripper, the Axeman of New Orleans and other serial killers

---

[1] The distinction between 'mass' and 'spree' murderers is blurred by time. The former set out to kill a number of people simultaneously, or close together, such as Fred Cowan, the Nazi sympathizer who killed four Jews and a policeman in New York (1977) or Michael Stone, a renegade Northern Irish loyalist who threw grenades at a Roman Catholic funeral in Belfast. The latter operate over hours or days, a geographically-related example being Michael Ryan, killer of the 'Hungerford Massacre'.

[2] Some of these killers are aptly named, the Green River Killer being a good example of geographically-limited serial killings, still unsolved. Another example of geographic bias is Lucian Staniak whose killings on trains or at railway stations in Poland helped characterize the killer, and yet there is no mention of geography in his nickname 'Red Spider'.

of 50 to 150 years ago suggest otherwise. This is reflected in the importance of geographic location: the use of Geographic Profiling and Geographic Information Systems (GIS) are modern developments understanding the relationship between humans and location that were well-known to Hans Gross in the late 19th century. Although he was writing from the perspective of an investigating police officer or magistrate, rather than a forensic scientist, Gross was completely objective in his advice. He saw no difference between the human- and natural-dominated landscapes, writing that

> There are localities which the Investigator must examine in the light of future events – hotels, public-houses, clubs and brothels, because of the brawls that may take place in them. Ponds and wells in villages on account of possible accidents by drowning, forests because of poaching and illicit felling.
>
> (Gross, 1891, p. 25)

Gross went on to consider the importance of both the location of industrial works (many of which use geological materials [mills, forges, quarries, brick or tile furnaces], as well as lines of communication. On the latter, his words of over 100 years ago are highly relevant to today:

> the Investigator has only to find and mark on the map all the roads and see whether they are correctly set down ... corrections will mainly show where a main road has degenerated into a side-track, ... or where a second-class road has been promoted to ... a main road.
>
> (Gross, 1891)

Gross recommends the same checking and re-mapping be done for buildings, wells, tanks, ponds, dams, etc. What he was effectively describing is the mapping of Ordnance Surveys, by walk- or fly-over, which is more fully developed in Chapter 4 on Remote Sensing. Gross was also very spatially-aware, noting (in the quotation that begins this chapter) that two witnesses both claimed a distance from an incident to be 2000 paces: when checked the distance was 450 paces, showing the need for absolute accuracy, especially when witness memory is to be relied on without re-visiting the scene. An adjunct to Gross's observations is that the tape measure and paper map of today's accident investigator are rapidly being replaced by laser range-finders and scanners and differential GPS systems. Thus measurement may become increasingly accurate, but mapped representations will always have a cognitive element, based on the bias of the mapper and their training, just as aerial photographs are subject to the vagaries of light and weather. Gross (1891) discusses some of the earliest types of crime scene map, and exactly these problems of representation. His over-riding theme is the same as ours: in mapping and describing a landscape, no feature (human or natural) must assume dominance until a reason for this is established. An integrated view of all features of the landscape must be taken, which is often easier for urban locations and buildings, because we are used to these regular geometric shapes, but more difficult for landforms, which are complex.

The spatial disposition of criminal data has always been of interest, but it is only with the advent of powerful computers that such data have realized a potential.

Spatial data are located on the surface of the Earth, and in some way this is what distinguishes them from other data and what underpins their usefulness for crime mapping. Typical questions include: where are particular features found? What geographical patterns exist? What changes have occurred over a given time? Where do certain conditions apply and what will be the spatial implications if certain actions are taken? The set of tools called Geographical Information Systems (GIS) have been developed to enable us to map and to analyse spatial information. The acronym GIS may refer to either systems or science. GI systems are essentially tools that have been developed to input, manage and analyse spatial data whereas GI science refers to the study of problems arising from the handling of spatial information in GI systems. The acronym GIS used in the current context refers to Geographical Information Systems.

GIS store two kinds of data: (i) spatial information and (ii) non-spatial attribute information, but it is the spatial component as well as the capacity to analyse spatial data that makes GIS different to other information systems and increases the applicability of GIS for crime mapping and forensic investigations. Several essential core texts are available which provide detailed information on the theory, methodological advances and analytical techniques of GIS (e.g., Burrough and McDonnell (1998), Heywood *et al.* (2002), and Longley *et al.* (2006)), and more specifically for crime mapping (e.g., Chainey and Ratcliffe (2005)).

Prior to the development of computer-assisted GIS tools such as ArcGIS™, two strands of geographically-related information regarding the investigation of life-threatening events had developed. One was the simple plotting, by spatial location, of crimes or deaths. An early important example is the work of John Snow, who in London traced the spread of cholera from plotting the time of disease contraction and location back to the source, a shallow water pump located on Broad Street (now Broadwick Street). How the infection came to be at the water pump is the matter of some interest, with a sailor arriving in the East London Docks as a likely possibility, again traced by mapping.

The other strand from which geographically-related research into criminalistics developed was in the 1920s when a group of sociologists based in Chicago (the 'Chicago School') suggested an ecological approach to the study of deviant and criminal behaviour and the urban environment. The ecology they referred to comprised distinctive neighbourhoods, in which the inhabitants shared a characteristic lifestyle, including the committing of, complicity with and suffering from crime. Shaw and McKay (1942) summarized the spatial and statistical aspect to this by dividing the city of Chicago into 5 zones of 2-mile width, radiating out from the central business area. Using male delinquency statistics from the juvenile court, they discovered a progressive decrease in the reported criminal offences with distance from the city centre. Further analysis showed that the very heart of the city (the commercial sector) had low levels of crime, and it was in fact what Shaw and McKay (1942) termed their 'transition zone', the area bordering the commercial heart and the nearby residential locations, which had the elevated crime rates. The transition zone is an area of high population turnover, where migrant workers arrive with little money and rapidly leave, creating social disorganization or preventing the development of a stable community. Indications of social disorganization include vandalism, illegal drug use and gambling, prostitution and racketeering: such behaviour flourishes because parental and social control such as people simply knowing one another, never develops.

Successor studies to those made by the Chicago School have found different spatial relationships to crime, but have generally confirmed that areas lacking social cohesion exhibit high rates of delinquency. Thus a transition zone in a different location to that bordering the commercial district of Chicago may show similar patterns, regardless of location. An example is a city with affluent sea-frontage, where juvenile crime is low, or perhaps a city with extensive river-side docks, where such rates will be high and occur in a linear, not concentric pattern. What the theory does not account for is the geographic location of organized or white-collar crime. When social or governmental policy intervenes, the simple Chicago model breaks down. Morris (1957) studied housing estates in Croydon (south London), where he found that the council policy of grouping 'problem' families together made for a pattern of delinquency dissipated from the simple model. The divisions of owner-occupied, rented and council-run housing, as well as access to transport and leisure facilities, have also been examined in terms of crime locations.

Two problems remain in our study of geographic location and crime: first, the reporting of crime causes major problems, with subcultures existing who suffer high crime rates coupled to a distrust of the authorities, and second, the low rates of serious crime that provide inconclusive spatial statistics. The second case is particularly acute because the Chicago School model suggests that it is environment that controls deviancy: this may be incorrect (Matza, 1964) or may only apply to certain types of crime. Many petty criminals are committing crime in order to uphold some form of social standing (access to money, cars, disposable goods) and are actually engaged members of society who abhor serious criminals (muggers, drug dealers, gangsters, pimps). It is then this latter class of criminal who may display far more complicated geographic patterns of criminal activity. Reporting levels as well as the low rates of serious crime have led to theories of geographic location and criminality being concentrated on deviancy and especially juvenile delinquency.

Organized crime and serious crime form two very distinct problems for the geographic analyst. Organized crime begins to fall into that bracket of social behaviour that is viewed from a political standpoint. Since the earliest wars between communities, hired warriors or mercenaries have been used to gain advantage. In the same way, large companies deployed mafia and other mob leaders, with state complicity, to break strikes and intimidate workers (Pearce, 1976). The striking miners of northern England would certainly have classed the British government's use of the police as organized crime against what they say as the legal right to strike. Governments, giving reign to criminal or paramilitary gangs who maintain some social order where the police or military could not ('turning a blind-eye') has the inevitable consequence of internal power-struggles and eventually internecine warfare or even genocide. Chambliss' (1978) conclusion when studying organized crime in the upper levels of Seattle society was that organized crime was a very essential element of maintaining a wealth and power base. Dissenters were often found drowned – a convenient accident.

Serious crime can be summarized as being the product of more individual activity or as being politically motivated, with the latter crossing over into the organized crime for financial gain. Political motivation for serious and organized crime often takes a Marxist perspective akin to that above, wherein crime occurs when a law is broken, and laws are made by the state (the puppet of the owners of the means of production) in order to exert social control. Again problems arise – the first being the perspective 'one man's terrorist is another's freedom fighter', the second being time. Strong cohesion within groups undertaking armed struggle leads to easy infiltration by competitor groups or government. A fragmented structure avoids this but inevitably leads to racketeering or intimidation for personal gain. Individual activity is rarely tackled by sociologists, being seen more as a product of psychological imbalance than that of society and location. Here we have two elements of the individual (not necessarily lone, but not part of a group) serious criminal: what makes them commit the crime in the first place and who may suffer (psychology, some element of opportunism), and when and where the crime takes place (geographic location, with some opportunism). At this point we have come full circle in this section – that the serious serial criminal (murderer, rapist, mutilator) often has a distinct geographic relationship to their crime. In the introduction we used this to demonstrate that geography or location is an important part of the serious criminal's label. We can now stress a second point, that the type of crime may also be spatially-dependent. Take the excessively simple, yet effective 'seafront' analogue (above) where pickpocketing and petty theft will be common, compared to car-breaking, which will occur in the back-streets to a busy sea-frontage.

Crime and location: 'Mapping and analysing crime data' (Hirschfield & Bowers, 2001)

In their edited book, Hirschfield and Bowers (2001) provide selected, representative examples of the use of crime statistics, especially in mapping criminal activity. In this section, we examine a few of the papers contained in Hirschfield and Bowers' book, which is a benchmark in the subject.

Turton and Openshaw describe the use and differences between their two computer programs, GAM (Geographic Analysis Machine, which detects patterns such as the location of diseases or repeat offences) and GEM (Geographic Explanation Machine). GAM and GEM allow the input of different variables such as population age, house prices, street length, etc. The results from each are powerful, providing quantitative information on how burglaries are high in well-off areas proximal to poorer locations; petty theft was concentrated around drug-dealing locations and the location of vulnerable elderly populations near to break-in 'hotspots'.

Costello and Wilks use GIS to examine journey times and routes to crimes. It is unsurprising that many offenders take short routes, indicating local activity, although older offenders will travel further. Exceptions may be where the offender is opportunistic, doing something else such as *bona fide* work, when a criminal opportunity presents itself. Some serious crimes have a historical element, where an offence or part thereof takes place where the offender used to live or know. This we have seen to be the case with Fred West (Sounes, 1995), who buried his lover Anna McFall and first wife Rena in fields close to areas he lived near before he started serial offending. The same pattern was seen with convicted English double-murderer, Ian Huntley, who took the bodies of Holly Wells and Jessica Chapman to a location near his father's home that he previously frequented.

Also in Hirschfield and Bowers (2001), Ratcliffe and McCullagh, as well as Bowers *et al.*, use databases with the time of crimes to examine repeat offences. The former use temporal GIS to show how similar victims or groups may be elucidated by the time the offence occurred. Chainey shows how a GIS can be useful in storing and analysing data from local communities, police and probation services to inform preventative measures such as the location of CCTV cameras. LeBeau takes this further, showing the potentials and pitfalls of mapping hazards to police officers who may be in danger: this is not dissimilar to Merrall's geographic analysis of the incidence of fire. Both LeBeau and Merrall are following the classical tradition, of using the geographic distribution of problem incidents in prediction.

The serial killer and location

Statistics once again cause debate among those gruesome practitioners of the serial killer story as to who is the most notorious (*www.serialkillermuseum.com*). For so many serial killers, the final death-count is never established, with newspaper revelations that many more cases may be attributed to a known killer than were at first thought. The converse is that with so many people going missing each year, maybe there are more serial killers or even individual murderers at large than suspected. The latter is rarely mentioned by the media: unsettling though the uncounted murders of the convicted serial killer may be, this is not as disturbing as the possibility of others, uncaught, at large in society (Cummins, 2003).

The police or government do not publicize repeated murder patterns prior to capture or conviction for fear of panic, incorrect assumption, or the generation of copycat killers. Thus we arrive at the conclusion that a comprehensive 'top ten' or 'most notorious' can never be established. However, in order to make some observation of the relationship between serial murder and location, an arbitrary list of those who fall into this definition (see above) is useful as a way of examining what impact geography had on their activities. Initially of course one of two geographic contexts will occur: the killer(s) either have a base or they move around. Thereafter their modus operandi (MO) may or may not be dictated by geography. The following, arbitrary list is from Lane and Gregg (1992) and serves this purpose.

- *Artieda, Ramiro*. All seven of his known killings were 18-year-old girls resembling Ramiro's former girlfriend. Thus psychology can be presumed to play a more major role than geographic location or opportunism. His first killing was motivated by greed. His later killings have some geographic element in being locations in Bolivia where Ramiro could operate.

- *Beck, Martha (Raymond Fernandez)*. This story bears some resemblances to the Fred and Rose West history of abused, self-loathing women whose sexual appetites were the impetus for predatory males with psychological problems (both Fernandez and West suffered head injuries prior to their crimes, as did Gacy (see below), whose MO however was different) to become out of control killers. Beck and Fernandez selected their 17 victims through lonely-hearts adverts, the geographic elements to these crimes being whether such respondents originated from certain communities and the circulation distribution of the newspapers.

- *Berkowitz, David*. The Son of Sam had a distinct geographic element to his killings of couples in cars on dimly-lit Bronx and Queens streets. The mix of psychology and geographic location is thus a classic example of how urban geography controlled what Berkowitz could do and when.

- *Bianchi, Kenneth (Angelo Buono)*. The work of these two cousin rapist-killers was initially ascribed to a single 'Hillside Strangler' who operated in Los Angeles by a series of opportunist killings of prostitutes but also by tricking professionals and students into accommodation they could access.

- *Birnie, Catherine and David*. These spree rape killers (their activity lasted four weeks and claimed four lives) based their activities around their respectable-looking home, where they enticed young women on pretext, allowing David Birnie a modicum of satisfaction in his voracious sexual appetite.

- *Bonin, William*. Most of the 41 'Freeway Killings' (aggravated homosexual rape and murder) were ascribed to Bonin, with some assistance from one main accomplice and two others. Bonin was the instigator in most cases, cruising the freeway and picking up hitch-hikers to attack. A linear pattern to his killings is the most notable geographic control.

- *Brady, Ian (Myra Hindley)*. The archetypal killer couple had a strong geographic element to their activities, wherein they picked up unsuspecting victims on quiet roads and streets in their van and either killed them at home or transported them to the now-infamous Saddleworth Moor near Manchester, where they were tortured, killed and buried. At least one victim (Keith Bennett) has yet to be found.

- *Bundy, Ted*. Bundy's main objective in killing was his choice of victim (young women), which controlled his activity more than geography, although the location of such young women (university campuses) did have a geographic element. The broader geographic location of killings is of note: when Bundy moved from working for the Seattle Crime Commission to study law at University of Utah, the killings ceased in the former area and began in the latter. The problem is one of database size and freedom of information: not many people move from Seattle to Utah coincidentally with a disruption in murders of one age/sex group. Finding that person is altogether a different matter.

- *Dahmer, Jeffrey*. Dahmer's victims were almost exclusively black and Hispanic young men, many of whom he dismembered and boiled the flesh from their skulls. Most of this activity occurred in Dahmer's apartment in Wisconsin, with the murder of at least one other victim in Ohio. In both locations, it was Dahmer's home that was the geographic focus of the murder and a victim type that dictated who was killed, as in the case of Bundy.

- *Duffy, John*. Known as the 'Railway Killer', Duffy's activities were often based on abducting women on trains or at railway stations, often in north London. This pattern of activity assisted an early use of psychological offender profiling, which as explained in this chapter can be compared to spatial patterns of crime and geographic profiling to increase prediction and suspect apprehension.

- *Dzhumagaliev, Nikolai*. The media name 'Metal Fang' belies a strong geographic basis for the seven known murders of tall, attractive women attributed to this respectable Kazakhstan builder. He would entice his victims on a riverside walk, rape, murder, dismember and then cook and eat his victims, the enclosed nature of river valleys giving him seclusion.

- *Evans, Donald*. When apprehended, police thought that Evans may be their elusive 'Green River Killer' (see page 125). However, Evans is the exact opposite of people like Bianchi or Brady in that after being dismissed from the US marine Corps in 1977, he wandered the length and breadth of the United States, killing 'at least 60' people along the way. Identifying the geographic pattern of killings thus had the same issues associated to Bundy.

- *Gacy, John Wayne*. Gacy's 15 known victims were largely heterosexual men whom he abducted in his car using chloroform, subjected to beating and rape and then killed, placing the bodies under his house or in shallow graves nearby. The geographic controls on Gacy's activities were travel distance from his home and likely abduction locations.

- *Gore, David (Fred Waterfield)*. Gore's serial attacks on young women centered on his home area of Vero Beach (Florida) and when apprehended verged on the spree type activity of the psychotic, in that he killed two women and held another hostage in a short period of time.

- *Lucas, Henry Lee*. Like Donald Evans (see above), Lucas was a drifter, travelling the United States, killing along the way, usually alone and for one year with the complicity of a pyromaniac (Ottis Toole) and his niece Frieda. It is possible that Lucas killed over 200 people.

- *Searl, Ralph and Tommy Ralph*. Searl was a hitch-hiker who killed 5 men over 10 weeks in Michigan, Indiana and Nevada, making him a moving almost spree murderer. Oddly his brother Tommy killed young women and all in the Kalamazoo area of Michigan, making him a geographically-isolated killer.

- *Sutcliffe, Peter*. The Yorkshire Ripper, an evangelical killer of 13 known women, had strong geographical associations to the areas of Leeds, Bradford and Manchester frequented by prostitutes. The geographical association was confused by a faked tape recording by a man from the north-east of England, focusing enquiries away from the area of both the killings and Sutcliffe's home.

- *Wuornos, Aileen*. Posing (and acting) as a prostitute, Wuornos went from serial killer to spree as she increased the rate and cold-bloodedness of her killings along the highways of Florida through the late 1980s. A victim-led killer with a strong geographic location.

All the crimes of our 18, arbitrarily-selected serial/mass killers can to some degree be defined in terms of geographic patterns. Eight (Berkowitz, Bianchi, Brady, Bonin, Duffy, Dzhumagaliev, Sutcliffe, Wuornos) had a strongly recurrent, geographic element to their activities, usually in the form of victim selection or murder location. Birnie, Dahmer, Gacy and Gore made their homes the focus to their activities, with other geographic elements to their killings determined by locations where they could engage victims. Bonin, Duffy, Dzhumagaliev and Searl (Ralph) exhibit a strong transport network-linked relationship with their choice of location for abducting individuals (whether it be highways, railways or riverside paths).

Table 5.1 is an attempt to summarize and highlight how geographic location in some way influences the killer, combined with psychology (sexual, financial gain) and opportunism. Evans and Lucas are the only arbitrarily-selected serial/mass killers to demonstrate no recurrent geographic feature (highways, railways, rivers) or focus (homes) to their activity, undoubtedly related to their lifestyles of drifting with no fixed abode. However even this type of activity can be addressed within a geographic framework in that a broad geographic scale is covered compared to the small or local scale geographic coverage of a home-centered crime. In a regional based crime pattern of serial/mass killers, the role of transport networks becomes crucial since travelling the length and breadth of a country such as the United States requires the use of trains, highways or freeways using mapped transport routes (as in the case of Ralph Searl whose method of victim choice was hitchhiking). The problem we set out at the beginning remains: murderers with one victim may have a strong random element, making comparative statistics meaningless. Thus to reduce our compared data to only those serial or mass killers with a number of victims, automatically introduces the problem of comparing killers with different psychotic problems. Nonetheless, even those who showed elements of wanting to be caught, nonetheless tried to avoid detection, or trusted tried and tested means of obtaining victims. Both these latter MO require geographic knowledge, which can be analysed and summarized over space and in time, most efficiently using GIS.

Case Study: The Shankhill butchers

The Shankhill Butchers were a group of extremely violent men linked to the Ulster Volunteer Force (a Loyalist Paramilitary grouping), but as Martin Dillon (Dillon, 1990) suggests, they were more like a group of serial killers than terrorists. The

Table 5.1 Illustration of the influence of location in criminal activity

Importance of location in criminal activity				
Serial killer	Area	Location of pick-up	Location of crime events	Body deposition site
Familiar or home-based location				
Artieda, Ramiro	Bolivia		known locations	
Berkowitz, David (Son of Sam)	Bronx and Queens	Urban roads	Known location	
Bianchi, Kenneth and Buono, Angelo	Los Angeles		Owned rented accommodation	
Birnie, Catherine and David			Home	
Brady, Ian and Hindley, Myra	Saddleworth moor, Manchester	Pick up from roads/streets		Familiar ground (killing and burial)
Bundy, Ted	Seattle and then Utah	Location changes with residence	Known location university	
Dahmer, Jeffrey	Wisconsin and Ohio		Own apartment	
Gacy, John Wayne		Pick up from roads/streets	Home	Home or near by
Gore, David	Vero Beach (Florida)		Home	
Searl, Tommy	Kalamazoo area of Michigan		Specific location based	
Sutcliffe, Peter (The Yorkshire Ripper)	Leeds, Bradford and Manchester	Pick up from roads/streets	Known area	
West, Fred and Rose	UK			Close to home; familiar locations
Transport network routes location				
Bonin, William	USA	Road network (freeway)		
Duffy, John (Railway killer)	North London	Trains or railway stations		
Dzhumagaliev, Nikolai		Riverside walkways		
Searl, Ralph	Michigan, Indiana and Nevada	Road network (hitch-hiker)		
Wuornos, Aileen	Florida	Highways		
Broad geographic scale				
Lucas, Henry Lee	USA	Broad geographic coverage	No fixed abode	
Evans, Donald	USA	Broad geographic coverage	No fixed abode	

group were led by Lenny Murphy, Robert 'Basher' Bates, and 'Big' Sam McAllister, who abducted Roman Catholics (or those they believed to be Catholics), rival loyalist gang members, as well as random people and savagely beat and killed them, usually by cutting their throats. Some of their murders were straightforward street killings, in some the victim was abducted, murdered in a vehicle (usually a taxi) and the body dumped elsewhere, and in some the victims were taken hostage and tortured before being killed. The bodies of these unfortunate souls were also often dumped in another location. A failed murder left a witness alive, whose testimony was key to convicting many of the gang in 1979, when 11 men were convicted of a total of 19 murders between them, and the 42 life sentences handed out were the most ever in a single trial in British criminal history. In his book *The Shankhill Butchers*, Martin Dillon (1990) said that his own investigations suggest the gang were responsible for a total of 30 murders. Murphy, the main leader of the gang, was released in 1982 and killed not long after by the IRA: Robert Bates was shot by persons unknown (probably not the IRA) in 1996.

The case of the Shankhill Butchers clearly includes spatial evidence which is (was) important is investigating the criminal activity. The victims were either killed in one location and their bodies dumped in another or taken hostage to be murdered and their bodies dumped elsewhere. The question is now that we have these spatial references, what can we do with them? Two options exist: summarize or describe the data as though they are not spatially referenced or take into account spatial location and assess the importance of spatial relationships between locations (Table 5.2). In analysing the geographic pattern of the Shankhill Butcher's killings, a number of factors have to be considered. First is (was) the territorial nature of west and north Belfast in the 1970s, in which nationalist (Catholic) and loyalist (Protestant) communities were clearly separated by physical barriers, by army and police patrols and by the indigenous community's own fear of the neighbouring population. This extreme territoriality broke down at the margins of the social divide – for instance the end of the Falls Road (nationalist) where it enters Belfast is not far from the end of the loyalist Shankhill Road. Thus at these nodes either member of the community could be found as they journeyed to and from their homes to Belfast, either for work or social purposes. Another factor in the activity of the Shankhill Butchers is

Table 5.2 Linking geographical reference with tabular attribute information

Abduction point					Body deposition site				
X	Y	ID	Date	Location	X	Y	ID	Date	Location
3338	3698	6	23/11/1971	Library St	3324	3699	6	29/11/1971	Urney St
3325	3697	7	25/11/1971	Brown Bear Pub	3330	3711	7	08/01/1972	Manor St
		8	29/11/1971	Windsor Bar	3305	3709	8	05/02/1972	Glencairn, Forthriver
3318	3700	9	08/01/1972	Burning Bush Ceylon St	3319	3699	9	25/02/1972	Esmond St
3329	3709	10	08/01/1972	Manor St	3328	3708	10	31/07/1972	Manor St

the sheer volume of shootings, bombings, riots and punishment beatings going on in Belfast at that time, leaving the police little time to maintain surveillance on these killings.

Abduction sites thus tend to fall into three groups:

1. Rivals and random unsuspecting Catholics were abducted close to the Butcher's centre of activity (the part of the Shankhill Road near Murphy and McAllister's homes).

2. North of there, the Duncairn–Cliftonville area was known to the gang and was a node where Catholics would pass through, and a place the authorities had difficulty maintaining surveillance.

3. East of these areas, on the west side of Belfast city centre, was a second node where again Catholics would pass nearby to the gang's territory and could be easily abducted.

Body deposition sites tend to be related to areas (1) and (2), probably from an opportunistic point of view, but also in terms of security: these areas were known to the gang and they were confident here. Area (3) was far busier, mixed, and had a high volume of through-traffic. Thus victims were either taken from here back to the locations the gang knew best, or to a housing estate well away from the abduction sites, yet again where the gang were confident of their surroundings, to leave their victims bodies. Several geographic implications stand out. First, that points of abduction satisfy two essential criteria: these were areas familiar to the abductors (territorial gang areas) and nodes of intersection with the Catholic population (public transport routes). Second, that deposition sites tend to be separate from abduction zones in territorial gang areas. Another important dimension to the analysis of the Shankhill Butchers case, alluded to above, is that in the 1970s this particular area in North and West Belfast could be mapped in terms of physical geographic areas (such as housing zones) but the perceived zones of territorial control and confidence were probably more important and how these overlap with public transport routes where random unsuspecting Catholics would be greatest at risk.

To investigate the Shankhill Butchers case some specific questions arise. Where were the territorial gang areas located in west and north Belfast in the 1970s in relation to the abduction and body deposition sites? Where were the nodes or zones of intersection between public transport routes and community areas? What were the shortest routes (with or without physical barriers) between the abduction and potential (or confirmed) body deposition sites? There are many cartographic mapping packages that fulfil many of the functions of GIS and could be used to map the locations of abduction and body deposition (Figure 5.1), but it is the analytical functions of GIS that allow us to link the graphic features with the descriptive attribute information and start to address the queries posed above. The GIS community has (under the auspices of the International Standards Organization (ISO) and the Open GIS Consortium (OGC)) identified core methods for testing spatial relations between geometric objects. Through implementing a structured

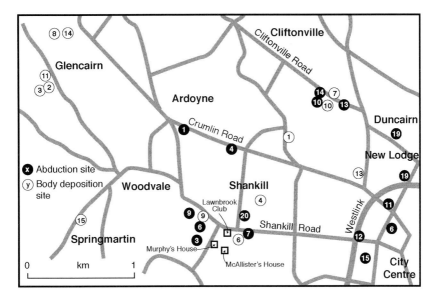

Figure 5.1 Location of the abductions and body deposition sites of the Shankhill Butchers in west Belfast, Northern Ireland

query language (SQL), spatial relations can be queried in GIS (as described in Longley *et al.*, 2001). Spatial relationships between abduction sites at the nodes or zones of intersection between public transport routes and territorial gang areas in west and north Belfast in the 1970s and body deposition sites can now be queried more fully in GIS.

5.2 Spatial data and GIS

Geographic information has always had a strong relationship to the map, or plan view of an area. The growth of GIS comprising plotting, storage, interrogation, comparison, manipulation and analysis of spatial data has been made possible with advanced computer technology. GIS are now used throughout society in demographics, map-making, planning, and resource and waste management, to name but a very few.

Most critical in understanding how a GIS works is the concept of related data – different data types in the same location, or relationships between data points. Previous to the development of GIS, such data had to be stored on separate maps, or on single, complex maps with two, three or more data types superimposed. GIS can be used to quantitatively relate different information spatially such as heights above sea-level, which when located by a spatial referencing system can be used to create contour maps, or more complex information captured on an aerial photograph or by remote sensing. Spatial referencing systems can be geographic coordinate systems (latitude and longitude), rectangular coordinate systems (for example, National Grid) or non coordinate systems (for example, postcodes). All spatial data must have one thing in common to be used in a GIS: they must be represented by digital information. Since we do not have the ability to represent accurately the real world in a digital format a model is used that can be considered as an abstraction or simplification of reality. There are various kinds of models or digital generalizations of reality that are used by different computer systems but two of the most commonly used in GIS are the vector model and the raster model. These allow us to represent in a digital format different types of spatial information or aspects of the real world that may be important in a criminal investigation. Vector digital data are composed of nodes (points) and arcs (lines). The nodes, joined by arcs, comprise features (for example, the edge of a road) or when connected may produce polygons (shapes bounded by lines). The vector model is often used to map discrete features such as artificial structures (buildings, towns), transport routes (roads, railways, waterways) and utility networks (water pipes, sewers), but sometimes also continuous line features such as contour lines. In the Shankhill Butchers case (above), examples of vector data would be the public transport routes in North and West Belfast and the boundaries for the housing subdivisions. The GIS may be used to delimit areas showing common attributes, thus automatically generating maps delimiting separate features (land use, geology, housing zones). The operator can then use a GIS to perform a number of analyses on the data, such as recognizing coincidences (for instance, where closely-spaced contours occur at locations of rock outcrops or the nodes or zones of intersection between public transport routes and loyalist (Protestant) housing areas).

A subtle add-on to the simple 'point or line on a map' coincidence, is the integration of areal data. This is achieved by using a GIS to divide the landscape into characteristic units (topographic, vegetation, soil type, housing type, social status or even perceived areas of territorial gang control or dominance as in the Shankhill Butchers case). Such cells or packets of land are recognized by a GIS as vector polygons, each of which can have attribute data attached to it.

The raster model in GIS is used to map continuous variables in the form of a grid of cells (pixels). The value assigned to each cell represents the attribute of that cell. In a digital elevation model (DEM, discussed in more detail below) cells in the grid represent height above some arbitrary datum (e.g., mean sea level). A raster model could also be used to map concentrations of pollutants where a cell value in this case will represent the proportion of some chemical in a soil. The human eye sees the 'connectedness' of the numbers, colours, etc., (such as in an aerial photograph) and yet GIS provides a quantitative measure of the distribution of these data with locational information. Such information may comprise information from human observation (e.g., at grid reference X, Y was noted) or be automated, most commonly from aerial/satellite information but also from geophysics.

Satellite image data that have been interpreted by a computer to produce a land use map can be input to the GIS in raster format. An example is land cover classification. Raster files can be manipulated quickly by the computer applying various mathematical functions, but they are often less visually appealing than vector data files, which can approximate the appearance of more traditional hand-drafted maps.

Spatial data may be captured in digital format from primary and secondary data collection. Primary data collection from remote sensing includes aerial photography, airborne and spaceborne sensors, which detect radiation for different parts of the electromagnetic spectrum, passive sensors (which sense naturally available energy) and active sensors (which supply their own source of energy to illuminate selected features) comprising radar and LiDAR (Light Detection And Ranging). Radar emits pulses of macrowave energy and LiDAR emits pulses of laser light. (Chapter 4 contains further details of remote sensing techniques.)

Critical to the development of GIS has been the rise in Global Positioning System (GPS) technology (Figure 5.2), which has enabled the primary capture of spatial data by the use of a GPS recording device. GPS data acquisition has become an integral part of geoforensic investigations because it enables the collection of elevation data from a search site for local scale topographic mapping and georeferenced samples to be taken for subsequent integrated GIS analysis.

Data used in GIS may also be derived from secondary data collection, for example conventional cartographic sources may be converted to digital format via keyboard entry, digitization or scanning. This may be important if historical data are relevant for an investigation. Scanning a conventional paper-based cartographic source results in a new raster-based (cell) representation of the map. A scanner, such as a drum scanner or a (less accurate) flat bed scanner, works by scanning line by line across the paper map and recording the amount of light reflected from the surface. However, some potential problems can occur when scanning: these include optical distortion,

Figure 5.2 Collection of spatial data in the field. (a) Setting up a base station for differential GPS (DGPS) data collection. (b) DGPS positioning of sample collection. (c) Samples taken and ready for collection. (d) Topographic and sample collection points

especially with flat-bed scanners, scanning of unwanted features and selection of appropriate background tolerances to achieve an appropriate balance between data identified as important for the search and background data. The process can be time consuming as editing is required to produce the final output. Identifying individual features of interest on a scanned map can be carried out by digitization, resulting in a vector representation of the features traced. Digitizing can be conducted on-screen by using the mouse cursor to trace features (for example, roads can be traced on a scanned aerial photograph) or using a digitizing table. Potential sources of error involved in digitizing a paper map need to be addressed because these will have an impact on any decisions based on the final output map. Potential sources of error in digitizing a paper map include errors in the survey upon which the map was based, distortion of the paper (a likely source of error if maps are not stored in climatically-controlled conditions) and user-derived error (for example, hand slips when digitizing).

Problems associated with secondary data collection, such as those addressed, have resulted in a preference towards primary data capture where greater control and targeting to specific need can be applied. However, as is the case for natural disaster management, the investigator or victim recovery team is not in control of where and when the crime or disaster takes place and is dependent on all sources of available data. In this case secondary data sources may be essential to provide valuable sources of material and information. The integration of historic data is a major challenge for the developers and users of geographic information: with the exponential increase in GPS accuracy maps become rapidly out of date, crude and ultimately as representational as landscape sketches. This is especially true for crime data, which may have the additional problem of intentional vagueness in location, for security purposes.

5.3 Spatial analysis within GIS

As it is impossible to sample every square centimetre, metre or even kilometre of the Earth's surface for every desired attribute, spatial analysis functions within GIS, using various interpolation techniques, may be used to estimate characteristics between sampling points based on where data have been generated. Surface analysis involves analysing the third dimension or the 3-D attribute of spatial data. Topography is a good example – if elevation data are known from a range of locations, GIS can provide a range of digital elevation maps, based on various mathematical functions to provide possible representations of topography (see Figure 5.3). The theory of spatial interpolation works on the principle of spatial dependence: values close together in space tend to be more similar than those further apart. With interpolation, observations or measurements in a certain area or 'neighbourhood' are used to estimate the value of a property at a location for which we have no data. The focus in forensic work may tend towards digital elevation data but surface analysis and interpolation are relevant in many other contexts and can be applied to any continuous data type such as soil properties.

Many different interpolation techniques can be applied in a GIS framework: widely-used techniques include Inverse Distance Weighting (IDW), fitting a spline function, trend surface analysis and triangulation, resulting in a Triangulated Irregular Network (TIN). Different interpolation techniques will be discussed in more detail in Chapter 6, but there are several surface analysis applications as part of a GIS that are of value to discuss further here. The uses of digital terrain models (DTMs) (or DEMs [Digital Elevation Model])[3] range from providing national databases of topographic information to enabling comparison of different terrain forms through statistical analysis. However, DTMs also provide source data for derived maps (such as slope, aspect, shaded relief). As part of landscape mapping and interpretation (as addressed in Chapter 2) with a view to reconstructing events at a scene, the investigator may want to find all areas in the terrain with a slope of more than 5°. The solution in a GIS would be to perform a slope calculation (and direction of slope or aspect). TINs as well as DTMs can be used to derive slope and aspect calculations. Map outputs displaying slope, aspect and change in slope are useful tools in partitioning the landscape into mappable areas for search and other purposes, as explained in Chapter 2.

Case study: Missing persons, West of Ireland

A scenario-based search method (Chapter 2, personal communication, Harrison 2007) was employed to investigate the disappearance of a teenage girl in the west of Ireland. Details of the last known actions of the teenager and the sighting of unusual activity by the last known person to have been in the company of the teenager led to an intensive search of an area of mixed bog and moorland. To facilitate a landscape domain-based search scenario for potential body deposition or burial (as described in Chapter 2), elevation data of the area were collected by differential GPS. These data enabled the creation of a DTM (Figure 5.3) and slope map (Figure 5.4).

At the time of the disappearance, the suspected body deposition site was accessible by vehicle down a narrow country track. However, vehicle accessibility diminished some 50 metres down the track and any further movement or transport of a body would have to be made on foot over rough moorland and bog. As part of mapping the landscape for areas to focus search efforts, it was useful to know which areas of the terrain would have been hidden from view from the farthest access point of the country track. The collected points which can be viewed from a particular location are called line-of-sight or viewshed maps. This kind of surface analysis has been used, for example, when a developer wishes to minimize the visual impact of a development or wind farm. The application in a forensic investigation may be the inverse: defining sites or areas of the topography that are not visible from a certain perspective (hidden areas for covert operations as suspected in the missing teenager case). The DTM and slope map were used to produce the viewshed map of the suspected body deposition site (Figure 5.5). The viewshed map enabled a feature based assessment of the terrain in this case the important features were hollows or hidden depressions that would have concealed covert activities of burial or deposition. However, another factor to take into account in the amount of

---

[3] We use DEM when we are explicitly discussing elevation, DTM when considering terrain.

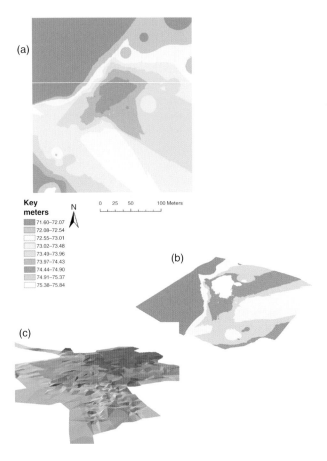

Figure 5.3 Examples of the type of digital elevation model that can be created by a GIS from topographic information: (a) 2D DEM; (b) 3D DEM; (c) and TIN

concealment possible is vegetation cover, in this case, bracken with interspersed saplings. Laser scanning (Figure 5.6, readers are referred to Chapter 4 for examples of terrestrial LiDAR) was used to capture the vegetation of the area and this combined with viewshed analysis provided a more complete picture of the degree of concealment afforded to covert activities in the area.

Discussion

Another surface analysis technique that has relevance to forensic investigations is what has been termed 'spreading with friction' or cost analysis. In a search scenario this technique allows the geoscientist to ascertain the cost of effort associated with walking from one point on a landscape (urban or rural) to another point. For criminal investigations it may indicate the most likely entry or exit route for a crime

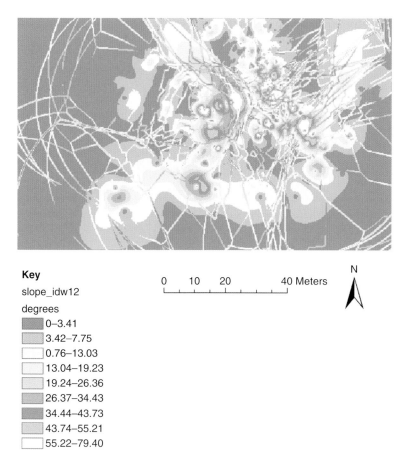

Figure 5.4 Slope map created by a GIS (using IDW with a neighbourhood of 12 points) from topographic information. Notice the circular artefacts of the IDW interpolation technique. The IDW interpolation technique is discussed in more detail in Chapter 6

scene. Two components will contribute to the cumulative friction value or cost of the route from the starting-point (adapted from Burrough & McDonnell, 1998). The first is distance, counted as cell steps on a raster map or in real units such as metres, etc. The second depends on the attribute properties of the cells traversed during the route.

For the crime scene scenario, the attribute properties differ for urban and rural setting. Urban scenes will involve access to alleyways and transport routes; rural attributes include elevation, land use, moisture content of ground (dry or marshland) or vegetation cover (grass or thick vegetation, especially bracken). As explained in Burrough and McDonnell (1998) 'the larger the value of the 'friction' attribute, the greater the accumulation of 'distance' when traversing a cell. The result is that the

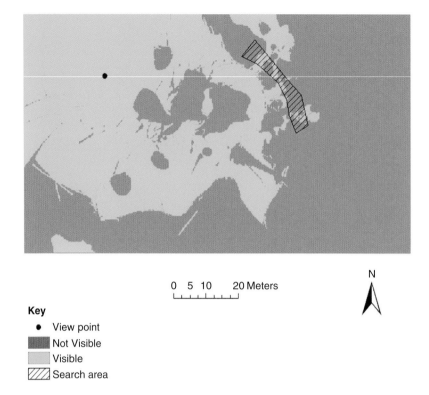

Figure 5.5 Viewshed analysis created by a GIS from topographic information showing areas of the topography that are not visible from a certain perspective

Figure 5.6 Terrestrial LiDAR image used to capture the vegetation of an area. This remote-sensing method (see Chapter 4, especially Figures 4.9(a) and (b)) provides an excellent example of the kind of digital topographic data that is ideally integrated into a GIS horizontal field of view, 15–20 m

effective spreading distance accumulates much faster where resistance is greatest, so that geometrically longer paths may be 'cheaper' ways to reach a given destination. In effect, the shortest exit route from a crime scene may not be the shortest route that involves crossing an area of marshland but a longer route over more even terrain. In the missing person case discussed above, the value of 'friction' across the mixed bog and moorland would be closely related to the moisture content of ground and vegetation cover resulting in higher 'friction' values for areas of wet bog and dense bracken cover. This type of information could aid the search for forensic evidence and help to target resources to certain search locations. This is especially relevant for searching for missing persons when a suspect has allegedly had to drag or carry a victim's body to a deposition site. In this case the suspect will invariably take the 'cheapest' way to reach the chosen destination.

Surface analysis derivatives from elevation data such as those discussed (slope and aspect calculation, viewshed and cost analysis) have been widely used in a range of applications in engineering and for military purposes (weapons guidance, pilot training), analysis of cross-country visibility, and hydrological and ecological modelling. The relevance to forensic investigations, however, has yet to be fully realized.

In many conventional maps the viewpoint is directly over the area of interest ('bird's eye view'). GIS opens up the opportunity of presenting data through a range of innovative technologies comprising interactive 'fly throughs' and 3-D visualizations where the angle of view, viewing azimuth and viewing distance can all be adjusted. This has the potential to increase the investigative power of the forensic geoscientist. Secondary attributes such as land-use maps and aerial photography can be draped or overlain over a digital elevation model. Shaded relief maps (using information from slope and aspect) can increase the convincing appearance of 3-D visualization to aid partitioning of the landscape into potential search areas. In this way, both the 'scenario-based' and 'feature-based' (M. Harrison, personal communication, October 2007) search strategies may be worked through to reduce time and cost on the ground and focus resources and provide optimum time to conduct intelligence led searches. It is worthwhile to state, however, a cautionary note that must be considered in any search strategy: information and outputs generated by GIS provide a representation of reality that will always involve some degree of generalization and simplification of the real world.

Cold cases

In Chapter 1 we highlighted how revisiting old cases may be used to inform current and future criminal investigations. Nowhere is this more pertinent than in GIS, which has immense power to assist in cases where location and timing are important. Some consideration of how GIS may have been used in cold cases follows.

Disappearance of Amelia Earhart. Although the major problem in tracking Earhart's movements was communication between her Lockheed Electra (with only a low-frequency receiver) and her guardship the USS Ontario, modern aircraft positioning by GPS now allows accurate tracking. The final location of Earhart's aircraft would have allowed rumours of her involvement (intentional or accidental) with the Japanese military to be established. The subsequent use of GIS would enable different layers of information – the recording of all US and Japanese boats, as well as non-combatants, storms and Earhart's plane – to be overlain, and thus coincidences or connections in space and time established.

Death of T.E. Lawrence (Lawrence of Arabia). On 13th May 1935, Thomas Edward Lawrence had a motorcycle accident: he died six days later in hospital, never providing his own version of events. Plenty of facts are known about the accident, with the specific distances from Lawrence's house (Clouds Hill in Dorset) to the scene, the precise timings of his movement down the road and the presence of two delivery boys who were involved as well as a mysterious black car, the true involvement of which has never been established. Plenty of other information about the events contradict: the only eyewitness was a duty soldier at nearby Bovington Camp who stated that Lawrence's motorcycle was travelling at speed far greater than the gear in which the damaged machine was locked after the crash. This soldier also stated that he saw the black car, which both boys denied. That Lawrence was a charismatic, politically unpredictable and possibly dangerous person is without doubt, even to those who ignore the glorification of the epic film starring Peter O'Toole. Thus conspiracy theories abound, making any test of witness accounts problematic (given that some people consider the authorities to have compromised some if not all these accounts). Nonetheless a critical part of the witness accounts is that Lawrence was periodically out of sight in dips in the road: it was within one of these that the accident occurred. As witness accounts conflict, a possible method of testing their observations using modern technology would be to create a digital terrain model of the road and surrounding area. Using GIS, viewshed analysis and 3-D 'fly-through' technology, a multitude of possible witness positions and lines of sight could be tested, in order to be sure that the timing of events and lines of sight were consistent.

Second gunman on the grassy knoll?. The yoyo-like recurrence of contradictions and conspiracy theories concerning the assassination of President John F. Kennedy on 22nd November 1963 hinge on a number of problems, some of them political, psychological, tactical, and as we shall see elsewhere in this book, forensic trace evidential. However, one central problem has always been the lines of sight and rapidity of fire between the bookstore repository where Lee Harvey Oswald was hidden and the unknown possible second gunman on the sloping area of Dealey Plaza, known since then as 'the grassy knoll'. Debates concerning events have largely centred on the reasons Oswald may have had for his crime, his subsequent death at the hands of Jack Ruby, and the evidence provided by amateur cine photographer Abraham Zapruder. The combination of Zapruder's film and terrestrial LiDAR scans of the extant area made into a 3-D digital urban model in GIS would, if such a cold case review by appropriate authorities be made, possibly disentangle these problems and provide clear views of what was possible in space and time on the day.

Case study: Locational GPS

The importance of accurate geographic location has been explained in Chapter 4, most especially with regard to GPS. From the examples above, location of the victim, possible offenders, suspect and proof of offender location are all seen as critical. A highly contentious use of GPS information arose in the trial of Scott Petersen, accused of killing his 8-month pregnant wife Laci and disposing of her body from his boat. He hired a truck not long after he was suspected and the police obtained permission to locate a GPS transmitter in the vehicle. The prosecution used his repeated visits to a location not far from where Laci's remains were washed ashore to argue an obsessive control over her possible discovery. This proved highly controversial in court, but was eventually admitted, proving to be one piece of evidence that persuaded the jury of a guilty verdict (Lee & Labriola, 2006).

Issues in Forensic GIS

The power of GIS is the ability of the operator to select a specific location and interrogate the databases about its origin, attributes and function. The map on the computer screen may thus be asked questions about the data behind its construction. Because GIS work is on the basis of X, Y, Z data (positional and attribute), information can be shared between anyone with similar data, regardless of computer system or software. This is critical in the mapping of criminal activity, especially that which crosses international and political boundaries where different data storage methods have been used. This does, however, raise issues concerning data quality and data storage. Ideally, all GIS outputs produced (visual maps and databases) should be supported with *metadata* (literally, data about data). That is, we should have information on data lineage – when and how data were collected and modified and by whom.

As the use of GIS increases in criminal investigations (as is the case for other techniques scrutinized under court conditions), the need to provide data lineage, with detailed records about the acquisition and processing of data will become even more important to facilitate repeatability by the defense. Data lineage should record details of data source, method of data capture, data models used, stages of transformations, editing and manipulation, and the hardware and software used. If dedicated data lineage recording software is not available, then it is recommended (personal communication, M. Harrison, October 2007) that a record of all data processing steps for each day should be saved (on CD or DVD), recorded and sealed, as is the passage of process for any other piece of evidential evidence. Although GIS may be used to carry out complex analyses or to produce impressive (3-D) presentations, the results may be worthless for criminal investigations if the data are subject to unknown and unacceptable errors.

The idiom *'garbage in, garbage out'* is often used to reflect a central concern whenever a model is used to represent the real world. Nowhere is this more imperative than in the use of GIS outputs in investigating criminal activity where error and uncertainty (in this case taken to refer to the difference between the real world and our representation) in data and map outputs is crucial for a sound evaluation of a crime scene or search area. Error may be introduced at various stages comprising errors in source data, errors in data encoding, errors in data editing and conversion, and errors in data processing and analysis. In terms of the currency of data utilized, we need to ask: are data up to date? If GPS data were collected for the investigation or available photogrammetry data or remote sensed imagery used, is a time series of data available and relevant? In terms of areal coverage – is it sufficient for the areal extent of the investigation? Consistency problems arise with map overlay and integration of different data types into a GIS including issues of compatibility of map scale and resolution between different data types to ensure that differences between maps are real and not a function of differences in scale of measurement.

Several sources of error may affect the production of DEMs, as for any other type of spatial data. This becomes especially relevant for error propagation issues because a DEM often forms the base layer over which subsequent layers of spatial information and evidence may be draped and integrated. As described earlier in the chapter, DEMs also provide source data for derived maps (slope, aspect) and subsequent surface analysis (e.g., viewshed analysis). As a result, errors introduced in the production of the DEM will impact on landscape mapping and interpretation: such errors can have disastrous consequences. Take for example the prediction (from elevation) of where floodwaters from flash storms or tsunami will move. With incorrect elevation modelling, the waters may inundate an area predicted to be 'safe'. Potential errors arise both in data collection (density of observations or measurements, positional accuracy, observer bias in measurement) and also from data analysis (choice of analysis model, method used for interpolation and artefacts in interpolated outputs [spurious sinks or local depressions]). Recognition of the potential for error and uncertainty in the use of GIS in forensic investigations does not negate its usefulness but should enable greater awareness of where potential errors can be introduced into a GIS and how errors may propagate through an analysis and thus aid the identification and management of such errors.

The UK holds the most comprehensive database of statistical reliability dealing with murder location and other subjects, which marries behavioural science with geographic location. Although examples of the use of this database are never made available in published form, the information is shared through law enforcement agencies worldwide. In this way, informative learning and valuable experience gained from decoding evidence from previous cases is used successfully to predict future criminal activity. An example of a geographically centred type approach is outlined in the next section.

Case study: Serial rapist and murderer Justin Porter (Las Vegas) by Canter *et al.* (2000)

The violence of a sexual assault and attempted murder in February 2000 shocked even hardened Las Vegas police. In what appeared to start as a break-in, a woman was subdued by beating, bound with duct tape and cord, sexually assaulted, further beaten and mutilated before the offender stabbed her again and doused the apartment and victim in accelerant and ignited. To try and ensure both final death and evidence destruction, the doors were barricaded to avoid escape. Amazingly, the victim released herself and escaped, making her a witness to a crime that was later linked to others. DNA and fingerprint evidence, followed by the bizarre murder of a Tibetan monk some five months later, showed that the Las Vegas police had a serial offender in their city, but without a positive identification or geographic base. The offender's mistake was to combine assault, sexual crime, murder and other offences with theft. The final location of associated stolen cars, other goods and a pram used to move a large television from one robbery gave police a geographic and temporal distribution of comparable crime in the city. Not all were proven as associated: those with positive fingerprint or DNA evidence were weighted, giving a consistent common area for these, as well as many other similar crimes. This target area comprised a number of low-rent apartment blocks. Further interrogation of the data allowed the officers to use a geographic centred analysis (such as Canter et al.'s (2000) Dragnet software) to predict a specific apartment block that housed a known offender – one Justin Porter. A visit to Porter's house by the police caused him to flee to his girlfriend's home in Chicago, where police apprehended him. This classic study in the use of plotting offences on a map, with information such as time/date, incident type, hierarchy of 'connectedness' (e.g., common fingerprints, DNA, MO) in a layered, GIS format provides both the simple triangulation or weighted centre of gravity solution with great effect: in this particular case almost leading police to Justin Porter's doorstep.

Other case studies from Canter (2003)

Canter (2003) makes a case for the type of geographic analysis outlined above (Las Vegas) being applied to both Jack the Ripper and Fred West. In the former case, certainly the late 1800s centre of activity (Miller's Court and Mitre Square, between Commercial Road and Whitechapel Road, in East London) exists as far as we can tell over 120 years after the events took place. In the latter case (Fred West), Canter makes a case for application of his model, which is reasonable given the acknowledgement of a modern-day complication in the form of the M5 corridor, linking Worcester to Cheltenham/Gloucester. The activity of Peter Sutcliffe (the Yorkshire Ripper) through the mid-1970s can be seen as a mix of both the focused activity (and thus analysis) of Justin Porter or Jack the Ripper and the more linear, transport-related MO of Fred West and Duffy and Mucahy (the Railway Rapist(s)). (Eggar (1999) gives a critique of Canter's work, especially related to the Railway Rapist(s) case.)

Canter (2003) develops the Las Vegas case study even further with two excellent examples that demonstrate how the conviction of Justin Porter was continued. The case of prolific and malicious telephone hoaxer Simon Wadland (jailed in 1992) demonstrates how, with a large database (he made 227 calls in 1998 alone) and limited information about victims being available to the offender, geographic profiling systems such as Canter's 'Dragnet' system and the sophisticated mapping technologies of various police forces produce spectacular results. Canter's map of the area west of Northampton, with Wadland's home at the centre of the area over which calls extended, is almost a bull's eye, elongated slightly north to south, possibly as an effect of the predominant road direction in that area. Wadland's MO was initially to look up potential victims in his wife's address book, call them on weekdays when alone and tell them he had kidnapped their husband or partner. To obtain a promise of their safe return, Wadland told his telephone victims he knew their address (which he often did), that accomplices were watching them, and that they had to self-mutilate. This was, he said in retribution for a previous prison sentence. When his wife's address book was exhausted as a victim supply method, he examined marriage announcements, located the bride-to-be and threatened them by telephone in the same manner. The lack of a digital telephone network at the time precluded tracing or locating the calls, which today can be done directly and by triangulation on mobile telephone networks. Wadland was caught after a victim managed to tape the call and his voice was recognized.

An even more successful application of Canter's system saved unknown amounts of ransom money, possibly the lives of uninvolved people and ultimately caught the offenders. The 'Mardi Gras' bombers left home-made explosive devices outside specific supermarkets and banks (apparently throughout London) and sent similar packages to individuals, all labelled 'the Mardi Gras Experience'. Their targeting lasted over three years, with ransom demands in most cases. A scheme was devised whereby the bomber could be caught: one supermarket chain was to 'fall into the trap' and provide a bank account, cashpoint card and PIN number. The idea sounded fine, until it was realized that there were thousands of cash dispensers available to the offender, making the prospect of monitoring each one impossible. Canter (2003) describes how the Dragnet software could be used to plot the locations of where bombs had been left previously: these showed an unusual hour-glass like distribution centred on Chelsea/Bayswater (west London) and suburban south-east London. This indicated one or more offenders, with two possible bases. Nonetheless, the highest concentration area (west London) was selected as an area for close surveillance, which paid off as the PIN number was used in this area, close to the centre of the Dragnet target: the offender's ex-wife lived within the south-east London focus, providing the answer to the hour-glass like distribution.

Selected examples using GIS

The flexibility, power and analytical uses of a GIS allow all manner of data (spatial and non-spatial) to be stored and analysed: moreover a GIS is ideal for the comparison and analysis of multiple spatial datasets, many of which are the key to forensic and legal enquiries. Common-sense tells us that when independently-derived data are compared and provide contradictory or complimentary indications, so levels of trust are raised or lowered. GIS allows these spatially located data to be overlaid and quantification of any comparisons to be made. For the types of information discussed in this book, such parallel datasets may be aerial photographs, compared to mapped geophysics, compared to soil samples. Conversely, results from spatially-collected samples, analysed separately may also be compared, for instance pH, particle size, mineralogy and pollen-counts. The problem with these datasets is that they are not truly independent, being often derived from the same sample: thus if compromise or contamination has occurred in one sample, the problem will affect all analyses. The spatial and geostatistical (to be more fully discussed in the following chapter) analytical tools available within a GIS can be used to interrogate the spatial datasets and provide information on outlier data that could be either the problem sample or the unique identifier, or by handling adjacent, yet separate samples taken for each type of analysis, creating vast datasets that are confusing to assess visually yet easily summarized and compared using a GIS.

The work of the National Institute of Justice (NIJ) in crime mapping

The NIJ has for a long time been at the forefront of using and developing crime data handling, storage and analysis mechanisms. Although French sociologists made some of the first spatially-oriented considerations of crime and crime type, it was the New York Police Department who pioneered what is familiar today as the 'dots on maps' approach that has led to the current use of spatial analysis in criminology. Hand-drawn maps and pins are limited in their flexibility, need space, can be quickly forgotten and become confusing. Nonetheless, they still have a place in criminal investigations as they allow time for the investigator to reflect on what is happening, and they often generate the 'eureka' moment, which, as they cannot be forgotten, cause the investigator to be idly passing and spot a (probably subjective) pattern not seen before. Using a GIS based approach does require the computer to be switched on and effort to be made in interrogating the databases. However, the opportunity for the 'eureka' moment has most certainly not been lost through the use of GIS: instead the opportunity may have been increased. Now the GIS operator may experience the moment of enlightenment as a previously unseen relationship is realized or a GIS output (digital or paper) will replace the hand-drawn map and enable a greater opportunity for subtle or long-forgotten spatially-based links to be made.

Perhaps the unrealized power of computer-based mapping and analysis tools is their use in cold case reviews or investigations that remain idle, only to suddenly reappear. The power of the GIS-based system is the ability to add other environmental or social data such as housing types, economic indicators, transport routes, search operations and DNA tests. Most systems are inter-operable and thus can be utilized by other agencies. Since the success of the New York police work in focusing police resources, predicting patterns of criminal activity and tracking major offences/problems, GIS has been used by many police forces throughout the world, with various in-house programs, commercial software and ways of using ready-made software being developed. Private developers of 'crime mapping' software often claim high levels of prediction in their outputs, suggesting that the police officer need only deploy their program in order to target resources and catch the next major offender! This is hardly surprising – the software developers are trying to sell (commercially or intellectually) a product, in a similar way to how Gross 'sold' forensic geology as overcoming tiresome enquiries. The problem of too many competing products is so great that in the USA the NIJ has set up the Crime Mapping Research Centre (CMRC: *http://www.ojp.usdoj.gov.cmrc*), whose mission is to conduct research related to the geographic analysis of crime, and, specifically, to support the development and evaluation of crime mapping technologies for criminal justice research. This latter aspect of the CMRC's work is crucial in respect of maintaining both quality control on the software products available, as well as encouraging the further development of successful analytical systems.

In the introduction to this chapter, the making of maps was taken as the origin of GIS and indeed the most required product of a GIS for criminal investigations may still be an up-to-date paper map. The essential role of maps (and geospatial data) in the early stages of search and rescue in disaster situations has been highlighted in a review document by The National Research Council (2006): 'Successful response starts with a map', and comparisons can be made between response to criminal activity and disaster management in the lack of control over timing and location. Similar challenges are faced in both situations: disparate sources of data collected for different purposes need to be quickly integrated to produce up-to-date data, maps and imagery.

Paper maps are easy to produce using a GIS and form a permanent output that is often the most effective means of communicating information. The usefulness of GIS, however, goes beyond the production of a paper map and includes the ephemeral output of visualization on-screen. This is an important part of the decision-making process that allows different scenarios to be tested and interrogated and may lead to a permanent map output for field reconnaissance. Commonly, a GIS operator is asked to produce maps that are 'familiar' to the users – yet the map is after all, a personal rendition of landscape, not an absolute. Scale, style and legend all dictate the final mapped output. The power of GIS is that the positional information will never change and in the future can be used to generate different forms of maps than we are familiar with today. Such maps are currently used to display, evaluate and quantify, water quality, hazard location, emergency response resources among many other features. The problem is that paper-based maps are abstractions of the real world; each map is based on a sample of selected features transposed to a sheet of paper with symbols to represent physical objects. People who use maps must know and interpret these symbols. Topographic maps show the shape of the land surface with contour lines: GIS offers the opportunity to create a DTM. This is only a model, and successive data points can be added to the database (sequentially) to provide ever-more refined images of ground surface. Thus GIS mirrors the historical development of maps themselves: through time we assume that maps have become ever-closer to representations of the 'truth' about ground conditions. In the same way, a crude collection of topographic or other information, input to a GIS will create a crude map of the data. GIS allows successive data-collection exercises to be added to the initial, creating evermore accurate representations of the landscape, spatial data, material distribution or social structure. Thus GIS mapped outputs can also be displayed as a time-series, showing how features of the Earth have altered through days, weeks or years, depending on the data collection time.

5.4 Use of Google Earth in criminal investigations

The success and widespread use of Google Earth and similar web-based digital Earth systems merits a mention in this context. There is no doubt that the increased availability of virtual globe systems as a media has fostered an increase in spatial literacy in the general public having both positive and negative consequences: the use of GIS in the legal system may be more readily accepted and better understood;

conversely the criminal will also have an increased spatial awareness and access to web-based spatial digital information. This necessitates a comparable spatial response on the part of the criminal investigator and underlines the crucial role of GIS in not only mapping criminal activity but also in analysing material to remain one step ahead.

Some readers may also be thinking why GIS operators and software are necessary when systems such as Google Earth are so readily available. The following aspects are noteworthy: Google Earth (and any similar virtual globe system) is not a GIS and as such cannot perform the full functionality of GIS (querying through SQL, analytical tools [buffers, etc.], quantitative analysis and cartography). Google Earth uses a composite of tiled images using processed and unprocessed data and provides no metadata on data currency, coverage, consistency or compatibility of image scale and resolution. In conclusion, although Google Earth and similar web-based digital systems have an increasingly important role to play, they cannot replace the flexibility, power and analytical functions of GIS in forensic investigations.

6
Scale, sampling and geostatistics

Truth will come to light;
murder cannot be hid long.

> Shakespeare, *The Merchant of Venice* (1596–8) act 2, sc. 2.

6.1 Scale and spatial resolution

The recognition and recording of scale, generally through the use of a coin or person in field sketches and photographs, is an implicit requirement for a qualified geoscientist. Criminal activity varies across space regardless of scale but this appreciation of scale by the geoscientist provides a useful means of analysing patterns of criminal activity and thus more effectively investigating crimes. The areal extent over which a crime takes place defines the scale of an investigation. Where a large areal extent is included and topography is important, the use of GIS in the production of DEMs and derivative maps (Chapter 5: slope, aspect) and other surface analyses (for example, viewshed analysis, Chapter 5) will involve decisions such as what spatial resolution should be used in data and map generation or has been used, if from a secondary source. These decisions will be based on a balance between representing accurately the property or area of interest and density and cost of available data (for example, elevation data acquired from spot heights or GPS and the collection of soil or water samples).

In simple terms, spatial scale can be described as dealing with the geographical extent of an area, whereas spatial resolution is a function of sample spacing and sample support (the size, geometry and orientation of an observation). Atkinson (2004) describes three situations that demonstrate the difference between spatial extent and spatial resolution: in a random sampling scheme, the spatial resolution varies locally; with gridded point data, the support is zero but spatial resolution will be determined by the point data spacing; in remotely sensed imagery (Chapter 4) the support (pixel) and sample spacing are fixed and approximately equal, thus spatial resolution can be seen as a function of support only. This is the reason why spatial resolution and pixel size are used interchangeably in remote sensing literature.

In the scenario-based and feature-based search methods, as defined by M. Harrison (personal. communication, 2007) and introduced in Chapter 2, consideration of spatial scale and spatial resolution becomes important for the integrated investigation of a crime or disaster. A scenario-based approach has implications for locational analysis of criminal activity, as each scenario is worked through, and so the spatial scale in terms of the geographical extent of the search strategy will vary. In a feature-based search method, physical and human anchor points such as windbreaks, a fork in a stream or key buildings will determine the spatial scale of the search and inform the spatial resolution of any invasive sampling strategy.

The difference between search strategies for urban and rural environments must also be taken into account when considering spatial scale and spatial resolution in the analysis of criminal activity for these different situations. Criminal activity in urban environments (such as in the example of the Shankhill Butchers gang murders in the 1970s in Northern Ireland, see Chapter 5) is generally based in certain areas (in this case territorial areas in North and West Belfast) and how these areas overlap with public transport routes. The areal extent of the abduction and body deposition sites (as in the Shankhill Butchers case) determines the spatial scale of the investigation. The spatial resolution of any sampling scheme employed will be determined by a scenario-based strategy: What was the mode of transport from abduction site, vehicle or by foot? Degree of activity at body deposition sites and opportunity for comparability of material (e.g. from car tracks, footwear or victim's clothing). In a rural setting where the areal extent of the investigation may be larger, topography and the production of DEMs (described in Chapter 5) may become important. The spatial resolution of the DEM and the necessary accuracy of the elevation values in forensic investigations will depend on the requirements for a particular scene of crime. It is important that the most suitable spatial resolution in areal photography and remotely sensed imagery is acquired so that it is 'fit for purpose'. For example, global scale analyses based on DEMs with 1 kilometer cell widths may be suitable for large scale search scenarios (such as the Boxing Day, 2004, Tsunami) but applications at a local scale utilizing data with a much finer spatial resolution, typically 1–10 metres, are more comparable to scene of crime operations. Fine spatial resolution DEMs (spatial resolution of 5–50 metres) are applicable for hydrological modelling and analysis of soil properties and as such may form part of the hydrology and geomorphology domain-based search operations described in Chapter 2. DEMs with a spatial resolution from 50–200 metres are defined as mesoscale or 'toposcale' and are required for broader scale applications such as hydrology, land-use and aspect-related microclimatic variations with associated vegetation patterns incorporating remotely-sensed data (Chapter 4). This spatial resolution of DEM may be especially relevant in search scenarios where a body deposition site is most likely to be in a certain vegetation or land-use type (vegetation such as bracken, peat bogs, etc.). In circumstances where overall elevation is more important than local variations in surface shape, mesoscale DEMs with a spatial resolution from 200 metres to 5 kilometres are more applicable. Macroscale DEMs (spatial resolution from 50–500 kilometres) provide generalized data and are applicable for broad-scale modelling or search scenarios.

6.2 Sampling for geological materials at urban and non-urban crime scenes

Sample number, method of collection and storage, location and context are generally agreed to be critical in subsequent analysis. In the various branches of geoscience, sampling is a basic skill, practised in ways developed for efficient and representative solution to a problem but rarely written about from the perspective of criminal investigations. Remote sensing and geophysics are both completely dependent on representative sampling, otherwise only ground-truthing can be used to test what the method has acquired, making the procedure otherwise largely useless. The main reasons why physical samples are rarely discussed in the literature may be twofold: first is the implicit assumption that the qualified geoscientist will have been taught appropriate sampling methods; second is the trust placed on analysis. Murray (2004), Pye and Croft (2004) or Ruffell and McKinley (2004) are three recent overviews on how *methods* not data *source*, *collection* or *integrity* are central to the current practice of the science. A third reason, specific to geoforensics, is historical. When soil or sediment was observed/recovered from a scene of crime by investigators, with appropriate training and/or support they knew a geologist of some sort was required. This meant the investigator could usually see the material involved, knew already roughly what it was, and the geologist could be deployed relatively quickly to recover material. Thus Murray's (2004, p. 116) method 'It is better practice for the evidence collector to examine a questioned sample first, for colour and particle size at least, and then to search for samples with similar appearance' could be deployed. Three problems arise: timing of the crime, sample size and lateral variation, and these are discussed in the next three sections.

6.3 Timing of the crime

A crime occurs in a location where transferable material exists yet where compromise to the scene(s) is likely (weather, construction, biological activity). No suspect is yet identified, begging the question of whether to sample the scene at all, or if sampled, on what basis? Locations of likely contact (Locard's Principle) and exit/entry routes are all possible, yet if the suspect is concerned about witness observation, they may actually avoid such places. How many samples (to representatively 'cover' the whole area) should be taken? Could maps of sediment, soil and loose material be made and samples of each taken? How representative are the maps (based on visual, pH, geophysical proxy observations)? The answers to these questions present some of the greatest challenges for those dependent on comparing/excluding samples for scientific analysis of any kind, and are the reason major research programmes into database development and sampling efficacy are underway at the time of writing.

6.4 Sample size

Issues concerning sample size have a similar history to sampling. Only in rare historical cases (Locard's work on Emile Gourbin, linked to the murder of his girlfriend by the face-powder under his fingernails) have geological analyses been used on what most practitioners would call 'trace' evidence (for example, a few micrograms, layers of dust or mud-spots). Thus once again, a brief look at Murray (2004), Pye and Croft (2004) or Ruffell and McKinley (2004) (overviews of geoforensics) shows that sample quantity is rarely debated, it being assumed the geoscientist will not have been employed unless suitable amounts of sample are already present.

Communication (in prison, in person, via the Internet) between those wishing to avoid criminal association, television programmes such as 'C.S.I.', books like this, and the popularity of 'forensic' courses, etc. have all contributed to a greater awareness of how to avoid transferring or preserving contact evidence. Personal experience has shown that often perpetrators do not know exactly why they are carrying out extensive clean-up operations, but being part of an organized criminal gang, or having previous experience, they know it can only assist their denial of a criminal act. Sophisticated terrorists have developed their own pre- onsite and post-operation procedures, including the sharing/swapping of clothing/weaponry/transport to confuse any DNA associations, taking above-standard forensic precautions at scenes and, most critically, following tight clean-up procedures such as washing or destruction of clothing and vehicles, rapid washing of hair, scrubbing of nails and disposal of all associated items. Nonetheless, Locard's principle pertains, and if indeed every contact leaves a trace, it is for the scientist to deploy methods suitable to detect that trace.

Sometimes the clean-up creates the trace itself: a domestic murder suspect in West Belfast in 2005 hosed his trainers following his movement along a muddy track. His washing was efficient, yet created mud-splashes on his trousers (secondary transfer) that only became obvious upon the material drying, allowing visual identification, sampling and analysis for comparison to suitable material from the scene. How to deal with small samples is considered in Chapter 8, yet there are repercussions for scene sampling. When Murray (2004, p. 116) takes his questioned sample (or description/photograph) to scenes of crime and alibi locations, he is either using the exclusionary principle, or comparing visually-similar material for further analysis. A mud-speck, barely visible to the naked eye is difficult to compare to a scene. It can be done with experience and caution but the tiny size of the sample and the hugeness of the scene will only provide ammunition for the defence, with analogies such as 'needles in haystacks'. Should the sample be so small as to only be recognizable using sophisticated analysis, then the task is indeed impossible: chemical/mineralogical analyses can rarely be taken to the scene for comprehensive sampling (see the use of Portable XRF in Chapter 4). Instead the same arguments as developed for timing of incident/sampling are pertinent: is common-sense about points of access, contact and transferability of material used? Or is an attempt made to take visually-representative samples using a proxy (colour, grain-size, geophysics, remote sensing), or an attempt made to collect a statistically-robust number of samples? In the absence (availability, time, access) of a proxy, McKinley and Ruffell (2007) suggest that as many samples should be taken as possible, and their representativeness tested geostatistically.

6.5 Lateral variation

The final sentence (above) will upset the close-minded geoforensic practitioner and challenge the open-minded. The possibility of a sample not being representative is of major concern: the innocent may be convicted or the guilty freed without charge on the basis of this assumption, if correct. Junger (1996) was the first person to challenge this assumption, taking 100 samples over a grid in a visually-homogenous location and comparing adjacent pairs, some of which were so similar that a court would be likely to rule them as indistinguishable. However, further apart, the comparability of pairs broke down. The work of Horrocks and Walsh (1998) and Horrocks *et al.* (1999), albeit on microbiological materials, demonstrated that soil or sediment could vary over a lateral distance of a few metres. McKinley and Ruffell (2007) used this approach in an urban location, where lateral variation could already be observed: their aim was to test the likelihood of a person being at a scene contacting the unique properties of that particular location. Like Junger, they conclude that the likely contact point, sampled by the scenes of crime officer, did not adequately represent the material found at the scene (gypsum-based plasterboard). A suspect in a fatal shooting was later apprehended and linked to the scene by the gypsum-based plasterboard on his footwear: the single police sample would not have demonstrated this.

The nature of the sample: Bull *et al.* (2006) on footprint cast samples

Bull *et al.* (2006) describe the following scenario:

> A young woman was out walking her dog in the early afternoon on a hot summer's day in a rural area in the midlands of England. The path she took crossed a small bridge, which forded a trout stream and ran parallel with [to] the raised embankment of a railway line. The woman was brutally attacked and dragged onto the railway embankment through a patch of thistles and it was assumed the attacker had attempted to lay her inert body across the railway track. The embankment proved too steep and the young woman was left at the bottom of the slope later to be found barely alive. She was taken to hospital where she died six days after admission. Tracker dogs at the scene followed a trail from where the girl was left, across the railway line into a field that had been ploughed [on] that very morning. Footprints could be seen tracking across the field. Plaster casts were taken of the footprints, primarily to ascertain the size of the suspect's shoes (with the intention of excluding the farmer as the originator ...).

A suspect was arrested and two pairs of his shoes and clothing were seized. The investigation centred on whether material on his footwear and clothes could be compared or excluded from material recovered at the scene. He denied ever being in the locations in question. The footprint casts provided adhered soil that would have been closest (originally) to the footwear of the maker: thus it was compared to both the soil from the field as well as that from 19 adjacent 'control' fields. Bull *et al.*'s use of binocular microscopy in identifying fibres, hair, as well as seven mineral/rock grains effectively excluded one pair of the suspect's shoes and the 19 control samples, the latter having shale pellets and limestone oolites when the scene, suspect and footwear cast samples did not. These results were confirmed by use of particle size and pollen analysis, which also provided some record of the previous history of the footwear that was not present in the scene or footwear cast material. The one matter not examined by Bull *et al.* (2006), which has arisen with other cases, is where plasterboard or gypsum/anhydrite (plasterboard mineral ingredients) occurs at the scene, yet is compromised by the use of plaster-casting. This requires even more sophisticated analysis in order to differentiate plaster types.

6.6 Use and misuse of statistics in forensic studies

When posed with the question of how to evaluate the degree of similarity and the significance of any similarity between evidence from a suspect and scene of crime, statistics provide a useful method to quantify or 'provide a number to' describe any similarity or dissimilarity between samples. This can be invaluable in a court of law when asked the question: 'How similar?' However, the inappropriate or misinformed use of statistics in a court of law can lead to important and genuine evidence being completely discredited. In the case of abduction and murder of a 19-year-old black male, Michael Donald, by members of the Ku Klux Klan in downtown Mobile, Alabama on 21st March 1981, evidence from 'similarities' in the chemistry of soil samples was invalidated due to the improper use of statistical methods and mistakes in the treatment of samples (Isphording, 2004). Although sound evidence did exist to link heavy mineral and clay mineral composition of soils at the crime scene with the samples taken at the defendant's residence, the inappropriate use of statistics provided an unfortunate opportunity for the defence to exploit Mark Twain's statement that 'There are three kinds of lies: lies, damned lies, and statistics!' and discredit the value of statistical analysis to the jury. Fortunately, to escape a possible death penalty the co-defendants (James 'Tiger' Knowles and Benjamin Franklin Cox) admitted involvement and testified Henry Francis Hays as the chief conspirator. A subsequent civil suit (in 1987) filed by the mother of the victim against the Ku Klux Klan returned a verdict of $7 million in damages and led to the demise of the United Klans of America (UKA) and Klan activity throughout the USA (readers are referred to Isphording, 2004, for full details).

A range of analytical techniques is available in forensic investigations to produce results to allow comparison between samples collected from a scene of crime and the defendant and victim. In response to this, a range of statistical techniques exist that can be used to test the comparability of scene of crime and suspect samples. Several of these statistical tests are tried and tested in the court of law setting and have become standard practice (for example standard (Pearson) correlation analysis). However the Ku Klux Klan abduction and murder case described above serves as a reminder of the pitfalls in the misuse of statistics (even in the use of those techniques accepted as standard practice). Several pertinent issues arise from a review of the use of statistics in forensic studies, including the number of samples required for robust statistical analysis; the usefulness of descriptive aspatial statistics; the important issue of comparing 'like with like'; addressing the problems of comparing related material; and the relevance of spatial statistics to address the spatial and temporal variability in nature.

6.7 Statistical sampling

Forensic geoscientists are interested in questions of comparison. How similar is the seized item from the suspect to material found on the victim or at the scene of crime? In an attempt to answer these questions a sampling scheme is deployed. A limited number of samples can be generally recovered from the suspect (clothing, footwear, vehicle) and similarly from the victim (if found). The number of samples that should be taken from the scene of crime (known or suspected) is more difficult to ascertain. It is useful to consider the aim of sampling in statistical analysis: we sample to make inferences about a population (say soil attributes, as opposed to people) based on information in a sample taken from the population. In forensic terms this provides a means of comparing samples from a suspect to the scene of crime and thus to the victim. We repeat, the term 'population' does not refer only to people but can be items or, for a forensic investigation, characteristics of an area of interest. However, the definition of a population, as it relates to individual forensic cases, is difficult to ascertain and no set of guidelines exist. The reference population (terminology as used by Aitken and Stoney (1991), also described as the background or target population) can be used to represent the scene of crime and the material taken from the suspect and victim form two separately sampled populations. However, as Aitken and Stoney describe (1991), the reference population and the two separately sampled populations may not be necessarily from the same population. This is the question that the forensic scientist needs to answer to ascertain whether samples taken from a suspect are from the same population as those taken from the scene of crime and the victim. Control samples are taken as standard good practice. These by definition should be from a different population to the reference population representing the scene of crime.

The usefulness of sampling is that we can generalize from the sample to make estimates or predictions about the reference population (or area of interest) as a whole. However, this poses questions that in forensic work may be exploited in a court of law: how reliable are such generalizations? Is the sample representative of the whole area? Samples can be misleading unless representative of the population (or area in the case of scene of crime). The most accurate sample would include the whole population (or complete coverage of the area) but this is unrealistic due to cost and time. Therefore samples should be chosen at random from the population or scene of crime. This means that every location within the population (or crime scene) will have the same chance to be the selected sample. However, in many forensic cases the samples may be chosen for the geoscientist in that the scene of crime officer collects the samples or there are key locations within the scene of crime where greater emphasis is placed on sample recovery (entry and exit routes). In terms of statistical procedures this is subjective sampling and introduces bias into the collected sample, which needs to be acknowledged in any subsequent analysis.

The next step for the geoscientist is to make inferences about the collected sample. Although the end goal may be to compare collected samples with those from a suspect's clothing, footwear or vehicle, standard statistical practice is to describe the characteristics of each attribute individually. The sampled data – once characterized statistically – can then be compared and interpreted. The value of using statistical analysis in forensic work is that we can describe sampled evidence by using statistical attributes or characteristics that distinguish between one individual object or measurement in the sample from another. Quantitative attributes or characteristics, where the values or outcomes are expressed numerically, comprise both *discrete* data where the possible values are clearly separated from each other and imply counting (e.g., number of plasterboard particles (McKinley & Ruffell 2007) and *continuous* data that vary continuously across an area and require measurement (soil properties). When faced with results from various analytical techniques used in forensic investigations, the value of taking time to examine and explore the data through descriptive univariate statistics cannot be understated. Descriptive statistics may be displayed in several ways – tabular (summary statistics: mean, mode, median, range, skew) or graphical (histogram and frequency distributions). As Pye and Blott (2004) suggest, this helps to identify groupings or different populations within samples and patterns of similarity between sample sets. These initial findings can inform further statistical analyses including selecting the most appropriate sample sets for statistical tests (e.g., Chi squared) and bivariate statistics (correlation analysis) and more complex multivariate techniques if appropriate (e.g., correlation matrices, multi-regression analysis, cluster analysis, principle component analysis). This initial exploratory step may also prevent the inappropriate use and presentation of statistical results to a jury at a later stage and subsequent discrediting by the opposition of valid and invalid findings alike.

Typical tabular summary statistics in forensic studies include the use of descriptors of central tendency (mean, mode, median) and measures of distribution shape (skewness, etc.). A histogram is the simplest method to present an overall picture of the distribution shape of a sample attribute. Straightforward descriptive information can be assessed from the histogram: Is the distribution shape peaked or rectangular? How many different peaks are there? Is the distribution symmetrical about some central point and so can be described as a normal (or Gaussian) distribution? This descriptive information about a sample attribute can then be used for comparison with samples from the suspect. The value of using summary statistics and histograms are that they are familiar to most people and can provide an uncomplicated comparison of the statistical characteristics of crime scene and suspect. Saye and Pye (2004) found the modal characteristics of sand samples to indicate a close similarity between sand grains retrieved from a glove and data from the England and Wales sand dune database (see page 166 for a case study of this incident). The median is less affected by outliers than the mean and therefore can prove useful to compare samples when the possibility of 'a rogue grain' or outlier could mask a potential similarity between samples from suspect and victim. Measures of variability are based on the scatter of data relative to a central value or sample mean (Swan & Sandilands, 1995). The coefficient of variation, or normalized standard deviation, is a statistical measure of variability that has been used effectively in forensic studies. Croft and Pye

(2004a) used the coefficient of variation during colour testing to measure variability between repeated measurement over time and also for variability between sample sets. Investigating element variability in small soil samples, Jarvis et al. (2004) found the coefficient of variation a useful measure to evaluate instrument measurement precision.

The end goal in any statistical analysis in a criminal investigation is to exclude all controls and provide a comparison between collected scene samples and those from a suspect's clothing, footwear or vehicle. The scientifically-driven forensic investigation seeks to objectively compare all scene, suspect or victim samples such that the exclusionary principle is maintained until only comparisons can be made. Ultimately the criminal prosecutor and forensic scientist may reach the same conclusion, and a case brought; either way, comparison and exclusion should have the same statistical tests applied. At times this will involve establishing the presence and strength of a relationship between two samples. An objective assessment is required that will provide a quantitative measure of whether a relationship exists between samples or attributes of samples.

The simplest method to judge whether any relationship exists is to plot a graph of one attribute against the other in the form of a scatter plot. This can prove to be visually effective in indicating a connection between samples; however, an objective measure of the relationship is needed to validate the subjectivity of the eye. The correlation coefficient is a measure of the strength of the relationship between sampled attributes that is independent of the units in which the attributes were measured. The correlation coefficient provides a quantitative measure of relationship between -1 and $+1$. A perfect direct relationship would produce a value of $+1$ whereas a perfect indirect relationship would be -1. A lack of any relationship would give a value of 0. In reality correlations will lie somewhere within this range. A high positive correlation is indicated by a correlation coefficient between 0.75 and 1, a moderate to low correlation by a value less than 0.75. Conversely the relationship can be indirect with a high negative correlation indicated by a correlation coefficient between -0.75 and -1. If when plotted on a scatter plot, the characteristics (grain size, mineralogy, etc.) of a sample taken from a suspect indicates a relationship with those from the victim and scene of crime and produces a high correlation coefficient, an association between suspect and crime is strongly indicated. However, if a moderate correlation is demonstrated (e.g., correlation coefficient of 0.5), then a relationship between samples cannot be ruled out. It must be remembered that material retrieved from the suspect, victim and collected from the scene represent only a few samples from the whole population.

Therefore, it needs to be assessed whether the correlation coefficient would be significant if it were possible to collect all possible samples within the population. To do this a hypothesis is set up that states that there is no relationship between the samples (another way of thinking of the exclusionary principle). That is, if all possible samples within the population were collected the correlation coefficient would be zero. Tables are available to find the distribution of the estimated correlation coefficient if the hypothesis were true. The estimated correlation coefficient value will vary around the ideal zero value. How much it varies depends on the number of samples or degrees of freedom. The calculated correlation coefficient value (0.5 in this case) can then be compared to the table for the relevant degrees of freedom. If the calculated correlation coefficient is significantly larger than the estimated value for the appropriate degrees of freedom, the hypothesis of no correlation must be rejected and there is a chance that a significant relationship exists.

Collected samples may have two or more attributes to analyse and a range of multivariate statistical techniques have been used in forensic studies (e.g., cluster analysis, principle component analysis and canonical variate analysis). In the work of Ruth Morgan and Peter Bull (2006, 2007) on data interpretation in forensic sediment and soil geochemistry, numerical data derived for two case studies from chemical analysis by ICP-MS and Dionex on 8 and 28 samples (respectively) were compared by hierarchical cluster analysis. Canonical variate analysis is a multivariate statistical procedure for treating the problem posed by simultaneous analysis of several sampling levels (Reyment, 2006; in geosciences this may relate to observations made over a stratigraphic sequence). Weight of evidence and Bayesian methods have been used in several areas of forensic analysis. Bayesian methods offer a natural tool for linking together many different individual and aggregate data sources. Small *et al.* (2004) adopted a Bayesian framework in a tracer approach to sediment provenance or 'fingerprinting' to access the influence of group contributions and the benefit of integrating the relationship between variables over discriminatory power alone. Readers are directed towards several key texts for further information on the use of statistics and Bayesian networks in forensic science (e.g., Aitken, 1995; Aitken & Stoney, 1991; Taroni *et al.* 2004).

6.8 Number of samples required for robust statistical analysis

One of the toughest questions that a geoscientist can be asked by a crime scene officer is how many samples should be taken. This is a very different question from how many samples are required for statistical robustness, but the compromise in both cases is usually as many as possible. Limitations to adequate sampling in order to obtain a statistically robust data set are generally determined by the individual circumstances of each scene of crime, but more generally include time pressure, the high cost of fieldwork, laboratory processing time and analytical costs. Small *et al.* (2004) stress the importance of adopting an appropriate sampling scheme to collect sufficient sample numbers to capture source group variability. 'A sufficient number of samples should be collected to ensure the quality of the data generated enables meaningful scientific statements to be made about the sampling target' Small *et al.* (2004). They conclude that the goal for forensic work is to gain an acceptable confidence level to obtain the legal outcome.

As heterogeneity increases, a greater sample size will be required to adequately represent the population as a whole. The larger the sample size the more confidence we can have that the sample mean will be an accurate representation of the mean for the population of the area of interest. In terms of statistical theory, the mean and standard deviation hold for any sample size but the approximation to a 'normal' or Gaussian shape requires the sample number to be sufficiently large. In a forensic investigation any sampling scheme deployed and associated uncertainty should be recorded and accounted for from the outset.

6.9 Comparing 'like with like'

One of the recurrent pertinent issues in the comparison of geological materials between scene of crime and suspect, regardless of analytical technique (e.g., soil chemistry, colour) used, is the need to compare similar material or 'like with like'. In geological materials this is most clearly seen in terms of particle size where failure to use data from the same size fraction in statistical procedures results in irrelevant and invalid statistics. For example, in the application of a tracer approach to sediment provenance or 'fingerprinting', Small *et al.* (2004) stress the 'importance of establishing "true differences" rather than apparent differences due to differences in particle size distribution'. Likewise, in colour tests of dried sand samples (desert and coastal), Croft and Pye (2004) found variation in reflectance curves from spectrophotometry for different size fractions. Moreover, conclusions from their findings suggest that the same size fraction should be used wherever possible and colour testing should be determined on bulk *in situ* material as well as on specific size fractions known to be characteristic of the particular mineral being analysed. As a result, greater confidence can be placed on any similarities and differences indicated by summary statistics of the data. One of the criticisms directed at the prosecution in the Ku Klux Klan murder case (above) was the lack of explanation for the discrepancy between levels of aluminium observed in samples from the defendants and from the victim. As Isphording (2004) states, a valid explanation was possible for the discrepancies if a size frequency distribution had been undertaken for each sample. This would have demonstrated that size analyses strongly affect the levels of metals that can be present: the lower levels of aluminium found in the defendant's samples simply reflected lower clay content in these samples and not a 'true' dissimilarity declared by the defence between the defendant's samples and those from the victim.

6.10 Addressing the issue of comparing related material

Forensic investigations, as in geosciences and other disciplines, often involve the use of compositions of data – multivariate data in which the components represent some part of a whole. Classic examples of compositional data used in forensic work include chemical analyses, geochemical compositions of rocks and sand, silt, clay compositions. Compositional data are usually recorded in closed form, summing to a constant (one if measured in parts per unit or one hundred if measured in percentages). As such they convey relative information: variables are not free to vary independently to one another. As Pawlowsky-Glahn and Egozcue (2006) explain, if one component increases in portion, others must decrease, whether or not there is a link between components. This has direct relevance to forensic investigations in that it creates a niggling uncertainty to the investigation that could be exploited in a court of law: distinguishing between spurious effects caused by the constant sum restraint and those caused by natural processes. Since the work of Karl Pearson (1897) and Felix Chayes (1971), the possibility of spurious correlation has been well documented (e.g., Aitchison, 1986; Butler, 1979; Buccianti *et al.*, 2006; Krumbein, 1962; Rollinson 1995). Compositional data generated in forensic investigations often comprise multiple variables. These relationships often show complex features such as nonlinear relations or mineralogical constraints. However, attributes in geosciences rarely exhibit 'normal' characteristics, hence transformation of data is integral to any further statistical analysis. A key question, which is as prevalent to forensic studies as to other research, is the appropriate use of transformation methods to maintain the integrity of compositional data and inherent constrained behaviour in multivariate relationships. Use of transformation methods, such as log-ratios proposed by the work of John Aitchison culminating in his seminal monograph in 1986, may maintain the integrity of geochemical data and inherent constrained behaviour in multivariate relationships. This is especially relevant for geochemical data and the multivariate geochemical datasets generated in forensic studies (e.g., Pye & Blott (2004) and Rawlins & Cave (2004)). Readers are directed towards the wide range of statistical techniques recommended for the analysis of geochemical data (e.g., Aitchison 1999; Davis, 1986; Rollinson 1995; Swan & Sandilands 1995; Webster & Oliver, 2001). The value of an appropriate method to investigate compositional data will reduce uncertainty in the manipulation of soil geochemistry data, honouring of inherent geochemical constraints and greater accuracy in the integration of the geochemistry data with other data. This will enable more meaningful interpretation of the nature of the geochemical variability in the geological, environmental and forensic inferences.

6.11 Spatial and temporal variability in nature

Geoscientists are all too aware of the spatial and temporal variability found in soils and sediments and moreover the different scales of heterogeneity in earth materials (Jarvis *et al.*, 2004; Pye & Blott, 2004). As a result, effort has been directed at the development of regional databases by governments (good examples occur in the USA, Canada, Australia, Russia [Omelyanyuk & Alekseev, 2001], and many European Union countries) as well as consortia (MacCauley Institute Soil Survey) and individuals (Saye & Pye 2004). These databases provide the opportunity to compare samples collected from a suspect to the database collection and then form a potential link to a scene of crime if contained within the area coverage of the database.

Dune database application to murder investigation, Lincolnshire coast, UK – Saye and Pye 2004

The discovery of a glove (at an inland roadside location) and a blood-stained shirt indicated evidence of a temporary body burial site in the coastal dunes at Chapel Six Marshes, Lincolnshire. The presence of sand grains and marine shell fragments trapped in the glove webbing pointed to a beach or dune environment. The particle size characteristics of the sand on the gloves were found to be similar but significantly different from the sand at the site at Chapel Six Marshes where the shirt was discovered, confirming this as a temporary burial site. The particle size characteristics of the sand on the gloves were then compared with data from the England and Wales dune database. The sand on the glove could not be sourced exclusively to Lincolnshire on the basis of particle size characteristics alone but combined with a comparison with the national dune sand geochemistry database, the Lincolnshire coast was demonstrated to provide the highest level of similarity. This enabled a higher resolution collection of control samples at approximately 50 metre intervals between the temporary burial site at Chapel Six Marshes and Anderby Creek (to the north of Wolla Bank). In particular, modal characteristics of the sand samples indicated a close similarity between glove and sand dunes. This was confirmed by optical microscopy and geochemical analysis and a short stretch of dunes, 500 metres from the known disposal site, was searched. No body parts were found at this site suggesting that if any had originally been deposited here, they had been later removed. Parts of two legs were later recovered from a location close to the Lincolnshire border and a partially composed torso was found adjacent to the A1 dual carriageway near Sawtry in Cambridgeshire. The hands, arms and head of the victim were never recovered.

Investigating spatial variability at the micro-scale: use of QemSCAN – Pirrie *et al.* (2004)

Spatial heterogeneity is present even at the micro- or pore scale. With the increasing use of small amounts of trace element material in forensic applications (see Chapter 8), the ability to examine and differentiate similarity and dissimilarity between microscopic amounts of forensic evidence becomes progressively more important. One of the pioneering analytical methods in this field is QemSCAN, an automated SEM (scanning electron microscopy) system fitted with up to four energy dispersive X-ray spectrometers, providing rapid quantitative mineral analyses. Consolidated samples, such as rock chips and ceramics, prepared as polished blocks and unconsolidated samples, soil and dust samples (mounted onto double-sided carbon tape), can be analysed at a user-defined pixel spacing. Each individual pixel is compared against a database of known spectra, identified and assigned to a mineral or phase name. Pirrie *et al.* (2004) describe how 'depending on the pre-defined pixel spacing and the particle size the system can quantify and map the composition of approximately 1000 particles, each 1–10 µm in size per hour'. As the X-ray spectra are collected the coordinates of each particle are automatically recorded. The QemSCAN system has a four-mode operation comprising particle mineral analysis, bulk mineral analysis, trace mineral search and field image scan. The fieldscan mode is of most interest in the present discussion of spatial heterogeneity at a micro-scale. Field scanning enables the spatial mapping of a rock fragment or ceramic block. The sample is divided up into a series of 'fields', each field is measured and the pixels assigned to a mineral species or phase. Adjacent fields are then 'stitched' together to provide a mineralogical or phase map of the scanned area. This scan mode has significant potential for quantitatively mapping spatial variability of geoforensic materials at a micro-scale.

6.12 Spatial awareness and use of spatial statistics: application of geostatistics

The role of this book is not to provide a detailed background on spatial statistics and geostatistics but to introduce the proposal that geostatistics can be of value in forensic work, and moreover that as spatial awareness in geoforensic work increases the tools to analyse the material need to include spatial statistics. The collection of samples or measurements at crime scenes presents a forensic investigator with questions such as how many samples should be taken and what sort (if any) of spatial sampling strategy should be deployed. Geostatistical techniques can be of value in addressing the issue of spatial sampling strategies at crime scenes and of informing decisions of how many samples are sufficient. In the search of an area for evidence relating to the deposition of a body, or other suspicious activity, a designated area may have been highlighted by forensic evidence and several 'hot spots' may even be identified by other intelligence such as a dog team. However, the whole of the designated area, which may be extensive, cannot be ruled out. In the case of no body (or similar) crime scenes, techniques are being developed including the use of isotope studies where samples taken from the crime scene are used for laboratory analyses in an attempt to provide information in the search for a body or obtain evidence of the past deposition of a body. This may equally apply to other hidden materials, such as chemicals. Excavation is a destructive, expensive and time-consuming process, which may not even be a viable option. The use of sampling and subsequent laboratory analysis can inform and provide greater clarity on determining specific areas to excavate and thus reduce cost, time and unnecessary destruction.

Decisions need to be made about sampling strategies and assessing the optimum number of samples required to sufficiently inform an invasive search. The most effective method of evaluation and interpretation of a spatial pattern of variation, which addresses any tendency to sampling bias, is systematic sampling in the form of measurement and collection of samples in a defined pattern or grid (Savvides, 2006). A nested grid system of sampling may be applicable in which the geographic extent of a wider spaced (large scale) sampling grid will be defined by forensic knowledge and 'hotspot' areas (indicated by dog teams, etc.) within this would be sampled in a more closely spaced (small scale) grid. However, with any sampling strategy there is usually a lack of complete knowledge about how the area varies in spatial extent. Interpolation techniques and geostatistical analysis can be used to make predictions at locations for which there are no data and provide a measure of uncertainty in the description. The outputs from interpolation and detailed geostatistical analyses, including variography, spatial prediction and conditional simulation, can provide further information on sampling strategies and better inform the search procedure.

Interpolation

Following the collection of samples, a hand-drawn field contour map may still be the preferred method by a geoscientist for an initial visualization of the spatial variability of an attribute at a scene of crime. This skill, albeit it in a crude sense, is applying the theory of spatial interpolation to convert the information collected from point locations to a continuous spatial map so that any initial spatial patterns can be observed. Interpolation provides a means of making predictions at locations where there are no measurements. Spatial interpolation works on the principle of spatial dependence: values close together in space tend to be more similar than those far apart. Commonly used interpolation methods include the following:

- *Theissen polygons*, where unknown values are given the value of their nearest sample;

- *Inverse Distance Weighting* (IDW), where samples close to the value to be predicted are given largest weights (their influence in making a prediction), whereas samples at a greater distance are given smaller weights;

- *Thin Plate Spline* (TPS), which in some respects can be described as stretching a sheet of rubber over points;

- *Kriging*, where the weights assigned to samples are obtained through modelling spatial variation using geostatistics.

Two of the most widely used interpolation methods commonly available in GIS packages such as ArcGIS™ (Johnston *et al.*, 2001), which have been used in forensic studies (McKinley & Ruffell 2007), include IDW and TPS. Both these methods require a search neighbourhood of nearest neighbours to be selected. Readers are encouraged to obtain further details of these methods from texts such as Burrough and McDonnell (1998) and Wackernagel (2003). There is a variant of TPS, termed Thin Plate Spline with Tension (TPST), where tension of splines may be adjusted to deal with extremes of values across an area. For example, in a large scale search scenario covering 50–100 kilometres, the topography is likely to vary across the areal extent of the designated search area to include flat ground (e.g., floodplain area) and areas of higher ground. The density of data collected (GPS readings for elevation or soil sampling) over this area is likely to be variable. In this sort of interpolation situation large gradients can be produced in data-poor areas producing overshoots or anomalously high or low edges to the curvature surface of the thin plate spline (Chang, 2002). TPST as a method overcomes this problem by allowing the tension of the spline surface to be changed so that it resembles a stiff plate or an elastic membrane (Mitáš & Mitášová, 1999). In the example of interpolating elevation data of a search area covering 50–100 kilometres, low tension can be applied to the spline surface to represent smooth topography (lower flat ground) whereas high tension can be used to represent sharp breaks of slope in higher ground. Using the ArcGIS Geostatistical Analyst software, the tension of the spline can be selected using a cross-validation procedure.

Use of interpolation in spatial mapping a scene of crime – McKinley and Ruffell (2007)

In an urban scene of crime one sample was collected by a Scenes of Crime Officer (or CSI: Crime Scene Investigator) in comparison to 112 gridded samples collected by the authors. In general the samples comprised urban dust but fragments of gypsum-based plasterboard were also found in samples. A hand-drawn contour map of the spatial distribution of the plaster material was used as an initial visualization. Spatial interpolation was used to convert the information collected from point locations (a count of the number of fragments of plasterboard in samples) at the suspect scene to a continuous spatial map so that any spatial patterns can be observed. The interpolated maps (Figure 6.1) were found to confirm and maximize information on the bimodal distribution pattern of plaster material indicated by the hand-drawn map. TPST was the preferred method of interpolation in this case and provided the best comparison with the smooth curves drawn by eye in the hand-drawn map. The IDW method of interpolation demonstrated a 'duck-egg' pattern around solitary points, which differed significantly from their surroundings. Cross validation enabled a statistical comparison between the use of TPST and IDW and provided robust validation for which interpolation method should be used. In this case TPST was found to produce a smaller cross validation root mean square (RMS) estimation error than IDW. Footwear tread samples containing plaster fragments could potentially be located to an area of the sampling grid depending on the number of plaster fragments. The single, CSI-collected sample allowed only minimal comparison to the 'suspect' samples. The grid of samples showed a wide variation, with samples from one particular area of the total sample area showing a greater comparability

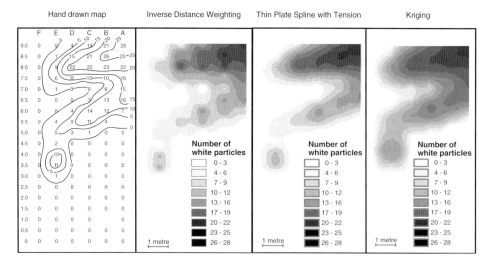

Figure 6.1 Comparison of a hand-drawn field contour map and interpolated maps using IDW, TPST and kriging: discussed later in this chapter (adapted with additional data from McKinley and Ruffell 2007)

to footwear samples, in terms of number of observed plaster fragments, than the remaining sample grid. The implications from this case are that where heterogeneity is observed or suspected at a scene, sufficient samples must be taken to map this variation.

Interpolation enables spatial patterns sampled by one set of measurements to be compared with spatial patterns of other spatial data. In the urban case discussed above, the spatial distribution of discrete data (number of plasteroard particles) was analysed. The presence of the plaster fragments as 'unique particles' (Sugita & Marumo 2004) or unique identifiers (Riverside Criminalistics Laboratory, personal communication. 2006) in this case led to the choice of variable. However, the spatial distribution of any transferable material (number of discrete particles or continuous variables such as grain size analyses or soil properties) can be analysed by interpolation and the spatial distribution then compared. The benefit of finding unusual particles in a forensic investigation is that they have a fairly restricted geographical distribution (Pye, 2004a). Once sampled, the distribution of unusual materials may have the potential to place a suspect to within a few metres of a scene.

Case Study: Spatial distribution of freshwater snail shells – work of Marianne Stam (US Department of Justice), (see Stam, 2002 or 2004)

In October 1999, a woman and her small children accepted a ride from a male acquaintance in California. He took them to an isolated location on the banks of the New River where he raped the mother, slit her throat, and attempted to drown her. He also hit her children with rocks. The victims escaped and hid for 30 hours in the river. The suspect was arrested only hours after the victims reported the crime. The suspect denied any contact with the victims and claimed to have not been near the crime scene. Soiled and wet clothing and shoes were found at his residence. Detectives also discovered scratches to the suspect's upper left arm and lower legs. The suspect claimed that these were from his girlfriend and from his work. All sexual assault evidence collected from the victim tested negative for semen; soil evidence was examined to determine whether the suspect was at the scene, or to exclude them, and to corroborate the victims' story.

The victim was able to pinpoint the area along the river embankment where she and the suspect entered the river during the attack. Soil samples were collected from the crime scene and the surrounding region for comparison to the soil from the suspect's clothes and shoes. Approximately 16 grams of soil were removed from the suspect's right shoe. Although other items of the suspect's clothing contained soil of similar appearance, the soil from the suspect's right shoe was examined in more detail because of the larger amount present. This soil was analysed and compared to soil from the crime scene using stereomicroscopy, polarized light microscopy, and X-ray diffraction. The soil samples were examined and compared as whole samples as well as sieved fractions using stereomicroscopy. The sieved fractions consisted of gravel, sand, silt and clay sizes. Portions of the silt-sized fractions of each sample were separated by density using bromoform into light and heavy minerals and were mounted in 1.54 and 1.65 refractive index oils respectively for polarized light microscopy. The heavy minerals were grouped according to their appearance:

colourless, green, or brown. The soil from the crime scene ravine and the soil on the suspect's right shoe were similar in Munsell Colour, mineral grain morphologies and gross content. Both soil samples also contained similar small white shells. The shells were identified as *Tryonia protea*, a freshwater snail species. Fossil shells (see Chapter 7 for a discussion of fossils) of this species were deposited within a natural lake that filled the Salton Sea Basin (part of which is now the scene of crime) during the last Ice Age between 10,000 and 50,000 years ago. These shells have a wide distribution throughout the Imperial Valley, thus reducing their significance for comparison. Polarized light microscopy revealed similar light and heavy minerals in approximately similar abundances in the samples from the crime scene and the sample from the suspect's right shoe. The minerals were not specifically identified, but were categorized according to their possible mineral types. Identification of minerals and further comparison of the soil samples was completed on the silt-sized fractions by X-ray diffraction indicated that the soil samples from the crime scene contained similar minerals to the soil sample from the suspect's shoe.

How significant was this similarity? Returning to the crime scene, the white deposits along the south side of the New River appeared to be limited in their distribution. They were mostly quartz mixed with the small shells. Based on the prevailing wind direction, their source could be large sand deposits to the west of the crime scene. The presence of these white deposits (with limited spatial distribution) in the crime scene soil samples and on the suspect's shoe added significance and helped corroborate the victims' accounts of the attack as well as place the suspect in the vicinity of the crime scene. In effect investigation of the spatial distribution of soil deposits demonstrated a limitation of similar characteristics to a small geographic area.

6.13 Geostatistical techniques

The urban case study (McKinley & Ruffell 2007) presented above is used to apply geostatistical techniques to model the spatial variation of the unusual particles (in this case, the number of fragments of plasterboard in samples). Kriging has been used to produce the mapped distribution of plaster particles shown in Figure 6.1. *Kriging*, as an interpolation technique, is similar in principle to IDW except that the weights assigned to samples are obtained using coefficients obtained from the variogram, a core tool in geostatistical analysis. Introductions to geostatistics are provided by Koch and Link (1970), Oliver and Webster (1990) and Burrough and McDonnell (1998), and the reader is encouraged to refer to these texts for further detail. Geostatistical analysis usually involves two stages: (i) estimation of the variogram and fitting a model to it; and (ii) use of the variogram model coefficients for spatial prediction (kriging) or simulation.

The variogram is used to characterize the spatial dependence of a property of interest and is a function of distance and orientation of that distance. It expresses, in the form of a graph (Figure 6.2), how variables vary according to distance in a certain direction. In simple terms, the variogram is estimated by calculating the squared differences between all the available paired values and obtaining half the

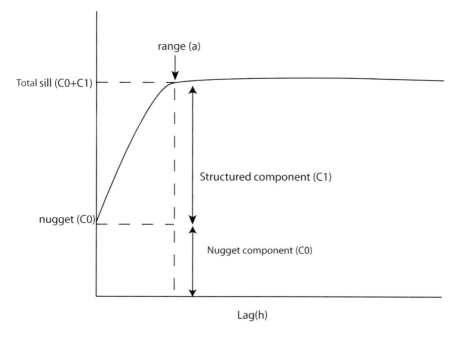

Figure 6.2 Parameters of a model variogram with a nugget effect

average for all values separated by that sample spacing (or lag). A mathematical model, usually selected from one of a set of authorized models, may be fitted to the experimental variogram and the coefficients of this model can be used for a range of geostatistical operations such as spatial prediction (kriging) and conditional simulation. McBratney and Webster (1986) provide a review of some of the most widely used authorized models in variogram estimation.

Use of geostatistical techniques in geoforensics

The variogram can be used to describe spatial variation in samples taken from scenes of crime or (body deposition) search sites. The distance between the pairs of samples is plotted along the horizontal axis and the value of the variogram (semivariance) along the vertical. The various components of the models fitted to the variogram can be related to structural features at the scene of crime or search site. The distance at which measurements become independent of one another is called the range of influence or correlation distance (Goggin et al., 1988). The range of the variogram can give information about the dominant scale of spatial variation in samples from the area. To enable the identification of directional variation, variograms can be estimated for different directions within a defined tolerance. The directional variogram indicates variation in spatial structure for different directions across the scene of crime or search site. For instance, if a linear feature or trend exists in a certain direction across the area then a directional variogram will indicate structure or maximum continuity in this direction.

Case study: Missing person (no body murder) West of Ireland – Harrison (2006)

In a conference abstract, elevation data were described from a mixed moorland, bog and agricultural area in the west of Ireland where police intelligence suggested human remains may have been hidden by a murderer. These data allowed creation of a DEM and the resultant topographic maps and 3-D visualizations allowed small water catchments to be defined. These informed the sampling of shallow groundwaters for carbon content and isotope analysis. Spatial variation in the distribution of the isotope data was characterized using an omnidirectional variogram of the dataset. The variogram model shown on Figure 6.3 is bounded since an upper limit is reached. The variogram characterizes spatial dependence in the property of interest (in this case carbon isotope content of shallow groundwaters). The variogram was estimated using Gstat (Pebesma & Wesseling, 1998) and fitted with a nugget and two spherical components. The spherical model is perhaps one of the most widely used variogram models. One reason for this may be that its form closely corresponds to observations in the real world: almost linear growth in semivariance to a particular separation distance and then stabilization (Atkinson & Lloyd 2007). The nugget effect represents unresolved variation and indicates a degree of randomness, which may be explained by a mixture of spatial variation at a finer scale than the sample spacing and measurement error (Journel & Huijbregts, 1978).

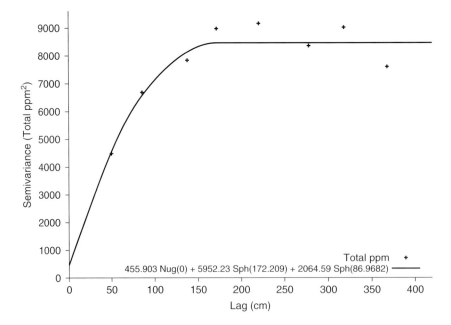

Figure 6.3 Example of modelled omnidirectional variogram using isotope data from a missing person (no body murder), West of Ireland. The variogram was estimated using Gstat (Pebesma & Wesseling, 1998) and fitted with a nugget and two spherical components

Different approaches are applied to fitting models to variograms with some geostatisticians preferring to fit 'by eye' to maximize the benefit of personal experience. In this example, a model was fitted to the variogram using the weighted least squares (WLS) functionality of Gstat. In WLS the weights are proportional to the number of pairs at each separation distance (Cressie 1985). Therefore, separation distances with many pairs have greater influence in fitting the model. The length of the variogram curve (distance h) in this case has been cut off at two thirds of the maximum distance between two sample points. The various components of the model fitted to the variogram can be related to the spatial structure of the carbon isotope content of shallow groundwaters at search site. The ranges of the variogram model give information about the dominant scale of spatial variation or correlation distance. In this particular instance, the ranges of the variogram model (1.72 metres and 0.87 metres) indicate a restricted geographical distribution of similarity in carbon isotope signature. The use of geostatistics through variogram analysis maximizes the information available on the spatial distribution of carbon isotope content of shallow groundwaters. The ranges of the variogram can be used to indicate the geographical area within which a similar carbon signature can be found.

Discussion on the use of variogram analysis in forensic investigations

If high variability in the spatial distribution of a property of interest (grain size, soil characteristic, etc.) from an area occurs over short distances, then potentially a sample from a suspect or victim can be located to a well-defined position with the sampled area. This last statement will undoubtedly provoke certain provisos. This approach is dependent upon a sampling strategy that adequately accounts for the spatial variability in the property of interest. The nugget value of the modelled variogram indicates a degree of randomness due to spatial variation at a finer scale than sampled spacing and measurement error. This needs to be accounted for in any conclusion drawn from variogram analysis. There is only one true variogram and the modelled variogram is only one possible solution and is dependent on several factors including the model used (spherical in this case) and user experience. The single, CSI-collected sample in the urban murder case (McKinley & Ruffell 2007) adopted a 'hit or miss' approach and allowed only minimal comparison to the 'suspect' samples. Experienced forensic practitioners may say that many cases have been successfully conducted using a tried and tested technique sampling only key locations in a scene of crime (for example, surrounding body or retrieved evidence, entry and exit points). If the objective is to compare samples from scene of crime with those taken from a suspect the question arises of how representative are the samples taken from the scene of crime? The use of geostatistics may provide a means to quantify how spatially representative a sample is. If a suspect can be placed within a metre rather than 5 metres to a scene of crime, this should help to address the issue of representative sampling.

Applying geostatistics to the example, discussed on page 171, of a sexual assault and attempted homicide on a woman and her two small children on the banks of the New River (Imperial Valley desert region, California (Stam, 2004)) the following observations are suggested. If a measured property is similar over much of an area and varies markedly only at the extremes of the area (at large separation distances on the variogram) then the property can be described as exhibiting long-range spatial variation (Atkinson & Lloyd, 2007). The small white freshwater snail shells (*Tryonia protea*) in the samples were found to have a widespread distribution and were therefore less significant for linking the suspect to the scene of crime. This would be represented by a large range value in variogram analysis in terms of density of snail shells (tens of kilometres if a sufficiently large area had been sampled). Conversely, a measured property that varies markedly over small distances can be said to exhibit short-range spatial variation. Variography of the soil samples in the vicinity of the crime scene would have demonstrated short range spatial variation in the form of a smaller range value (correlation distance) for the presence of a mixture of the white sandy deposits and the light brown silt deposits. Although a successful conviction was achieved and the victims' story corroborated, the inclusion of geostatistics in the investigation may have provided a quantitative measure of the limited spatial distribution of soil characteristics at the scene and decisively linked the suspect to the location. In the discussion above one of the key requirements is the implementation of a spatial sampling strategy. This returns to questions posed before in this chapter: how many samples should be taken for a representative sample and how to assess the optimum number of samples required to sufficiently inform an invasive search? Coefficients of the variogram models can potentially provide information on spatial variability in a search site for the variable measured. This has implications for sampling design and could inform future sampling strategies for no body search sites. The use of variography can help to address these questions in that the nugget value (Figure 6.3) indicates unresolved variation (and measurement error). A high nugget effect suggests high variability in the property of interest over short distances and so sampling at a finer resolution may be required to minimize the nugget effect and adequately capture the variable characteristics of the area. However, as stated before with any sampling strategy there will always be a lack of complete knowledge about how the area varies in spatial extent. Difficult or remote terrain, available resources and cost may all be contributing factors to the limited information available. The interpolation techniques described before (IDW and TPST) were used in the urban murder scene to make predictions at locations for which no data were collected. The coefficients of models fitted to the variogram can also be used for spatial prediction (kriging) and conditional simulation as discussed below.

Kriging can be used to generate predictions on a finer grid than the original sampling scheme. In effect, kriging can be used in areas where data coverage is not as dense as one would like. Thus simply taking numerical values, of say sediment softness for the location of reworked soil, will provide a guide or prediction to where an object may be buried. There are many varieties of kriging, the simplest of which is called simple kriging. To use simple kriging it is necessary to know the mean of the property of interest and this is modelled as constant across the region of interest. In practice this is rarely the case and so a more widely used variant of kriging tends to be ordinary kriging, which allows the mean to vary spatially. Kriged predictions can be generated using the coefficients of the models fitted to omnidirectional variograms, as a first step to investigate spatial patterns in the values across a scene of crime or a search site. The kriged outputs may indicate zones of high and low values and directionality in these zones across the sampled area.

Case study: Urban murder scene of crime, relating to McKinley and Ruffell (2007)

In the urban murder crime scene (McKinley & Ruffell, 2007; Figure 6.1) ordinary kriging was used as a comparison with the other interpolation methods. The kriged output was most visually comparable with the TPST interpolation and bears a resemblance to the hand-drawn contour map produced by the geoscientist. The advantage of using kriging in this instance, is that spatial dependence using coefficients from the variogram have been taken into account.

Case study: Missing person (no body murder) West of Ireland

Zones of high and low isotope values are indicated but the kriged output map (Figure 6.4) has a smoothed appearance, a common feature of maps derived using kriging. Two anomalies were indicated through this approach, in places consistent with criminal behaviour. This informed survey locations for cadaver dogs and the use of ground-penetrating radar.

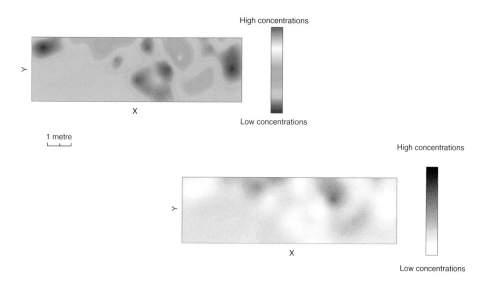

Figure 6.4 Kriged output map (ordinary kriging of a missing person (no body murder) case, West of Ireland. Kriging was conducted using algorithms supplied as a part of the Geostatistical Software Library (GSLIB; Deutsch & Journel, 1998)

Discussion on the use of kriging in forensic investigations

Kriged predictions are weighted moving averages of the available sample data and as such represent a smoothed interpolation. Smoothing, by definition, means that the predicted map cannot be correct and thus cannot provide a 'possible reality' for the scene of crime or search area. This should not negate the usefulness of using kriged maps as a visual approximation in forensic investigations and indeed as an interpolation method: its theoretical basis sets it apart from other interpolation algorithms (such as IDW). However, as Atkinson and Lloyd (2007) conclude the smoothed interpolation means that 'the spatial character of the output map is guaranteed to be different to that of the original data'. The kriging variance is a measure of confidence in the predictions and is a function of the form of the variogram, the sample configuration and the sample support (the area over which an observation is made, which may be approximated as a point or may be an area; Journel & Huijbregts, 1978). The kriging variance is not conditional on the data values locally and this has led some researchers to use alternative approaches such as conditional simulation (discussed in the next section) to build models of spatial uncertainty (Goovaerts, 1997).

Conditional simulation. Conditional simulation (also referred to as stochastic imaging in some texts) is not subject to the smoothing associated with kriging. Therefore simulated realizations do represent a 'possible reality' of the scene of crime or search site, whereas kriging represents a smoothed interpolation of the sampled data. Possibly the most widely used form of conditional simulation is sequential Gaussian simulation. With sequential simulation, simulated values are conditional on the original data and previously simulated values (Deutsch & Journel, 1998). In other words, simulated values honour the values at their locations and simulated values are added to the data set. This means that the values available for use in simulation are partly dependent on the locations at which simulations have already been made. Simulation allows the generation of many different possible realizations. This enables the uncertainty in the spatial prediction to be assessed and may be used as a guide to potential error in the construction of a map of the search site. Sequential Gaussian simulation is discussed in detail in several texts (for example, Chilès & Delfiner, 1999; Deutsch, 2002; Deutsch & Journel, 1998; Goovaerts, 1997).

Case study: Missing person (no body murder) West of Ireland

In contrast to the smooth kriged output (Figure 6.4), the sequential Gaussian simulation realization appears visually noisy but does not seem to be entirely random (Figure 6.5). Spatial dependence, indicated by variogram analysis and illustrated in the sequential Gaussian simulation realization, may point to the presence of structures present. In the case of a body deposition search site, this information was used to inform an invasive search. The two anomalous locations were surveyed by two types of ground-penetrating radar system, and by cadaver dog. GPR failed to indicate any subsurface disturbance or grave, yet the dog indicated close to a coincident location to one anomaly. Further search near this location failed to

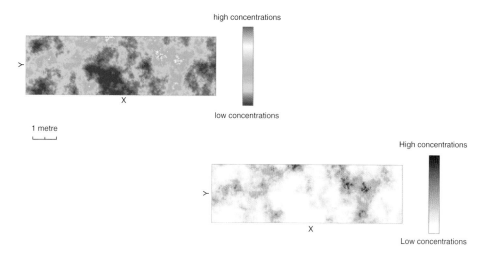

Figure 6.5 Example of a SGS realization using isotope data from a missing person (no body murder), West of Ireland. SGS was conducted using algorithms supplied as a part of the Geostatistical Software Library (GSLIB; Deutsch & Journel, 1998)

discover human remains, yet the isotope, topography and dog indications showed that some anomaly existed. This may have been a false-positive or the remains may have 'returned to earth', in the light of the missing person never having been found.

We hope that the advantage of this method should be clear in that the target is better defined and the important more subtle values on which the search is based are visually highlighted. Sequential Gaussian simulation can provide a means of generating a simulated map of a search site from a limited number of measurements and the generation of many different possible realizations that allow uncertainty in the spatial prediction of the property sampled to be assessed. To reiterate the concept of metadata in the use of GIS, any mapped output needs to be accompanied by information on the sampling strategy employed, any transformation methods applied to the data, the interpolation and geostatistical techniques used and the associated uncertainty. As Small *et al.* (2004) conclude 'when looking at data portrayed as a surface it is necessary to be aware of the processing steps that have been applied to the data from the initial sampling scheme to interpolation methods applied'. Only then can the pitfalls experienced in the misuse of aspatial statistics be avoided and the exploitation of Mark Twain's statement that 'There are three kinds of lies: lies, damned lies, and statistics!' not be applied when geostatistics are used in a forensic investigation or presented in the court of law.

6.14   GIS and geostatistics

One of the main advantages of using GIS in forensic investigations (discussed more fully in Chapter 5) is the opportunity to incorporate a large range of data sets in different data formats into one system as long as they are spatially referenced. The

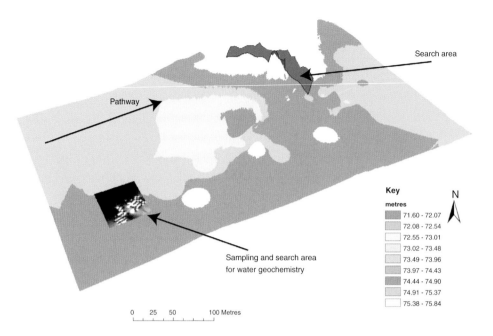

Figure 6.6 Integration of spatial data using GIS and geostatistics demonstrated for the missing person case, West of Ireland

appeal of GIS is the ability to overlay these spatially-located data and make an informed assessment of any comparisons.

The use of GIS in the integration of spatial data is demonstrated for the Missing Person case from the West of Ireland (Figure 6.6). Differential GPS was used to collect data from the mixed moorland, bog and agricultural area and a digital terrain model created. This allowed a sector or topographic domain approach to be used at a scale far finer than usual in geomorphology. Small water catchments were then defined for sampling of shallow groundwaters for carbon content and isotope analysis. The elevation data have been used to produce a DEM over which an interpolated surface of water chemistry of the sediment (isotope data) has been overlain. This provided useful 3-D visualizations of the search area and anomalies of interest to be identified (in this case potential body deposition sites, but it could equally apply to illegal waste, etc.). In this example geostatistical functionality with GIS (kriging through the Geostatistical Analyst in ArcGIS™) has been used to produce a combined output for search purposes.

Atkinson and Lloyd (2007) discuss the complementary role of geostatistics within GIS. Several key issues, pertinent for the integration of geostatistical functionality within GIS in forensic applications, arise from their conclusions: the proper handling of scale in the integration of a range of data sets of different format; limited statistical constraints on the use of GIS operations such as overlays and buffering compared to geostatistics (although recognition of uncertainty has grown in GIS in recent years (Zhang & Goodchild, 2002)); and the increasing availability of geostatistical tools within GIS without the pre-requisite of geostatistical training. In the particular example coefficients from the variogram (produced in Gstat) shown in Figure 6.6 were input into the Geostatistical Analyst in ArcGIS™ to produce the kriged output, however settings in ArcGIS allow the user to produce a kriged map without analysing the variogram.

Another important issue relates back to the inherent smoothing involved in kriged prediction maps. If the goal is to compare interpolated surfaces of several spatial data layers of forensic evidence then it must be noted that the maps provide only smoothed predictions, are thus different to the original data and do not provide 'possible realities'. Each kriged prediction will depend on the specific choice of variogram model and estimated coefficients, and the neighbourhood selected. As Atkinson and Lloyd (2007) stress geostatistics is a set of tools that can be used to perform a wide range of spatial analysis tasks but is not useful for all problems. In particular the object-based view of the world is beyond the scope of geostatistics. This is particularly relevant for the forensic geoscientist as tasks such as partitioning of the landscape into topographic or hydrological search units, etc. are object-based with clearly defined boundaries. The search area in the missing person case was geographically-limited by road and dwellings, an impassable bog and open farmland. Within the search area, individual locations were limited again by views, inaccessibility (thick trees, bog, river) or rock (insufficient depth of soil for burial). This domain approach to create a priority scenario of search areas is object-based and most easily described in GIS by a vector model. However, sampling spatial variability (of soil, sediment, water, etc. characteristics) of the scene is integral to this approach. Atkinson and Lloyd (2007) suggest that a combination approach, comprising a nested object-based and random function (RF) model of geostatistics, may provide the most appropriate model of reality.

Although unfortunately, the missing person has never been found, the methodology in this case describes an innovative combination of GIS in traditional landscape interpretation and geostatistics in hydrological chemical analysis and strongly supports the supposition by Atkinson and Lloyd (2007) for a 'nested object-based and RF model' as the most suitable approach for forensic applications.

7
Conventional geological analysis

'It is simplicity itself,' he remarked, chuckling at my surprise – 'so absurdly simple that an explanation is superfluous; and yet it may serve to define the limits of observation and of deduction. Observation tells me that you have a little reddish mould adhering to your instep. Just opposite the Wigmore Street Office they have taken up the pavement and thrown up some earth, which lies in such a way that it is difficult to avoid treading in it in entering. The earth is of this peculiar reddish tint which is found, as far as I know, nowhere else in the neighbourhood. So much is observation. The rest is deduction.'

'The Sign of Four' by Arthur Conan-Doyle.
This version: Conan-Doyle (1988)

The description and analysis of rocks and sediment has developed over the past 300 years and was one of the first applications of geology to criminal cases. Jackson (1962, in revising Gross, 1891) cites Locard, Giesecke and Popp as three pioneers of the use of soil and sediment in comparing suspects to locations. Geologists use their skills mainly in the description of rocks, physical geographers work more with sediment and soil, and engineers work with both, as well as 'rock-like' human materials (e.g., concrete). The same descriptive tools are used for each, being modified for the type of material under examination. Consideration of the type of material (soil, sediment, sedimentary, igneous, metamorphic, human origin) is the first stage in analysis, so that we can deploy tried-and-tested methods and appropriate descriptions. This may seem elementary, but sometimes the nature of a substance is never established, requiring tiresome analysis with no precognition. Manufactured products are so highly developed that they can mimic natural materials very effectively. In a recent survey of building stones used in Belfast, the origin of a limestone façade to a well-known fast-food outlet needed to be determined. A number of geologists visited the restaurant and noted the sediment layers, with algal laminations and even some fossils, concluding that this was a cut and polished limestone building stone termed Cotham Marble by manufacturers. What was not known was that this multi-national restaurant could afford to import manufactured facing slabs, made by placing lime, water, sand and pebbles (including oyster shells) in tanks to set and then cutting them into slabs, later to be polished facades. In this case it was the lack of destructive analysis that precluded correct identification of the material, a problem common in geoscience where samples are precious or in small quantity. In such cases, it may be that methods and descriptions developed for one material offer advantages when applied to others in terms of ease, cost and accuracy. An example is the analysis of concrete, usually undertaken using civil engineering methods of determining hardness, tensile strength, etc. When the source of some concrete needed to be established in recent criminal cases, such tests were useful for comparison, but provided no information on ingredients that indicated the manufacturer's supply.

Initial descriptions of rock, sediment and soil rely on simple, unbiased observations that will be used to make a preliminary classification. This includes size and shape of both the fragment and particles within it, overall colour and again colour of included particles. These features may be described regardless of knowing the rock, rock type or sediment/soil classification, but an initial evaluation of rock classification (igneous, sedimentary, metamorphic) determines some of the terminology used. For instance, the size and shape of constituent particles may be described from pure observation, even though these might most commonly be crystals in an igneous rock and grains in a sedimentary rock. Colour may be assessed by visual comparison to a chart of known values (for soil, the Munsell Colour Chart) or by digital analysis of RGB values following scanning or photography. For *in situ* soil and rocks, structure is all-important because it often provides information on origin. Igneous rocks may contain structures indicative of a once-molten state; metamorphic rocks of when the rock has been heated or deformed and sedimentary rocks of particle movement or fossil content. In most cases these initial descriptions provide a crude classification of the kind of material so that a focused collection of comparison material is possible. Occasionally, such brief description is all that is required to establish exclusion of samples: when investigators seize sediment or soil from scene and suspect, they see only 'mud', but the geoscientist will easily exclude granite-bearing sand from basalt-bearing glacial till, such can be the simplicity of exclusion. More often far greater depth of analysis is required, first to establish beyond reasonable doubt the crude observations; second because some materials, although of different origin end up appearing indistinct, and third, although some exclusion (above) beyond reasonable doubt may be simple, comparison or less straightforward exclusion require far greater accuracy. A natural example may be a rock type such as basalt, which may be of Carboniferous age (300 million years old) and from Germany that looks identical to basalt of a few years old from Hawaii. A second example occurs when materials are highly processed, such that lime made from Silurian limestones in Estonia (410 million years old) ends up looking the same as lime made from Cretaceous chalk (90 million years old) from southern England. This simple example belies the fact that the non-professional geologist cannot take this book and use it in the collection and analysis of geological materials: all we can do is demonstrate how geology *may* be used, by a properly-trained professional.

7.1 Elementary analysis of rocks

Rocks are classified by their mode of origin:

- *igneous* ('fire-formed'): comes from surface or subterranean volcanic and related activity (e.g., the rising of magma);

- *sedimentary*: originally laid down as sediment and subsequently hardened by cement and compaction;

- *metamorphic*: may be derived from igneous or sedimentary precursors, but retains evidence of textural or mineral change by heat, pressure, or both.

Establishing which of the three elementary rock types one is observing assists with the application of later terminology: igneous rock textures refer to crystal size, shape and distribution, whereas sedimentary rock textures refer to grain size, shape and distribution. Thus similar observations apply to all three rock classifications, although their individual characteristics are different. Within each group, simple divisions occur. These are outlined in the following sections, as they will become important in many of the case studies that follow.

Igneous rocks

A preliminary examination establishes the size and orientation of crystals (if present). Few or small visible crystals implies rapid cooling of magma and thus maybe a volcanic or similar origin where rapid cooling may occur. Large crystals suggest passage of time to allow crystallization, possibly in a magma chamber. The colour of such rocks is important, but can be misleading. Although, igneous rocks that contain light-coloured minerals (quartz, feldspar, muscovite mica) will be light-coloured and dark minerals (hornblende, pyroxene, biotite mica) in igneous rocks will produce darker-coloured rocks (Figure 7.1), fine-grained mixtures of minerals can produce dark or light variations that do not reflect the colour of the constituent minerals!

Sedimentary rocks

Preliminary examination will reveal whether suspected sedimentary rocks contain particles, or have no particulate matter and thus may be precipitates such as salts or some kinds of limestone. Particles may be of derived rock material (cobble, gravel, sand or silt grains) or fossil or eroded crystal debris. Precipitates may be of lime (limestone, dolomite) or other salts (e.g., sodium chloride). Establishing whether the constituent material within a suspected sedimentary rock has been derived from the fragmentation, erosion, transport and deposition of particles (clastic), or by the precipitation of calcium (or other) salts by biological or chemical processes, aids in a

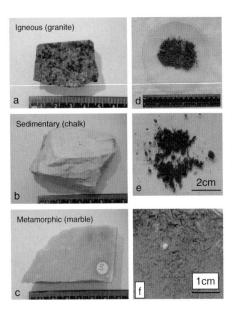

Figure 7.1 Examples of well-known rocks, classified by their type (igneous, sedimentary, metamorphic), and examples of some common soils. d = brown earth, e = black earth, f = urban soil from waste land

preliminary division of sedimentary rocks into clastic or chemical/biochemical. The clastic sedimentary rocks comprise conglomerates, breccias, sandstones, siltstones and mudstones; the (bio)chemical sedimentary rocks comprise limestone (Figure 7.1), dolomite, rock salt.

Metamorphic rocks

As implied by the name, these rocks have been altered or changed in some way by the Earth's natural tectonic forces of heat and pressure. Prior to the change, they may have been igneous, sedimentary or even other metamorphic rocks: the clue to their origin lies in deciphering what original minerals or fabrics remain, and how they have been altered. Good examples include marble (originally limestone, often metamorphosed by heat (Figure 7.1)) and slate (originally mudstone or shale, metamorphosed by pressure and often heat combined). With increasing heat and pressure, the origins of the rock become obliterated, the final stage of the process being complete melting (and creation of igneous rock) to be perhaps followed by uplift and erosion to form sediment and sedimentary rock. As common precursors are igneous or sedimentary material, key features of metamorphic rocks are both the constituent minerals as well as fabric. The movement of pressurized and heated fluid, especially water, through rocks during metamorphism is often key to the types and locations of change. Hydrothermal activity is a fundamental process in the formation of many precious ores and minerals. Therefore, knowledge of such rock textures is key to understanding ore-bodies: many a mining scam has been uncovered because of the impossible juxtaposition of ore and rock host, uncovered by geologists with specialist training in this area.

7.2 Hand-specimen analysis – case studies from Murray and Tedrow (1991)

Rocks

This study involves a case where a Canadian liquor store imported scotch whiskey in wooden crates, only to find the bottles replaced by limestone blocks of the same weight. These were shown to various experts in both the country of origin (United Kingdom) and Canada, who all agreed that the specimens were Carboniferous limestone from central England, close to the point of export. A suspect was identified as someone who worked for the exporters, had access to the quarry and was known to take rocks away.

Soil

Murray and Tedrow (1991) describe a 'comparison' case study, now a benchmark in many investigations involving vehicles in collision that have subsequently departed the scene, either as hit and run, or in order to destroy the vehicle. In the hit and run incident they describe, clumps of soil fell from the car's bumper during the collision with a pedestrian. The soil contained minerals known to be common in Missouri (a lead mining area). Vehicles passing from this area to the scene of crime were checked and a suspect apprehended: the remaining adhered soil compared to those at the scene as well as to roadway materials from Missouri, where the car had been. Murray and Tedrow's (1991) case study of identifying the theft location of stolen cacti by comparing material adhered to the roots in the imported location to that at the vacant holes (theft site) is a similar comparison. Lee *et al.* (2002) describe a similar problem with the theft of stolen palm trees.

Less soil was available for comparison when a soil-covered body was found in a refuse sack (above ground) at a police shooting range in the eastern USA. Nonetheless, this is still a hand-specimen case study, because the quantity of material allowed simple classification and definition of areas nearby with such soil, which was of a type common to brackish environments. An area of reclaimed landfill on an estuary turned out to be the previous burial site, where a mother and daughter had murdered and buried the husband/father, only to be so disturbed by the smell of decomposition that they exhumed the body and left it in the only remote location they could travel to – the police firing range!

Murray (2004) suggests a sequence of geological analyses for the standard comparison of questioned (often suspect) materials to location (Figure 7.2). He also compares the sequential methods of analysis recommended by different practitioners, most especially the work of Ritsuko Sugita and Yoshiteru Marumo of the National

188 CONVENTIONAL GEOLOGICAL ANALYSIS

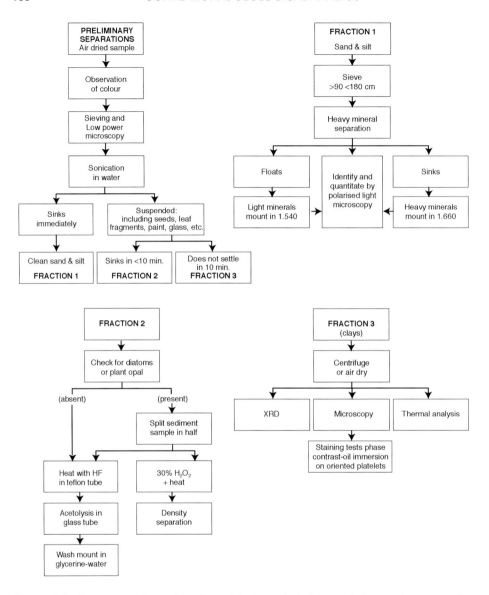

Figure 7.2 The succession of sediment/soil analysis for exclusionary/comparative forensic investigations recommended by Skip Palenik. After Murray (2004, after Palenik)

Research Institute of Police Science, Tokyo (Figure 7.3). As we shall see in Chapter 8, these 'best practice' analytical sequences are based on a reasonable sample size being available, when in some cases this type of work either cannot be done or may cause such compromise/biased results (low numbers of particles for statistical testing) that only selective parts of the examination succession can be undertaken.

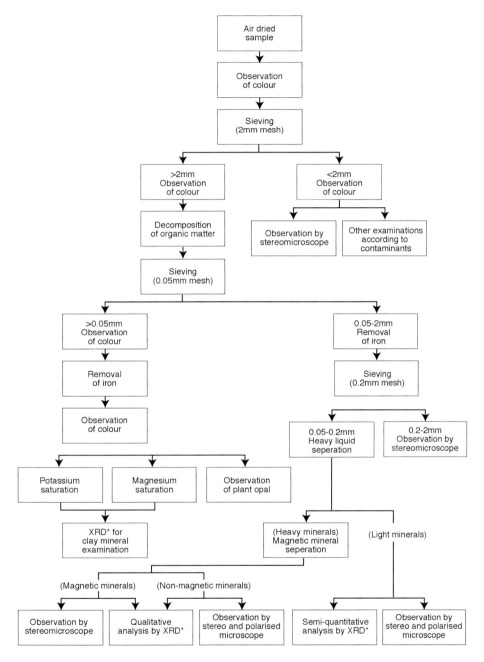

Figure 7.3 The succession of sediment soil analysis stages recommended by the National Research Institute of Police Science, Tokyo (work of Sugita and Marumo, 2004)

7.3 Sediment analysis

Context

By far the most common material considered in forensic work is soil and sediment. We may recall from Chapter 2 that soil has the ability to sustain life, which sediment may also do, but for reasons of rapid deposition, reworking or harsh environments, sediment may be capable of supporting life but has not yet done so. A common misconception in forensic geology is that purveyed by stories of extremely gifted and possibly intuitive practitioners such as Oscar Heinrich (the 'Wizard of Berkeley', see Block, 1958), who could take two sand or soil samples and make positive or negative comparisons of scenes and suspects. This can of course be done, and modern-day practitioners such as Peter Bull (Oxford University) do such work. However, theirs is based not on intuition but on a background database and experience in examining many thousands of samples and many millions of sediment grains (in this case, quartz grains: see Bull & Morgan, 2006). Thus their work relies on a database and thus a context.

A far more common context that can prove crucial in soil and sediment work is the field location of the sample. Thus the primary context of any sample is both its large-scale location on a map and in a stratigraphic succession, as well as micro-scale lateral variation and stratigraphic layering. This context applies equally to suspect or victim samples: we may not know their original location, but their location on an article of clothing, on a car or in footwear tread is the same context of spatial and stratigraphic variation, that must be noted. In Chapter 2 we saw the modern-day replication of the pioneering work of Georg Popp and Hans Gross in how spatial variation in sediment type can be related to the layers (or microstratigraphy) built up in a person's shoes as he or she traverse this landscape. In court, the comparison of one sediment or soil type from a shoe to one type from a scene of crime will always force a question from the defence of how variable soil is. The combination of soil types in footwear tread microstratigraphy renders such questions as increasingly irrelevant. The other reasons for championing the importance of field location as context to an investigation include understanding the scene in a more holistic manner, as well as making the final laboratory analysis easier. Demonstrating lateral variability in soils and sediments through costly and time-consuming analysis of hundreds of soils of varying texture, colour and fossil content complicates comparison of materials that for all intents and purposes look similar. Murray (2004) champions taking the 'questioned' sample (or its description) to the known or potential scene of crime and alibi locations and making a direct comparison, or more critically, excluding those materials that are not visually-comparable. When no questioned sample (or of sufficient volume) is available, or when the scene has yet to be identified, sufficient samples to reflect the limits of the location's natural spatial and stratigraphic variability must be taken. This evaluation can stand the test of a law court's examination if based on good scientific protocol. This is made easier in environments where 'natural' processes dominate, as it is here that the field-based Earth scientist is trained. The urban environment holds altogether different challenges, more usually related to trace evidence or analysis of industrial materials, as described in other chapters.

Initial observations of field context include:

- Field notes of lithology (mineralogy, composition, colour), texture (grain size, grain shape, sorting, fabric, stratigraphy, geographic and stratigraphic boundaries), structure (bedding, soil structure), fossil and organic content.

- Photographs and drawings. These are often best completed as accompanying presentations, avoiding the questioning of one without the other. Current practice sees digital photographs burnt onto CD, with a date stamp.

Of over-riding importance in assessing soil and sediment variation is the relationship between lateral extent (the plotting of soil/sediment types on a map) and vertical, or stratigraphic relationship. The observation of a sediment type in a suspect's footwear or vehicle tread that occurs frequently as a lower layer to an area's stratigraphy will automatically limit contact with this material to either exposed layers (river undercuts, quarries) or digging activity, focusing search operations or implying an activity that may be suspicious. Allied to the semi-quantitative descriptions of texture is the more statistically robust analysis of grain (or particle) size. Because particle sizes cannot be influenced by changes in colour or biology, their variation is often seen as more robust in comparison than other methods of soil/sediment comparison. The user-friendly graphs of particle-size variation makes comparison of many samples simple and understandable by the courts. Problems may arise when suspect/victim samples were derived by a different 'pick up' mechanism to the field control or scene samples. Some fabrics and tyre treads may be prone to the adherence or 'sticking' of certain particle sizes, or of inadequate or incomparable sample quantities: a size analysis of maybe 10 particles in a speck of material adhering to a suspect's clothing will not be statistically comparable to the analysis of 1000 particles (or more) from a virtually-unlimited field sample.

A primary consideration when examining soil and sediment is the size of the particles. This may be further split into the maximum, minimum, average and proportion of grain sizes. Initial evaluation of grain size by mechanical or optical means depends on comparing similar amounts of material, and thus efficient means of splitting and weighing dried samples is critical. Sieving of samples is not efficient if grains adhere to one another, thus salts, organic matter and moisture must be removed. Sieves with laboratory-standard (and tested) sizes must be used in sequence and the captured material weighed. Depending on the number of sieves used, this will give a crude particle size distribution. The advantage of sieving is that the samples can be subsequently analysed, or the sample re-assembled: the disadvantage comes in comparing questioned and control samples of different amount, because individual grains assume a disproportionate significance in smaller samples. Automated particle size analysers deploy light, X-rays and most commonly lasers in order to make measurements. Laser light is projected and sometimes converged by lens onto a rotating solution containing the grains that are of interest. The angle of beam diffraction caused by the grains is measured and equated to particle size. The method is a very powerful discriminant: some sophisticated devices use slightly more sample and post-analytical recovery is problematic (Malvern Mastersizer). Other devices are more flexible, with dry and small samples being used in a non-destructive manner (Coulter Granulometer). As with sieving, sample preparation (deflocculation), testing and repeated measurement achieve better results.

Another method of estimating sediment particle size is settlement in solution, the theory being that large, heavy particles fall to the base quickly, resulting in a spread of particle sizes. The main problem with this method is the highly variable density of the minerals (or their proportions) that make up sediment grains, resulting in small, heavy particles falling quickly to be mixed into large particles. The method does have other applications in sedimentary analysis in both soil density measurement and the study of heavy minerals (by separation in liquids of specific gravity: Petraco & Kubic, 2000). The density gradient column was a commonly-used soil comparison technique throughout the 1950s and 1960s. The method deploys two separate glass tubes, usually 30 centimetres long and 1 centimetre in diameter, filled with layers of liquids with successively changing densities. When disaggregated soil is added to each column, the particles that are of the same density as the fluid remain, separating the soil. The resulting banding in the columns makes for easy comparison, but as Chaperlin and Howarth (1983) have demonstrated, soils from very different locations can give visually similar results. The method is now rarely used in forensic analysis as a result of their work. Problems occur when small, heavy particles mix in with lighter, larger particles. The settling medium may also cause contamination issues: separation of each 'layer' in a tube may be impossible.

A number of spin-off techniques have arisen from the basic principle of sedimentation analysis that are more rigorous than the simple visual display of the gradient column. Sedimentation analysis takes an altogether different starting point, being that various materials were moved and deposited by different currents and agents of transport, each of which will have left evidence of its activity. The process of splitting a sediment into its constituent parts allows the sedimentologist to begin reconstructing this history. Grain size, shape and density thus become only the initial classes into which we further group the materials within the sediment. Dispersal by chemical agents is especially useful when analysing those materials that flocculate, or because of their atomic makeup, bond to each other. Wet (by running water through the samples) or dry sieving has the advantage of allowing the investigator to make cursory visual examinations of each size fraction as they are produced (wet or dry). Contamination is an issue in sieving, especially if limited amounts of fine-fraction material, or rare (unique indicator/particle) are to be used. Sieves must be cleaned and scanned for embedded particles after each analysis.

Following context – detail

Field-based analysis of sediment/soil distributions, backed-up by laboratory analysis, is essential in understanding the gross variation of transferable material at (for instance) a scene of crime. Many of the methods used to describe, exclude or compare samples can be deployed at the macro- or 'hand specimen' scale or at the micro- or trace evidence scale. Thus some of the methods discussed here will also be mentioned in Chapter 8 on trace evidence). Although effective, preliminary assessments of particle size distribution are crude: akin to saying that all lorries of a certain size can be allowed across a toll-bridge, regardless of what they are made of, carrying or their weight. The same goes for soil and sediment particles: size and gross variation aid to our understanding of the scene and likely variation, *they do not tell us what the material is made of*. For this, our first analysis is optical magnification (handlens or microscope) and comparison to known standards. This may be completed satisfactorily by examination of particle sizes (hence sieving) or by making an acetate peel or cut thin-section of the specimen. These both allow the passage of transmitted light, which if polarized can give indicative colours for certain minerals, and thus begin to take the crude analysis of size and shape to makeup. Sieved portions may be examined and photographed or placed on a computer scanner. Peels are best made by immersion of a flat surface in acetone and then adherence of acetate: peeling will extract a layer for microscope examination. Many minerals are optically so similar that further analysis or treatment is needed for diagnosis. The latter includes staining with agents such as alizarin red, which picks out minerals that otherwise appear similar. To obtain the sort of quantitative data useful to courts, a count of the numbers of different minerals in each sample, peel or thin section must be made, quantitatively (using a point counting method) or qualitatively (using comparison charts, see Figure 7.4). Such analysis and comparison has been used many times in courts of law by specialists in this field such as Skip Palenik (Palenik, 2000). Such peels and thin sections can be further subject to analysis by irradiation by X-rays or by observing their cathodoluminescence. These all provide greater differentiation of mineral types in each specimen and become

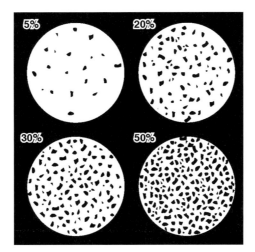

Figure 7.4 Cartoon microscope or handlens views of soils/sediments with known, scattered proportions of black fragments (could be minerals, asphalt, paint). The trained geologist can take such views and compare them visually to real samples, and estimate to within a few percent the proportions of specific fragment species. Redrawn from Tucker (1988). Field of view in each case, typically 0.5–1 centimetres

relevant when certain minerals are apparent. Hence the need for specialists as analysis progresses: often an experienced all-round geologist will be sufficient to enable an initial assessment of a scene or soil/sediment comparison. However, at some point a specialist in certain landforms, minerals or soils will be required, enabling the best assessment.

Further specialist analysis

Two main avenues are available to the forensic geologist following optical examination and comparison (hand specimen and microscope) of soil, sediment or rock. These are to determine either the mineral content (e.g., quartz, mica, clay) or the chemistry (e.g., silica, iron, magnesium, potassium) of contained particles. Although chemistry and mineralogy are related, they are not the same: silicon oxide for instance can take different mineral forms, from quartz to tridymite, or may not have a crystal structure at all (glass). Thus a mineralogical analysis can predict with some certainty the chemical makeup of a rock or sediment, just as knowledge of the chemical makeup can be used to determine likely molecular arrangement and thus infer mineralogy.

Mineralogical determination

For coarse-grained soil and sediment (particles above silt-grade or 62 microns), the optical properties of many minerals in a thin section under a polarizing microscope allow mineral determination to be made. Some minerals are opaque (blocking the passage of light), occur as such fine fraction (generally below silt grade or of the clay-size fraction), or are stained or mixed with organic matter, iron oxide, etc. that occludes light. These non-optical fractions of soil and sediment are unfortunately common: many soils are a mix of clay-grade minerals and organic matter. These material types, or the fine fractions of mixed-grade material, are best suited to mineralogical and chemical analysis.

A standard and robust method of determining mineral content is X-ray diffraction (XRD). The technique uses a focused X-ray beam that is directed at successive angles onto a layer of material. The method is particularly useful when studying fine-grained materials too small for easy identification by standard microscopy. The focused X-ray beam will be partly transmitted and partly diffracted by the molecular layers in the sample. The angle of diffraction is dependent on the spacing of the molecules (d spacing): a simple equation (Bragg's Law: $n = 2d\sin$) calculates the d-spacing of the crystal (measured as 2-theta) from the angles of the incident diffracted beams. A diffraction is recorded by a detector and recorded as a peak on a chart (a diffractogram or XRD trace: see Figure 7.5) of diffraction angle versus number of diffractions per angle (or part thereof) of beam incidence. The number of diffracted X-rays per increment of time (set by the operator) depends on the number of times the incident beam has irradiated a material with that molecular spacing. Conventional XRD can be used to scan from as low as 1°, with 2° to 5° being the convention. The position of peaks on an XRD trace can be compared to known standards to give mineral or crystalline substance determinations.

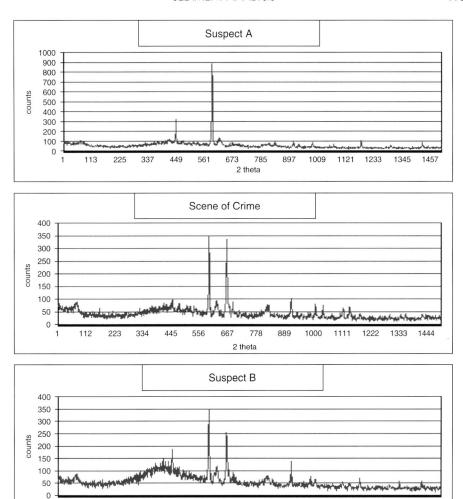

Figure 7.5 X-ray diffractograms. An X-ray beam is directed at the sample from a range of angles. The layers in any crystals present diffract the beam, and the resultant beam angle can be detected and related to the mineral. The dried soil samples were scattered on a glass microscope slide: this provides very crude and non-quantitative estimate of mineral contents. In this case, these samples formed part of a multi-proxy study (this evidence alone would rarely come to court), but XRD was used as a screening tool, there being limited samples

There have been a number of attempts at generating quantitative XRD data. Two methods based on bulk (2 grams or more) samples are considered by Ruffell and Wiltshire (2004): the first is popular in the US and France, and uses ground samples with a known standard added (a possible problem in forensic casework as this could be portrayed as tampering with a sample). The second, as advocated by

Steve Hillier of the MaCaulay Institute (Aberdeen, Scotland) freeze-dries the sample. Quantification of the clay-fraction only, or of trace sample amounts (milligrams), has also been considered by sedimentologists (see Tucker, 1988: the methods favoured require less original material (depending on clay content). As XRD is most suited to the analysis of small amounts of soil, we shall deal with this method more fully in Chapter 8.

The percentage minerals in a rock provide a key discriminant in comparison. Optical methods of estimating mineral proportions depend on the methods above – visual estimate, point-counting and image analysis. These all depend on the minerals being visible when significant proportions of many rocks contain minerals that are either opaque to transmitted light or are too fine-grained to allow significant passage of light. Opaque minerals include many of the important ores, and thus estimates of content can be the subject of the criminalist when fraud is suspected. Many muds and soils comprise significant proportions of nearly-opaque clays, making optical mineralogy by petrographic microscope incomplete. In such cases, the material may be subject to examination by scanning electron microscope (SEM). The higher resolution of the SEM allows the crystal shapes of the optically-indeterminate fraction of a sample to be determined, making the method a perfect complement to optical mineralogy (for silt, sand and granules). The SEM can be fitted with a secondary device (probe) to measure the chemical component of an individual grain or crystal. The probe comprises either a wavelength dispersive (WDS) or energy dispersive (EDS) specrometer, the latter being most commonly used in the analysis of gunshot residues and geological materials.

The key to the method is the position of the specimen and the varying angle of incidence of the X-ray source. This creates changing diffractions, related to varying mineral molecular spacing that can be related to known standards. Problems with the XRD method include different minerals possessing similar molecular structure and non-standard (thus possibly incomparable) preparation methods. The former problem can be overcome by further preparation of the samples to alter specific crystal structures (e.g., hydration or heating) as well as detailed examination of the diffractogram. Failing this, chemical analysis and examination by SEM (see Chapter 8) may resolve the problem. Sample preparation is the subject of great debate in the literature, with proponents of specialized grinding machines, freeze-drying, exacting preparation methods all championing their method, often with one aim, the production of as quantitative data as possible from the sample. This of course has great relevance to investigations of how comparable (or exclusive) two or more samples may be: as mentioned for particle size analysis (above) the production of simple comparability statistics for judge and jury can be highly effective. The scientific reasons for complex statistical analysis are not questioned in court, when for all its foibles, a simple 1 to 10 scale of comparability (1 being poor, 10 being excellent) can be so persuasive. An alternative to the never-ending quest for higher and higher quality data extraction from the XRD method is instead to use it as one mineralogical in a multi-proxy approach where grain shapes, chemistry, biology are used.

Light microscope analysis

A common procedure when wishing to determine the mineralogy of grains in rock is to examine the material by transmitted polarizing light. This light causes definitive colours to be produced that allow greater discrimination of grains than by simple grain surface (textural) observation. Such polarized colours are difficult to establish in whole grains unless a known reference is used and the grains immersed in oil of known optical characteristics. The grains can however be washed and further analysed. A more precise method of determining mineralogy is by thin section analysis, whereby a rock, or resin-impregnated (commonly Canada Balsam) soil or sediment, is cut and then ground to 20–30 μm thick, allowing the easy passage of transmitted light when crystalline substances are encountered. Both methods are used conjunctively with a description of the grain size, shape and distribution. Thus common observations may be made that the quartz grains (determined by mineralogy) are mostly rounded (determined by shape analysis), while the mica grains (determined by mineralogy) are mostly angular (determined by shape analysis) (Figures 7.5 and 7.6).

Providing percentage counts of grain sizes, shapes and mineralogy is notoriously difficult. Two common methods have been developed, both of which are deployed in comparing sediment for forensic analysis. The first is point-counting, whereby a thin section is mounted above or below a tiny grid or graticule that allows the operator to divide the slide into small sections. Then a mechanical microscope stage is deployed and within each square the number of a certain grain type (round, mineral, size) is counted. Ribbon counting uses the same principle but along a line instead of throughout a square. Averaged counts are deployed and experienced operators frequently come within 5 % of each other's estimates. Image analysis by computer package has advanced the method considerably. Software similar to that used in GIS (Chapter 5) is used, which automatically identifies materials of separate class and can calculate their total area in a given slide. The method works as well as the programming that the operator provides, which at some point becomes pointless as operators can spend as much time training the image analysis software as doing the work themselves! The advantage of image analysis is when many hundreds of samples need to be described; the disadvantage is that when individual (unique) particles, textures, etc. are encountered, the automated process will not recognize them. Point-count, image analysis, mechanical sieve and laser particle size analyser measurements have frequently been compared. Again, the automated procedures remain superior for grain-size analysis, whereas shape and unique particles have been better determined by optical assessment. A conjunctive approach (as always) is best.

The percentage makeup of particle size, shape and mineralogy is naturally a friendly use of data for court. The most frequently used method of displaying such information when comparing materials would be the bar-chart or histogram. These and the cross-plot or scatter-graph remain useful tools when two samples are compared with regard to one variable. As we have seen, however, sediments

Figure 7.6 Thin sections of sandstone granules (~1 centimetres diameter each) found lodged in two suspects' boots, and from the scene of a murder. These were created by extracting the granule, cutting it and embedding the flat side in resin on a microscope slide. The upstanding side was then cut away and ground to 20 um thick (thin!), such that normal and polarized light can pass through the minerals; when viewed under a microscope at 10×, 20× or even 100× magnification, the grain shape may be observed and their mineralogy determined. In this example, we see how the sandstone granule in Suspect A's boot comprises subangular quartz grains 'floating' in a clay-mica matrix. Sandstone grains from the scene comprise subrounded quartz grains, many of which are compacted into one another. These samples came from the same case as the X-ray diffractograms in Figure 7.5, and show the use of two comparative analyses, if based on the same fundamental materials

and soils have many more than one variable: many being a mix of sand, silt and clay, requiring some form of expressing three variables, the most common being a triangular (ternary) plot. A more sophisticated method in the quantitative analysis of particle size data is to use conventional moment statistics in comparison: these can be used because most grain-size distributions will achieve a normal (or Gaussian) distribution when plotted on an arithmetic scale. One of the most effective means of showing non-geoscientists what we mean by different particle size fractions of a sample displaying different features is to avoid bombarding them with graphs, but simply show photographs of the different sediment size fractions, as demonstrated in the following case study.

Case study: Murder associated with drug-dealing, Belfast, N. Ireland

This case is the subject of an inquiry, and thus details cannot be given.

In the early hours of a late September morning, 2004, a farmer awoke to see smoke issuing from a vehicle parked on a track within view of his home. He called the fire brigade who doused the flames and then moved away as a body was noted beside the vehicle. The deceased was identified as a minor drug dealer and fence for stolen goods, known to the authorities for a variety of offences. His girlfriend had reported him missing earlier that morning. She claimed he had gone to meet some associates (not known to her) at 11pm the previous evening.

A number of the deceased's known contacts were visited, and one person in particular claimed to have just arrived home from a late shift at work, which was strange given his vehicle was cool and he had washed. When arrested, this suspect was compliant until some shoes he had hidden were seized. He claimed that the mud and grass on these shoes was like that on the pair already worn – from his own garden. Although minor amounts of material remained on the hidden shoes, nonetheless a range of analyses were carried out. These produced an array of results that excluded the alibi material from his garden and included material from the hidden shoes and body deposition site (now established as a different location to the murder, from discharged cartridges found elsewhere).

In court, the various experts who worked on the materials demonstrated their results and were extensively cross-examined. For instance, the X-ray diffraction expert decided to display four diffractograms, one from the worn shoe, one from the alibi garden, one from the hidden shoe and one from the body deposition site (Figure 7.7).

He anticipated being asked about the numerous (over 100) other samples analysed from these items, scenes and the murder scene. Instead the defence lawyers bored the jury with long questions about the XRD method. The only expert not to get such a rough ride also chose to risk being questioned about all the samples and the apparently selective nature of their exclusion/comparison by also showing just four samples. These were the same four items as used by the XRD witness, only in this case dried, sieved to four fractions and photographed in their bags (Figure 7.8). On close examination, there are problems with the exclusions and comparisons made from selective images such as these.

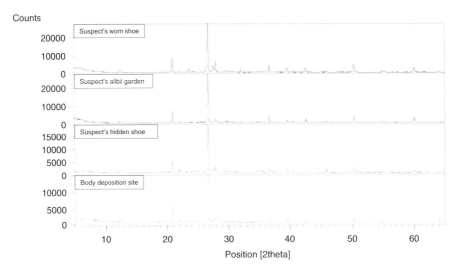

Figure 7.7 X-ray diffractograms from four items selected from over 100, used by the prosecution expert to demonstrate the nature of the material and subtle differences between four key groups of sample

Sedimentary particles

Grain shape is most commonly referred to in terms of roundness versus sphericity and angularity. Flatness is also used. Quantitative methods of measuring how close a grain is to a perfect sphere have been developed, although these are time-consuming when many grains are to be measured. More often, a standard chart of angular to round spherical grains and sphericity to roundness plots are used for visual comparison. Both allow a consideration of the texture of grain surfaces, a characteristic that can depend on mineralogy (cleavage, strength). Throughout the 1960s, Krinsley pioneered the use of grain surface textures (especially quartz) in defining the original or dominant environment from which that material came. His work was especially successful in determining quartz grains that had spent the last period of their time prior to deposition in lake or sea shores, in wind-blown (aeolian) places, in areas of harsh chemical weathering (tropical locations) and glacial locations. By classifying these types, Bull and Morgan (2006) have shown that with careful division of classes by geographic location, the percentage of quartz grain surface types in a given sample can be used to provide a likely original location, or number of locations. Their work is laborious, but shows great potential for limiting search and alibi locations.

Textural, structural and mineralogical determination for forensic analysis is by no means confined to sediments and soils. In many cases, bedrock has not fully broken down, and thus granules of the parent rock reside, with the capability of transfer. In these cases, identification of the parent rock is essential as this may provide a clearer indication of its source than the inherently mixed and weathered nature of sediment and soil. In order to provide as unbiased comparison as possible, the questioned

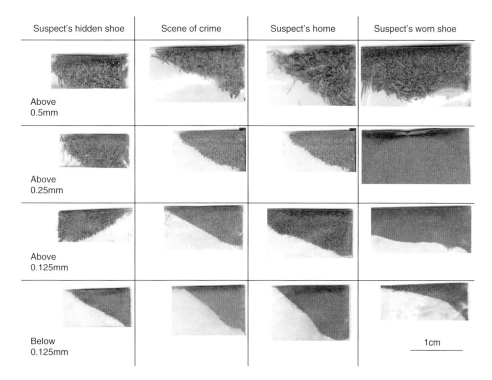

Figure 7.8 A demonstration of the simple method of showing the similarities/differences between selected samples. This display was backed up by similar analysis of over 100 other samples, each analysed by XRD, plant macro-morphology, coarse fraction thin-section petrography and scanning electron microscopy. Note the variation even in samples supposedly from the same source (suspect's hidden shoes and scene of crime)

and control rock fragments must be described using the same standard protocol of colour, grain (crystal) size, shape, rock structure and mineralogy. These will allow a preliminary classification and, if possible when many grains are present (most critically in the questioned sample), variation to be determined and any unusual or unique minerals or fabrics be discovered. Thus the common igneous ('fire-formed') rocks of granite, basalt, rhyolite, gabbro, solid lava (pumice, obsidian, etc.) as well as many of the common metamorphic rocks (slate, schist, gneiss, marble) may all be described, classified and variations determined by reference to the above features.

Case study: Igneous rocks

Vigo Granite (and analogous types) occurs in the mountain belts of northwest Spain, where it is widely-quarried for export as ornamental stone. The rock is relatively homogenous, with few cracks and only minor variation in grain size or mineralogy. Areas with more dark (mafic) minerals form elongate clots, but otherwise the homogenous nature makes this a mechanically-strong material that cuts well and polishes to provide high-quality flooring and building cladding. The quarries generally produce 2–3 metre square or tabular slabs of varying thickness (up to 50 centimetres) for export: these are later cut by the importing stonemasons for flooring, cladding, grave headstones, etc. Thinner slabs are produced in the quarries, but these suffer the possibility of breakage. Thicker slabs are strong and thus even when they fall over, may not break. The granite quarries often sell to an exporter, who can arrange for cheap passage on a ship. The slabs may reside in the quarry, the exporters storage or the docks for many months prior to shipment, where again they may move to and reside in one or more storage points. The blocks are stored at import and moved on to the various stonemasons.

There was some surprise at the Belfast docks where such granite was imported when a thick slab fell from a forklift truck and broke, revealing a series of powder-filled tubes (Figure 7.9(a)). Suspicion aroused, the authorities and police determined the powder to be very pure cocaine. Somewhere between the cutting of the slabs and the last import to the docks, the drug importers had drilled and retained in sequence and orientation a series of rock plugs from the end of the block (Figure 7.9(b)).

A rock drill had then been used to drill into the slabs. These were possibly cleaned, filled with packed cocaine, and most remarkably, the plugs re-installed in their original orientation. These were glued into position such that the rock fabric continued from the slab, across the edge of the plug, into the plug fabric and out again, with only a thin line to separate the two. The ends of the granite were of rough texture (not polished), with much dust on them making the plugs difficult to observe.

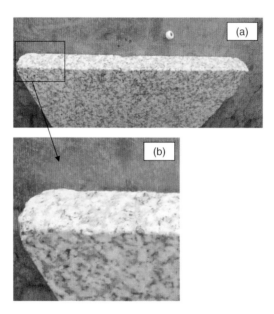

Figure 7.9 (a) The broken slab of Vigo Granite, with the hollow tubes that were filled with uncut (pure) cocaine. (b) The intact Vigo Granite. Close inspection of the left hand side shows some circular marks – the ends of the plugs outside of the cocaine-filled tubes

A number of problems faced the authorities: which route had the slabs taken in coming to the port? How many other slabs had cocaine hidden in them? Where were the slabs going to go? Close examination of the means of transport provided some answers to the first question. The slabs were protected by rough, pallet-type crates, the bases of which scrape the floor of wherever they are unloaded. Examination of some of the crates revealed pollen and marine microfossils (see page 221) of a warm, Atlantic origin. Rock fragments were determined as limestone found commonly in specific locations between northwest Spain and northwest France. These, compared to the various shipping points indicated by all the possible ship's logs, indicated storage in a specific location consistent with the soil, microbiology and geology in the pallet 'dirt'. A second microbiology-rock assemblage was found, but the provenance not determined. Later investigations showed this to be in northeast Europe.

Interviews began with dock-workers in both locations. Next the question of whether other slabs had cocaine in them. Initially it was decided to X-ray (Chapter 4) the slabs, until an easier method was found, as the thickness was too great for X-ray penetration. For ballast and tonnage calculations the docks have abundant accurate weighing machines and trucks capable of moving large, heavy materials. The specific gravity of granite is well-known, the dimensions all given and unadulterated granite was found as a cross-check: those slabs that were too light were seized and subjected to high-resolution scanning by 1 GHz GPR (ground-penetrating radar) pipe detector: see Chapter 3). Hollows were mapped out and those with elongate tubes taken apart and the cocaine recovered. During examination of the slabs, various tests were conducted on them: it was considered that some had been marked with a UV-fluorescent pen: unfortunately this was not the case. Later intelligence showed that each slab had a bona fide numerical marker on it: the 'cocaine' slab's numbers had been passed on by covert means, allowing the importers to identify 'their' materials. Some piles of high-quality, broken granite were later found in an illegal landfill, suggesting that this method had been used successfully before.

Some evidence for rock and sediment being used as a smuggling medium comes from anecdotes of North African rock slabs that have highly-valuable fossils in them being bought by reputable people in Europe. These are then split, hollows drilled out, and wrapped, high-grade marijuana resin introduced. The split rock is re-glued: glued valuable fossils and their host rocks are common in collections, thus not arousing suspicion. The similarity between rock, soil and sediment samples on X-ray machines to marijuana resin makes differentiating the two impossible, as many soil scientists and geologists who have been searched in airports can confirm!

Case study: Sediment – story of the Japanese incendiary balloons concluded (nearly!)

The importance of meteorology was introduced in Chapter 2, where the account of Mikesh (1990) was summarized. A Japanese military campaign in World War II used some 9,000 automatically-detonated incendiary balloons to attack the United States. The balloons were released from Japanese beaches and carried westwards by the jet stream to eastern North America. The balloons were self-regulating, releasing hydrogen to lose elevation when warm and sandbags to gain elevation during cold periods. Limited damage was caused by the 1,000 or so balloons that traversed the Pacific (apart from the death of a family of six who accidentally ignited one) and their descent as far east as Michigan caused concern. The US Geological Survey were asked to determine the origin of the ballast sand recovered from balloon crash sites, as this was a likely comparison to a launch site. The combination of sand particles, microfossils, unusual minerals and absence of coral precluded both North American beaches as well as Pacific warm-latitude islands, leaving only Japan as a likely suspect launch site. Pre-war publications on the mineralogy of Japanese rocks and sediments demonstrated a similarity to beaches on the northern coast of Japan. Following the campaign, it was determined that balloons were being launched from three sites, one of which was identified by the geologists and a second one close by.

Case study: Particle size analysis and the work of Blott *et al.* (2004)

Blott and co-workers provide an excellent overview of how the modern generation of particle-size analysis machines (laser granulometer) work, with two case study applications. Previous to the invention of these devices, to determine particle size, sedimentologists separated fractions through a series of sieves with set mesh sizes (wet or dry), separated the sediment by settling in liquid, or used other automated devices. The use of laser diffraction measures clay, silt and sand particle sizes rapidly and with high reproducibility. Strictly speaking the sample is not destroyed, but more wasted as the water suspension medium is ejected after analysis. Thus the sediment could be captured, although many devices use large quantities of tap water that would require settling and may have become mixed/compromised in the analysis process.

To date, few works have considered the conjunctive use of the laser diffraction device in analysing small or irreplaceable samples. Small sample volumes can be catered for by many of the devices available on the commercial market. In support of this technique, Blott *et al.* describe a hit-and-run incident where a vehicle travelled across a roadway, hitting 'banks' on both sides before also hitting two pedestrians, one of whom later died. The car was driven away and later seized and mud retrieved from the nearside and offside tyres. These were analysed by a number of methods, including laser granulometry for particle size evaluation. Blott *et al.*'s statistics show a very convincing set of comparative figures (mean, mode, size fraction percentages) for the front offside tyre and the offside verge (bank) and the rear nearside and nearside verge, as would be expected if the vehicle was the same as involved in the hit-and-run.

In their second case study, drug traffickers caught in coastal southern England were linked to a location in the Netherlands from where a boat carrying their consignment departed. Significantly, the particle size distribution of the sand samples did not compare with the English location the traffickers were arrested in, denying any alibi they may produce that suggested the sand to be English not Dutch.

Historic case study: Murder by arsenic of Francis Blandy

This section describes the famous case of Mary Blandy in the mid-1700s (Bleakley, 1905). Mary was the daughter of a wealthy English lawyer called Francis and the lover of a recently fallen from favour Scots-born marine lieutenant and son of the 5th Lord Cranstoun, named William Henry Cranstoun. Francis Blandy grew tired of Cranstoun's continual presence and banished him from his Henley-on-Thames home, denying him the opportunity of marrying his daughter and securing a £10,000 inheritance (some hundred millions of pounds in the present-day equivalent, depending on choice of conversion).

From his refuge in Scotland, Cranstoun sent Mary (his co-conspirator) 'Scottish pebbles and powder with which to polish them', the powder being arsenic that Mary added to her father's tea. The powder would not dissolve however and he refused to drink the tea, but a maid (being less fussy) did, causing terrible illness. Having established a successful smuggling method, Cranstoun sent a second package of pebbles and powder, which Mary added to the old man's gruel, which killed him. The gritty texture of the gruel was noted and a physician sent for. Being suspicious, he interred Mary and successfully analysed the residue. Mary Blandy was sentenced to death and hung in Oxford on 4th April, 1752.

Although probably not the first case of using geological materials to smuggle illicit goods, Cranstoun and Blandy were early practitioners of what has been a common method for the covert movement of drugs, explosives, precious artefacts and gems, and stolen industrial or military materials; that is, the hiding of materials amongst or inside rocks and sediment.

Cold cases – opportunities missed?

As mentioned in Chapter 1, it seems easy to use hindsight and suggest methods of investigating past crimes using present-day technology. There is, however, a point to such thinking: although we cannot anticipate future technological advances, the careful measurement of features and collection of evidence allows us to revisit such cases, often with success. We can then learn from these past cases: below are two famous examples where geological materials were noted, but not used, or could maybe still be used in criminology.

Lindbergh baby kidnap and murder. Yellow clay was noted on the floor besides the window by which the kidnapper entered, as well as on the ransom-money suitcase. Lindbergh himself noted the similarity to the clay soil outside his New Jersey home, but would the scene and garden clay compare using modern analytical methods? Would other materials be captured in the scene clay that recorded the previous environmental history of the footwear?

Janie Shepherd murder. In April 1977, two boys found the decomposed, clothed body of Janie Shepherd, an Australian living in London, in Hertfordshire. Her mud-splashed car had been found 10 weeks earlier in London, and analysis of soil and microfossils by a team from University College London had suggested that the mud originated from areas in Oxfordshire, Hertfordshire, Wiltshire or Surrey. However the mud on the car did not match the area of Hertfordshire where Janie's body was found (northeast of St Albans). To this day, the origin of the mud on the car has apparently not been established, and thus a possible murder scene remains undiscovered. Predictive work (pioneered by Heinrich, practised for many years by environmental profilers such as Dr Patricia Wiltshire and recently published on by Rawlins *et al.* (2006))that locates a possible scene (excludes all others) could well have provided a more specific search location for where Janie's car had been, possibly the murder scene, or close to a perpetrator's home.

Chemical analysis

Most methods of chemical analysis are destructive in some form or another, requiring crushing to a laboratory standard, digestion in acid or conversion of the sample to plasma by heat. Non-destructive chemical analysis, however, can be achieved by X-ray fluorescence (XRF) and neutron activation analysis (see the PINs technique for assessing weapons of mass destruction in Chapter 4). Nearly all the chemical analysis methods have been deployed at one time or another in the comparison of samples thought to be associated. Space precludes us from a full description of the science behind each method: we refer the reader to the cited work in each case study, and also for general background to Fairchild's chapter in Tucker (1988). We have assumed that those reading this book are doing so to widen/confirm their knowledge or for application to their field of work. The following case studies therefore demonstrate how the method was applied, whether it was fit for purpose and replicate effective experiential learning. They do, however, have the unfortunate effect of narrowing the reader's vision of what the method *could* be used for as opposed to what it *was* used for.

Atomic Absorption Spectroscopy (AAS) is a widely-used and very reliable method of chemical analysis that uses the absorption of UV or visible light to measure the concentration atoms following vaporization in a flame or graphite furnace. The amount of light absorption relates to the amount of atoms in a sample: sophisticated AAS devices use laser light that additionally allows atomic fluorescence measurements. The method generally requires 2 grams or above of sample material and the range of elements detected can be limited by the type of device used.

XRF is the emission of fluorescent X-rays of a known spectra from a material that has been excited by bombardment with high-energy gamma or X-rays. XRF is used in many types of chemical analysis, especially where reasonable amounts (2 grams or more) of otherwise difficult to assess (e.g., homogenous) material. Geology, engineering product analysis and manufacturing (glass, ceramics) use XRF widely.

Inductively-coupled plasma mass spectroscopy (ICP-MS) requires the production of sufficient ions and electrons in a sample to make the gas electrically conductive. This is achieved by subjecting the sample to rapidly-alternating magnetic fields in quartz-holders. The result is a plasma or fireball – in which argon and free electrons are present. The number of free electrons can be used to determine the composition of the sample. ICP-MS and its counterpart ICP-AAS (atomic absorption spectroscopy) are super-sensitive and can be used on much smaller sample sizes than AAS and XRF. Many investigators, however, still favour these latter methods over ICP, which unless sufficient, representative samples are taken, can be prone to over-sensitivity, emphasizing the chemical content of an unrepresentative sample. As Jarvis *et al.* (2004) point out: 'instrument measurement variation can play an important role in controlling ICP data quality. However, replicated sample preparation is perhaps more critical if a true estimate of uncertainty is to be made'.

These comments could be equally applied to many of the trace material analytical techniques mentioned here; choice and location of sample, preparation and analytical conditions can individually, or collectively, create more variation in result output than occurs through natural change. Hence standard preparation is essential, backed-up by multi-proxy analysis, in which ICP is an excellent contributor, providing many element abundances and thus diminishing problems of natural homogeneity/variation as key chemicals may provide unique identifiers in the same way as unusual minerals, human fragments or biological remains. The measurement of trace elements, generally by sequential extraction, is of great benefit in characterizing soils (Cave & Wragg, 1997).

If the reader is now overwhelmed once again with geological acronyms, refer back to Fairchild (in Tucker, 1988), or consult many of the generally reputable web sites (Wikipedia can be trusted in the first instance for simple explanations of common analytical methods).

Two further techniques deserve mention: Fourier Transform Infrared Spectroscopy and Near-infrared spectroscopy, essentially methods of measuring the atomic makeup from using induced radiation, in this case infrared (Cox *et al.*, 2000). The methods are used in many soil and forensic laboratories, yet do not always appear in the standard list of techniques in the multi-proxy armoury (Bull *et al.*, 2004; Croft & Pye, 2004b; Rawlins *et al.*, 2006).

Case study: Atomic Absorption Spectroscopy (AAS)

Abbott (2005) once again provides us with a clear and straightforward example of mining fraud. AAS is the traditional technique for analysing the chemical makeup of rocks and ores. A professional assayer testified in court that he made his own mental adjustment, based on a secret formula, in order to correct the values provided by AAS! These values invariably gave greater precious metal contents than the original data. Abbott explains that one does not even need such a devious mental adjustment – should iron be incompletely removed prior to analysis, then the levels of platinum-group elements can be erroneously high. Abbott recounts a published case where a US nickel was analysed with incomplete iron removal, the results showing high levels of palladium, platinum and iridium – all very rare and expensive metals.

Case study: XRF by Bergslien and others

Bergslien *et al.* (2006: *www.geolsoc.org.uk/pdfs/FGtalks&abs_pro.pdf*) give an excellent set of examples of the use of portable (field-based) XRF in forensic and associated applications such as tracking pollutants in the environment to their source (screening soils for toxic metals, for instance in a local school playground). In

Chapter 4 we used their work in a field-based case study, but the method can equally be applied to the rapid screening and detailed quantitative analysis of laboratory materials. The technique can be used in a variety of environments, has minimal sample preparation requirements, and analysis takes a few minutes. These characteristics make both portable XRF and non-quantitative laboratory XRF ideal tools for use at crime scenes where rapid, non-destructive screening of materials may be necessary – as Bergslien et al. (2006) write:

> Field portable XRF uses either an X-ray tube or radioisotope as an excitation source to irradiate samples. The incident X-rays interact with the samples' atomic structure by displacing electrons from their inner shells, leaving vacancies that are filled as outer shell electrons release energy to fall into new ground states. The energy released will be an X-ray equivalent in energy to the energy difference between the two shells. Since each element has a characteristic arrangement of electrons, the X-rays released by such transitions will be unique to that element, allowing its identification. By comparing the intensities of X-rays from an unknown sample to those of a suitable standard, elemental composition can be quantified ... Forensic geological applications of FPXRF include distinguishing between mineralized tissue, such as bone or teeth, and inorganic materials of similar appearance. Mineralized tissues are a calcium phosphate, often called bioapatite, which closely resembles naturally occurring hydroxylapatite. Preliminary results using FPXRF, sometimes paired with X-ray diffraction (XRD), demonstrate that bioapatite might be clearly differentiated from inorganic materials, such as Plaster of Paris and powdered apatite. Both methods are non-destructive and can be used in concert to determine both elemental composition and crystalline structure. The bioapatite samples tested thus far have significantly lower strontium and iron levels than geologic materials and had no detectable barium, bismuth, chromium, manganese, sulfur or zirconium. An operator skilled in use of the FPXRF instrument may also provide significant help at the crime scene in identifying composition of personal effects, including gemstones and other metal jewellery. One of the most significant advances of the technique is the ability to obtain usable spectra in as little as 6 seconds, allowing rapid sequential analysis and sorting of materials at the scene.
>
> Bergslien et al. (2006)

Their method promises to have great value in the screening of materials, especially where a first indicator of the source of a questioned sample must be made. For instance, debris is observed on a suspect car underside, parked in a city. A recent crime that may be associated took place well beyond the city. Hours of arduous CCTV examination will likely show the movement of the vehicle. However, all urban samples currently analysed show comparable elevated lead, iron and zinc to rural samples, making a quick decision possible: high lead levels do not preclude the car from having been in the countryside, they do diminish the possibility at the time.

XRF can also be used in a similar mode in the laboratory. A farmer in Somerset (England) found a fist-sized, smooth metal object on the floor of an infrequently-used barn. He also noted a hole in the roof that he had not noticed before. The reinforcing opinion that the object was an iron-nickel meteorite developed as ablation (wind-erosion) flutes were noted on the object. Staff at Queen's University Belfast cut the object at one end to obtain a smooth surface: no characteristic meteorite structure was noted (Windmanstetten structure) but nonetheless the cut end could be placed under the XRF and irradiated. No nickel was found, and the object was excluded as likely iron-smelting slag. No quantitative measurements could be made, nor were they needed. How the hole came to be the farmer's barn we could not ascertain!

ICP-MS (conjunctive) case study: Pye (2004a) on tracing the origin of a body found on an aircraft

Pye (2004a) shows how the abundant red mud found adhered to the boot of a person found dead on an aircraft allowed a multi-proxy approach to the analysis and predicted provenance of the soil (and thus person) to be undertaken. The body was that of a presumed illegal immigrant to the United Kingdom, found in the undercarriage of a large aircraft at Heathrow Airport (London). The plane had landed at over three locations since its original point of departure, giving the investigators some measure of possible origin: the man had no other means of identification on him. The conjunctive approach used sediment description (including colour measurements), whole-rock and clay-fraction X-ray diffraction, palynology (see page 232) and most importantly for this study, ICP-MS. Combining the results, Pye shows that the initial conclusion was that the red soil came from a wet tropical country, limiting the search. Comparison samples from possible airport grounds showed Ghana to be the only realistic comparison, the other sites having been excluded.

Geological and engineering analysis of concrete used in the covert burial of a murder victim

In late 2003 a woman was reported missing by her husband after failing to return home from her daily visit to a nearby town. He alerted friends and the police and a missing persons record was made. Friends and family throughout Ireland and the United Kingdom were alerted, as given her age it was possible she had become confused and decided to visit them. The river was also searched in case she had fallen in. By Christmas, her case was very serious and thereafter considered a suspected abduction and likely murder.

On the same day as her disappearance, the fire brigade were called to deal with a vehicle fire in a small settlement some miles from the missing person's home. The fire had been started intentionally, by person or persons unknown. Living at the house was a person with a previous criminal conviction for abduction and rape: this person had not wished for the fire brigade to be called, for reasons unknown. The burnt car was seized, searched and stored by police, given the fact that the fire was started intentionally and thus an insurance fraud suspected. Given the coincidence in time, a known sex offender and rough location, the car was further examined and the owner became a suspect in the disappearance. The car interior was badly-burnt but DNA evidence was recovered from materials associated with the car, leading to the suspect's arrest in March 2004. A search, focused on the suspect's house, revealed

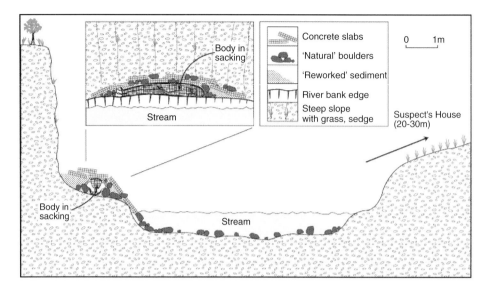

Figure 7.10 Cartoon sections through the body deposition site

the body of the missing person on a small ledge next to a stream under 100 metres from the suspect's house. Her body had been tied up in agricultural sacking and covered with stones and concrete slabs (Figure 7.10), three of which were of 'slab' like proportions (7 centimetres thick, 60–70 long and 30–40 wide).

The stones were recognized as coming from the stream-bed, but the concrete had no immediate source, prompting interest in this material. Where had it come from?

A further search of the suspect's house revealed many concrete slabs in the garden. Two areas had broken concrete, suitable for the making of 'crazy paving' piled up and in many sizes from breakage. Thus it was not possible to bring the concrete from the body deposition site to the garden and make a direct 'jigsaw puzzle' type fit: in addition, biological material associated with the victim would then be introduced back to the suspect's house, possibly falsely incriminating him or other persons. Two approaches were taken to testing whether the concrete from the grave site could be compared to that from the suspect's garden. First, samples were taken of each location. Second, the area surrounding the body deposition site was searched for control concretes, especially of a 'slab' (tabular) form that would be useful in covering a body. Five slabs were recovered from the suspect's garden (two of which fell apart), five from the stream (another one of which fell apart) and five controls from surrounding lands. Ideally, a similar number of each would be useful to compare, but the investigator is not in control of what the protagonist has done, or the locations of debris scattered about the countryside, thus the original three body deposition site slabs were analysed.

A *multi-proxy approach* to the analysis and comparison of the concretes was decided as the best approach. All slabs were described and photographed: this showed some variation, as the garden slabs were fresh, body site and some controls were algae-covered and others had soil adhering to them (Figure 7.11).

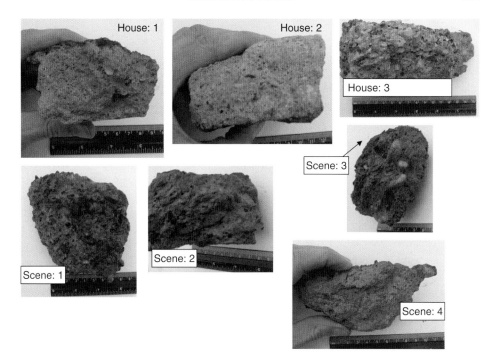

Figure 7.11 Replicated photographs of broken blocks from the body deposition site and suspect's garden prior to control materials being found. Note how the different environmental history of each fragment has affected its external appearance. The body deposition site is a stream, so blocks are algae-covered; the garden is dry and thus the blocks have no algae

Thus ten cut surfaces were created, photographed and described. These showed some comparison between body site and garden materials (Figure 7.12). However, such comparisons are subjective, and although convincing to a jury, may not always be accurate.

Ten thin-sections were cut from each block, mainly to provide better information on the constituent mineralogy, but also to show how thin-sections varied through the concrete (Figure 7.13). Common civil engineering methods of assessing concrete would be to test for strength (tensile, shear) and analyse chemically. The problem faced here was that some of the blocks had been in a river, probably for up to four months and possibly more, and others (the controls) had unknown origins. Thus strength tests may have provided false-negative results.

Chemical analysis of the concrete also faced this problem, although 'skimming' outer surfaces removed visually-weathered areas. XRF was then deployed on these surfaces. The remaining material was disaggregated in acid under controlled conditions and replicate experiments (to test the effect of the acidization process) were run on other controls. Consistent results showed that after sieving and drying, the fractions were representative. Rather than attempt to explain particle size distribution curves to the jury, the method (above) of photographing each size fraction was

212 CONVENTIONAL GEOLOGICAL ANALYSIS

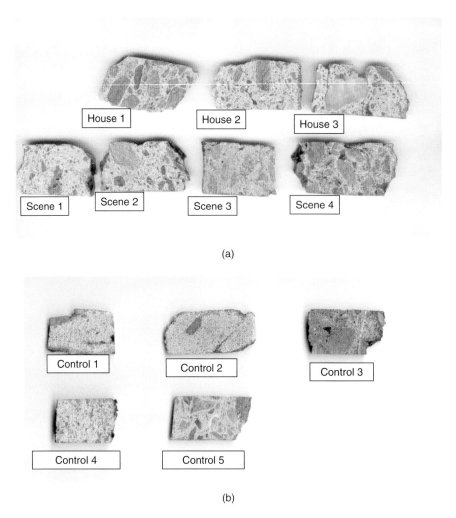

Figure 7.12 (a) Example cut surfaces (of the 10 made) of the house and body deposition site blocks. (b) Cut surfaces of control blocks found using standard search methods (see Chapter 9), emanating from the body deposition site outward. Each block is ~3 centimetres long

used, which provided a strong visual exclusion/comparison that replicated the more intangible curves (Figure 7.14).

Six visual geological tests were thus completed:

1. hand specimen description of the uncut blocks;

2. cut section description;

3. thin section visual appearance (pattern recognition);

4. petrographic analysis from thin-section;

5. particle size fraction from visual appearance;

6. petrography of the size fractions.

Automated particle size analysis provided additional data. The challenge was to provide some measure of comparison, when most of these data rely on subjective visual comparison (pattern recognition), subjective choice of grains to be counted (petrography) and colour/texture comparison (particle size visual appearance). Various published methods of exclusion/comparison description were considered. It was found that in the analysis of concrete, comparing weathered to fresh material is not commonly undertaken. Thus the most simple method of communicating complex geological exclusion/comparison was chosen, wherein for each test a 0–5 score was established, with 0 having no test results to compare, 1 having one result, 2 having two tests to compare, and so on. This created a possible 'best comparison' score of

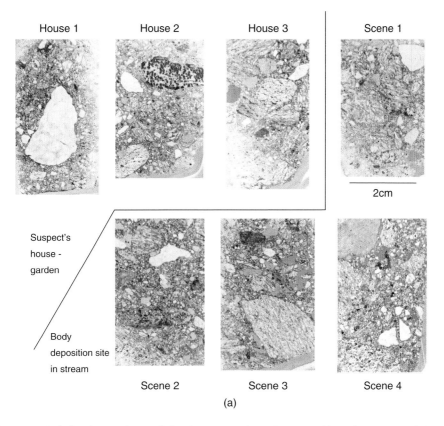

Figure 7.13 (a) Thin sections of the house and body deposition site blocks. (b) Thin sections of the control blocks found during searches of surrounding property. Each thin section is ~3 centimetres long

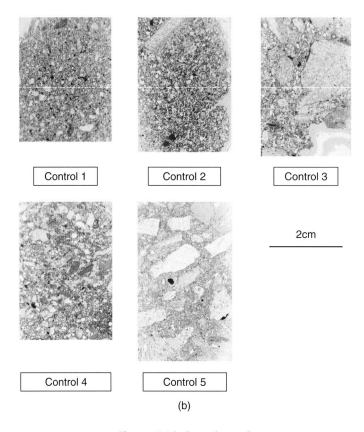

Figure 7.13 (*continued*)

30 visual geological tests, each scored 1–5: Table 7.1). An internal check of House to House and Scene to Scene comparisons established some efficacy in the method. However, a second interpreter of such data would likely obtain different results. For this reason, the score from 30 was compared to the automated particle size results (Table 7.1), with some interesting variations to the expected comparison/exclusions. House exhibits 2 and 3 repeatedly compared to Scene samples 3 and 4, and yet the visual estimates (score from 30) started to contradict the automated particle size results when the 50 % confidence level (actually at about 16/30) in visual estimates was reached.

The first notable problem was Scene 2 to Scene 4, which showed poor visual comparison when cut sections were examined, but with visual examination of the size fractions, showed increasing comparison to the automated particle size analysis. A similar pattern could be observed with Control 3 to Scene samples 4, 2, 3 – with particle size (visual) making a better comparison to automated particle size analysis. Control 3 did not appear until 15 previous comparisons (of 48) had been made.

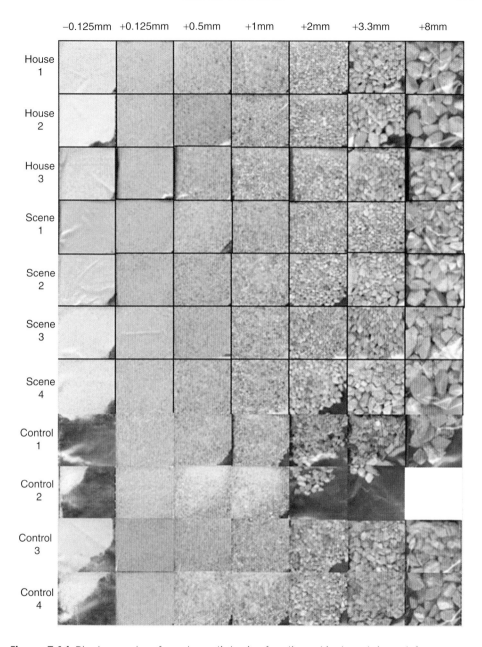

Figure 7.14 Photographs of each particle size fraction, dried and sieved, for presentation to the jury. Each square is 1 centimetres across

Table 7.1 Comparison of methods used to analyse concrete found above a clandestine grave, at a suspect's house and from surrounding area (controls)

	1. hand specimen	2. cut section	3. thin section	4. petrography	5. particle size (visual)	6. size fraction	score/30	Particle size r^2
House 1 – House 2	4	3	4	3	5	3	22	0.97
House 2 – Scene 4	3	3	4	3	4	4	21	0.85
House 3 – Scene 1	3	2	4.5	4	5	2	20.5	0.98
Scene 3 – Scene 4	1	5	2	3	4	4	19	0.83
Scene 2 – Scene 3	2	1	4	3	5	3	18	0.9
House 2 – Scene 2	2	1	4	3	4	4	18	0.8
Scene 2 – Scene 4	1	1	2	3	5	4	16	0.97
House 2 – Scene 3	2	2	3	3	2	4	16	0.66
House 1 – Scene 2	2	2	3	3	2	2	14	0.69
House 3 – Scene 2	2	2	2	3	1	4	14	0.55
House 3 – Scene 3	2.5	3	3	3	2	0	13.5	0.61
House 2 – House 3	1	4	2	3	0	3	13	0.33
Scene 1 – Scene 2	5	1	4	3	0	0	13	0.48
Scene 1 – Scene 3	4	1	4	3	1	0	13	0.57
House 1 – Scene 4	3	2	2	3	3	0	13	0.73
Control 3 – Scene 4	0	0	2	3	5	3	13	0.92
House 1 – House 3	1	3	2	3	0	3	12	0.21
House 1 – Scene 3	2	3	3	3	1	0	12	0.59
Control 3 – Scene 2	0	0	0	3	5	3	11	0.91
Control 3 – Scene 3	0	0	0	3	4	3	10	0.88
House 3 – Scene 4	1	1	2.5	3	1	1	9.5	0.51
Control 3 – House 1	0	0	0	3	1	5	9	0.59
Control 3 – House 3	0	0	2	3	1	3	9	0.54
Control 3 – Scene 1	0	0	2	3	0	4	9	0.47
House 1 – Scene 1	2	1	2.5	3	0	0	8.5	0.19

Table 7.1 (continued)

	1. hand specimen	2. cut section	3. thin section	4. petrography	5. particle size (visual)	6. size fraction	score/30	Particle size r^2
House 2 – Scene 1	2	1	2.5	3	0	0	8.5	0.29
Control 3 – House 2	0	0	0	3	2	3	8	0.7
Scene 1 – Scene 4	1	0	2.5	3	0	0	6.5	0.45
Control 1 – Scene 3	2	0	0	0	3	0	5	0.75
Control 1 – House 1	3	0	0	0	1	0	4	0.58
Control 1 – House 2	3	0	0	0	1	0	4	0.56
Control 1 – Scene 4	1	1	0	0	2	0	4	0.62
Control 1 – House 3	1	0	0	0	0	2	3	0.15
Control 1 – Scene 1	2	0	0	0	0	1	3	0.12
Control 1 – Scene 2	2	0	0	0	0	1	3	0.48
Control 2 – Scene 1	3	0	0	0	0	0	3	0.11
Control 4 – House 3	0	0	0	0	2	1	3	0.69
Control 4 – Scene 1	0	0	0	0	3	0	3	0.71
Control 4 – Scene 3	0	0	0	0	1	2	3	0.52
Control 4 – Scene 4	0	0	0	0	0	3	3	0.16
Control 2 – House 1	2	0	0	0	0	0	2	0.037
Control 2 – House 2	2	0	0	0	0	0	2	0.011
Control 2 – Scene 2	2	0	0	0	0	0	2	0.01
Control 2 – Scene 3	2	0	0	0	0	0	2	0.01
Control 2 – House 3	1	0	0	0	0	0	1	0.1
Control 2 – Scene 4	1	0	0	0	0	0	1	0.08
Control 4 – House 2	0	0	0	0	0	1	1	0.07
Control 4 – Scene 2	0	0	0	0	0	1	1	0.25
Control 4 – House 1	0	0	0	0	0	0	0	0.04

These latter points were damning of both the methods of analysis and comparison when selected data were extracted; the geological data compared well to the chemical analysis. Furthermore, when the location of each 'House' sample was plotted on a map, those closest to the body deposition site (Scene) were found lower in the pile of slabs on the body. Those from furthest away were found on top of the pile, implying the suspect covered the body with the closest-to-hand slabs first, and then with time and effort found slabs further away to continue the covering.

The accused was found unanimously guilty and his record of abductions, rape and lewd conduct made public: he is currently one of only a few people in the United Kingdom sentenced to life with no opportunity for remission.

Conjunctive use of petrology and geochemistry in an engineering/environmental pollution case by Hower *et al.* (2000)

Major pollution of the Tennessee River occurred in 1996 when massive quantities of coal slurry were released in the space of a few hours to days. The cause was a pressurized blow-out from a mine portal in Virginia, the result was a massive fish kill, the destruction of valuable irrigation water and unquantified associated environmental damage. The problem faced by investigators was tracking the origin of the slurry: two likely sites were considered possible and the river slurry was compared to fine-grained materials associated with each. A combination of petrology and geochemistry established the makeup of the slurry/fines, with chemistry being the best exclusion/comparison tool (specifically Zr and Y contents). The slurry was successfully traced by these elemental characteristics back to one mine and its associated plant, the coal-seam they worked also having unusually high Zr and Y contents.

Specialist geological analyses

Gross's 1906 version of *System der Kriminalistik* (*Criminal Investigation*) introduces what is still one of the main uses of the forensic geologist – the detection of fraud. Gross begins his section on fraud (page 396) with the still-common, present-day practice of making flint (and other conchoidally-fracturing, hard rocks such as porcellanite) axes and arrowheads: indeed, many archaeologists expert in dwelling reconstruction can re-create authentic-looking Stone Age materials. The specialist archaeologist can often tell the difference, given the common use of modern tools in 'knapping' the flint core. Likewise, trace geological evidence such as an absence of weathering, soil and presence of modern fibres can provide exclusionary tools.

Primitive copies of Egyptian sculpture are identified by the glass paste and siliceous clay used to mould them. An effigy of Rameses (King of Egypt), apparently in black basalt from the region, was sold in the late 1800s for 100,000 francs. Further tests, however, revealed it to be made of black schist from Antwerp. The geologist reader will be appalled by this story: basalt is rarely foliated or showing a preferred texture, schist always so. However, the Antwerp Schist used in this case is more correctly a meta-basalt/dolerite and when polished shows no easily-observed lamination.

Pottery is more easily faked as the materials and firing mechanisms are largely the same, yet firing can destroy many of the modern trace particles seen on flint objects. Modern inclusions can be of assistance: glassware is easily faked yet the crystallization of ancient glasses cannot be easily replicated. Gem fraud and the tracking of the illegal movement of gems is a major crime in itself, and due to the massive financial gains can often be the origin of associated crimes through to murder, hence some forensic science services have specialists in gemmology (Small, 2006).[1]

Towards a best practice methodology – the multi-proxy experiments of Croft and Pye (2004b)

In their abstract and preamble, Croft and Pye (2004b) outline the opportunities and limits of their work. Their experiments were conducted on soil samples only, but as we have seen, soil is the weathered product of rock and as such the use of multiple methods can be equally-well applied to sediment, soil and (as in the analysis of concrete [above]) manufactured 'geological' materials. A second, enlightened point comes in their introduction:

> The range of techniques available to the geoscientist [and thus the investigator] is very wide but not all have yet been applied or tested in a forensic context ... No single technique can provide a unique characterization of a soil, sediment or rock, but used in combination the results provided by several techniques can provide good discrimination with a high degree of significance.
>
> Croft and Pye (2004b)

The statistical test of their discrimination and significance are explored using simple and impactful cross-plots that perhaps underlie the word 'significance' in this context, and that is one of the law courts. Croft and Pye use the methods on which they are experts: colour, particle size, stable isotopes and bulk geochemistry. They sampled bulk soil from four locations and wore a variety of footwear at these same places. The extent to which their four methods could separately and conjunctively be used to compare the footwear with each location were then discussed.

The results of two or more independent (see below) methods that provide the same comparison/exclusion are easily grasped by judge and jury. They also provide the expert witness latitude when the opposition lawyers seize upon the fallibility (and all scientific methods are fallible) of their one favoured method. Two expert witnesses cannot enter the box/take to the stand at the same time, which forces either the long process of recalling witnesses (can be disadvantageous in a jury-based court) or an examination of results, *in the whole*, as they should be.

---

[1] A famous example is the 1977 murder of Pinchos Jaroslawicz in the diamond district off West 47th Street, New York. The mutual trust between diamond traders in this area was shaken by the brutal murder and was compounded by their use of handshake- and cash-based agreements, with no paper-trails to assist who was making transactions to whom (Baden & Hennessee, 1991). The story is one of a terrible failure in police search techniques, the victim having been trussed up in a foetal position and hidden in a tank beneath a work space, remaining there throughout a failed police search.

A better legal and scientific criticism of the multi-proxy approach is the extent to which analyses are truly independent of each other: in response to Croft and Pye's work, Bull *et al*. (2004) cite the work of Brown *et al*. (2002) who use pollen and soil analysis techniques separately in a murder investigation. Their main tenet is that the geological ingredients of soil are all inter-related, and thus mutually dependent, a concept vindicated by studies of the influence that particle size has on soil/sediment geochemistry: the two are related and thus are not independent comparative data sources. Bull *et al*. (2004) make the point that multiple analysis of the geological content of soil or sediment will of course tend to provide good comparisons, because the various methods used are essentially analysing the same ingredients, just by different means. Mineralogy is often related to grain size; chemistry to mineralogy; and major to minor element geochemistry. This is good in a way, as numerous 'geological' analyses that internally agree at least show the geology is the same in two samples. A check of geological analyses is to consider an independent variable, such as footwear or vehicle tread analysis or establishing the water chemistry, biological content (e.g., microbiology) of samples. The microbiological content of a soil is not necessarily independent of mineralogy/geochemistry because this depends on how 'local' the pollen, fungal, bacterial content is to the soil. If transported microbiological elements are few (making for improved spatial resolution, see Chapter 6), soil makeup may still be related to micro-organisms. Luckily, indigenous and transported microbiological debris is common in many soils. Croft and Pye's initial point about soil analysis can thus be related to the concrete case study (above): where no more independent variable exists such as microbiology or human evidence (footwear, blood, DNA) then the individual ingredients in any comparative material provide the next test of a multi-proxy approach. The conjunctive use of microbiological and geological materials currently provides the best method of comparing or excluding soil, although the future use of multiple geological and biological methods must be the next logical step, followed by analysis of water contents and other materials or features captured by the minerals and organic remains. Before we proceed to examining how far these techniques are likely to proceed, some examination of geological and fossil biological materials is desirable to bridge the gap between geological–ancient biological (fossils) and biological evidence.

7.4 Fossils and microfossils

In the above sections, the reasons why geoforensics is focused on all solid Earth materials (igneous, metamorphic, sedimentary rocks, sediments and soils) is given, partly because all these materials are involved in the different forms of criminal enquiry, and partly because igneous and metamorphic rocks are the common precursors to soils, sediments and sedimentary rocks. Nonetheless, the fact remains that many cases and thus the bulk of publications on Earth sciences and forensics have soil and sediment as a central theme. The macro- and micro-scale organic component of soils and sediments vary, but are nonetheless always an important component. In soils (as we observed in Chapter 2), the ability to support life is inherent in what a soil is. In sediments and sedimentary rocks, fossils (macrofossils and trace fossils, both observable by the naked eye) are one of usually 8–10 (hand-specimen) descriptive elements that provide a classification and thus means of exclusion/comparison. If we then include microfossils (not visible to the naked eye) that are very often present in sediments and soils, then the importance of these organic components in our exclusionary/comparative principles becomes even greater.

Most dictionaries define fossils as 'anything dug up', which is unsatisfactory for our purposes, in that a fossil is a record of past life, usually preserved below ground. Thus fossils can be of any age from months or years to millions of years and can occur in sediment, ice, sedimentary rock or solidified lava and some metamorphic rocks. They can also be large (dinosaur bones), very small (algae) or the tracks or burrows of organisms (trace fossils). The most common fossilized remains are of the organism's hard parts, either its shell or skeleton. However, exceptional preservation of soft tissue, such as Siberian mammoths in ice or Californian mammals in tar-pits, do occur. In historical, industrial archaeological and forensic contexts, the difference between a fossil and a buried dead organism becomes blurred: if the extent of petrifaction, mummification or mineral replacement is used to differentiate, further problems are introduced. How much replacement or mummification must occur for the dead organism to become a fossil? There are further complications – some organisms such as calcareous algae make their skeleton from what is effectively limestone and furthermore can become dormant, with calcareous overgrowth when buried, only to come alive when exposed to fresh sea-water and light. These are not exceptions: desiccated wheat grains from Egyptian tombs have been successfully germinated after thousands of years (they had original organic material preserved), yet World War II Japanese and American soldiers, buried in the coral sand of Pacific islands, have had all their skeletal content re-mineralized.

Thus age means little when discussing fossils, but longevity of preservation seems more significant. The biologists have sought a clever way around this problem by introducing the term *sub-fossil*, or we suppose nearly a fossil. Fossils used in a forensic context fall into one of these brackets:

- *geological remains*, which are inert particles (e.g., re-mineralized organic remains: true fossils as the layperson would recognize);

- *sub-fossils*, which are generally recent (years, tens of years, possibly hundreds of years old) and are dormant, undergoing change of chemistry through desiccation, mineralization, etc.;

- *dead organisms*, whose only post-mortem change is desiccation, soft-tissue putrefaction and cell-wall degradation.

As indicated in Chapter 1, the 'geo' in this book's title, and the volumes of published work on non-fossil biological materials, precludes significant coverage of the latter. Some peculiar attributes of sub-fossil, or preserved by burial or incorporation into a preserving medium, require consideration: organic materials in human remains are one obvious example that, because geological methods of analysis are used, are included. The use of macrofossils as comparative evidence is limited compared to microfossils. However, geological criminal cases and fraudulent activities involving fossils are abundant.

Macrofossils

The use of microfossils as trace geological evidence on suspects and victims (see below) is well-known: in Chapter 6 we saw how the spatial distribution of fossil snail shells assisted in the conviction of a rapist and potential murderer. Macrofossils (easily visible to the naked-eye) are of interest both to those using conventional scientific methods in the analysis of crime, as well as palaeontologists whose fossils become the objects of investigation themselves. Because of the scientific and financial value of many fossils, analytical techniques have been developed that maximise information gained from specimens that can be translated to forensic work. The most obvious example would be fossil human remains, whose careful diagnosis bridges the pathologist–anthropologist gap (Spindler, 1993) and provide the forensic scientist with excellent proxy studies for anthropology. The question of the age of the remains dictates whether criminal proceedings occur (most especially whether suspects may still be alive) but the method of establishing victim age, sex, race and manner of death remains the same for a 10,000 year old bog man from Denmark as for a recently-buried murder victim.

The analytical treatment of materials as well as humans also translate from palaeontology to forensics. The involvement of wood in serious crime is of particular interest because as soon as a tree is felled, cut, dried and treated, its decay is slowed down, much like a fossil. In addition to standard biological methods of analysing wood (beyond the scope of this work), palaeontological and palaeoecological methods (radiocarbon and dendrochronology) may be used in conjunction with cut-marks or staining types and patterns. As early as 1906, Georg Popp described the common association of wood with crimes: sawdust packing of illicit goods and old-style safes; wood splinters from breaking and entering; weapons and accidents involving wood.

Case studies using fossil wood and plants

Gross (1906) describes an early case. 'A country house was entered by drilling round the special safety device of a window: the framework of the window was made of pinewood and painted white. A locked rosewood cabinet was then forced by drilling round the lock....' A number of valuable items were stolen. The suspect was identified from his efforts to dispose of the stolen goods. A brace and bit recovered from his home had fragments of white-painted pine and preserved rosewood in it: grass fragments and seeds were recovered from his boots that corroborated the wood analysis.

Conventional wood identification was crucial in the Drummond Murders (1952). Sir Jack Drummond (a famous nutritionist), his wife Ann and their daughter Elizabeth were found murdered between Digne and Villefranche on the French National Route 96 ('La Grand Terre') while on their August holiday. A chip of wood recovered from under the dead girl's head compared to the wood and damage on the butt of a US carbine owned by a local family of co-conspirators in the crime (Mack *et al.*, 1986, p. 154).

Comparison of wood cut-marks analogous to the methods archaeologists would use in determining past building, warfare, hunting and agricultural practices are best known to criminalists by the Lindbergh kidnapping case. The kidnapper(s) abandoned a home-made wooden ladder, used to gain entry to the Lindbergh's New Jersey home and snatch their 20-month old son (later found dead in adjacent woods) in March 1932. Regular, distinctive cut-marks on the wood indicated its origin in a timberyard with a slightly defective circular saw. Detectives visited numerous yards until they found the source, which was then used to locate purchasers of wood and eventually a comparison to the cut beams in a suspect's attic, from which the ladder was constructed.

Wood is also commonly analysed in cases of fraud: for example, Topham and McCormick (2000) describe how they used dendrochronology to age violins without recourse to destructive radiocarbon dating.

Case study: Fossil wood and mica assists search for a murder victim

Fossil wood lay behind the case study discussed in the above section 'Geological and engineering analysis of concrete used in the covert burial of a murder victim' (on page 210). To summarize the background again, in late 2003 a female from a border town in the northwest of Northern Ireland went missing. On the day she was reported absent by her husband, a car fire was reported at the house of a known sexual offender some miles away. No connection was initially made, given that the missing person had relatives in Dublin and London and may have decided to visit them. Nonetheless the partly burnt-out car was seized and two months later, when the missing woman did not return, a murder investigation was launched. Lists of likely suspects were produced, including the owner of the burnt car. His vehicle was examined and, apart from the exterior being suspiciously clean (where unburnt), 1 centimetre-diameter lumps of mud with grass and other material were found embedded in the vehicle chassis. The mud contained rounded mica – an unusual shape for this easily-broken mineral and one associated with slow-flowing rivers, streams and ponds (Figure 7.15(a)). On further analysis of this mud, a 0.5 centimetre diameter piece of wood was extracted (Figure 7.15(b)). This was analysed and suggested to be a hardwood. Furthermore, the possible types of hardwood in the area were considered, with only a few species being likely.

A search of databases for the area indicated likely places where stands of possible trees grew. A second question arose: how would a section of such wood be transferred onto this vehicle? Analysis of the wood suggested it had been cut in a crude way, most likely by a large hedge-cutter and not a saw or broken. Thus mature (on account of the likely size of the original wood fragment and visible tree-rings) hedges with characteristic hardwoods were a possible source. The last known location of the missing person was established and routes to the suspect's home mapped and examined. Only one route had recently-cut hardwood hedgerows: this roadway was frequently covered in mud and silt from flooding of an adjacent slow-flowing river.

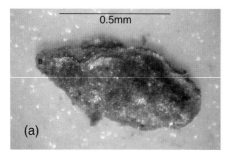 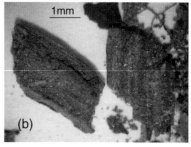

Figure 7.15 Materials extracted from 'mud' caught in the underside chassis of a burnt-out car belonging to a suspect in a no-body murder investigation. (a) rounded mica; (b) fragments of wood, suggested to be cut hardwood

This route led directly to the suspect's house. A later search of this property recovered the decomposed remains, covered with concrete slabs, of the missing person in a stream next to the suspect's home. Whether the suspect passed this particular wood- and fluvial silt-strewn road, or another, and whether his passage was connected with the crime, is not known.

Green (2006) has reported on the work of wood specialist Peter Gasson of the Kew Gardens Jodrell Laboratory: 'There was one case where someone was battered to death.... A rolling-pin was thought to be the murder weapon and a fragment of wood was found in the bathroom where the suspect was thought to have cleaned up. It turned out the fragment was the same kind of pine as the rolling pin, so we could help identify the likely weapon'. Green (2006) does not explain how the wood was compared, but reference to works such as Farmer *et al.* (2007) suggest that microscopic structure, used conjunctively with 'jigsaw' matching, would be rational in this instance.

Case study: Graeme Green on work at Kew Gardens and the 'Torso in the Thames'

Quoting newspaper articles is avoided (where possible) in this text, due to their inaccessibility compared to scientific publications. The infamy of this case, however, and the excellent article make a summary desirable. Green (2006) describes how Kew Garden's expert Hazel Wilkinson sifted through the gut contents of a torso, named 'Adam' after recovery from the River Thames in 2001. She found over 15 different types of plant material, including calabar beans (probably not fossils in the true sense), whose sedative and paralyzing properties are used (for example) by Nigerian witch doctors in black magic. This initial observation confirmed an early police theory that a ritual killing was being investigated. Subsequent analysis of the isotope content of the bones from the body also pointed to a Nigerian source, which the UK police used to launch an overseas investigation and appeal for help.

Fossil fraudsters

The Piltdown Man (see the 2004 talks of Boylan, Brook and Stringer [separate authors] in *http://www.geolsoc.org.uk/template.cfm?name=HOGG0954895486*) is an early example of the seriousness of fraud involving the 'remains' of former life on Earth. Faked human remains continue to be a problem in science and in illegal trafficking of such materials (Wienker *et al.*, 1990). Fossil hoaxes and fakes are among the most famous and widespread geological frauds, together with claims to have discovered gold, diamonds or oil. Hoaxes and fakes involving fossils continue to be both committed and discovered, the feathered Chinese dinosaurs being revealed as fakes in 2000 (Monastersky, 2000). Ross (2004, see *http://www.geolsoc.org.uk/template.cfm?name=HOGG0954895486*) conjunctively used geology, palaeontology and detailed microscopic investigation to track the work of Victorian fakers who inserted modern insects into fossil amber. They hollowed out small cavities, placed the realistically-looking aged insects in, and used resin or the heat-annealed amber extract to seal the modern insects into the ancient materials. These archived materials caused some problems for years, because they indicated a recent fauna was alive long before other fossil records of their appearance. Ross demonstrated how the limits of the excavated amber could be observed microscopically and how the modern insect simply could not have existed at a time before it had evolved!

The most incredible fossil fraud of recent years involves the work of Professor V.J. Gupta (Panjab University), described by Talent (1989). Gupta's initial deviousness came to light when the amazing coincidence of fossil faunas 600 kilometres apart in the Kashmir Himalayas was remarked upon: this led to the realization that these faunas were identical to those found near Buffalo, New York, and that photographs of some of the specimens, from these different locations, were indeed the same fossil. This was only the start: Talent (1989) goes on to show how this discovery reflected Gupta's 25-year-long history of faking, stealing and making fraudulent claims. What is most embarrassing is that Gupta published two papers in the journal *Nature*, and persuaded many international scientists to co-author his works, often following dubious circumstances at conferences where he would produce superb fossil material from his pocket.

Such fossil fraud continues to the present-day and will go on: Rowe *et al.* (2001) describe the Archaeoraptor forgery, the 'missing link' between dinosaurs and birds, shown by them to be a forgery, based on CT scans of the skeleton, which showed where fossil dinosaur and bird materials had been joined by the forgers.

Chemistry of fossils – Trueman's applications to anthropology

Because geologists and palaeontologists have used the chemistry of fossil materials (especially calcareous or phosphatic shell and bone) extensively in their studies, it is only natural that the bulk, trace and isotope geochemistry of human teeth and bones be analysed in the same way. In parallel with this, archaeologists, always keen to extract what information they can from human and animal remains from ancient sites, have used similar methods to link materials to places, infer dietary habits and examine pathology.

The mineralogy and chemistry of bone, teeth and hair (roughly, in order, the commonest non-flesh elements of humans and animals to be preserved) is well-known from medical investigations and have powerful uses in forensic studies. Trace element and isotopic composition, which vary more and can be related to environmental factors when the bone etc. was formed, have a wide usage in criminalistics, disaster victim identification, genocide studies and the effects of pathological pollution. The literature on this subject area is extensive (especially when archaeological, anthropological and medical publications are considered), and so information coverage will be necessarily selective, such as microbiology (below), largely because this topic is right at the limits of what we have considered 'Geoforensics', being equally valid to archaeological, anthropological or biological forensics.

Trueman (2004) considered the geochemistry of bone, examining chemical changes in time. When human bones are found during excavations or erosion, law enforcement agencies are informed, yet their actions will largely depend on the age of the material. Anthropologists can often give an estimation based on visual examination (bleaching, rootlet infiltration), but to provide an age in years requires lengthy and often problematic radiogenic (uranium or carbon decay) measurements. Trueman (2004) shows how the internment of bone causes rapid and definable geochemical changes that can assist in assessing bone age. Trueman's main thesis centres on the removal of bones, in his studies because of the sale of such artefacts, but with obvious implications for those wishing to conceal acts of murder and genocide. The 'passing off' of recent bones as archaeological remains is fraudulent and surprisingly lucrative, as well as being deeply upsetting to relatives and possibly causing problems of human health (bones and associated remains of the recently-diseased can carry the pathogen for many years, one of the worst problems being smallpox).

A further use of bone geochemistry may be to compare bones to a possible source location: although this is possible using major and trace element data, the carbon, oxygen, strontium and other isotopes found in such material have been shown to have great discriminatory power (Pye, 2004c, gives a summary of the main isotopes useful in forensic studies). The use of isotopes in this context comes from the inheritance all organisms derive in isotopic composition from their environment. Major influences include food, water, maternal milk and nutrients. The problem that isotopes can help solve is the accidental discovery of bones, teeth and sometimes hair, nails and skin. When conventional methods of identification are exhausted (missing persons lists, facial features, dental/fingerprint/medical records, DNA) for a region, a wider search of such databases is required.

The problem with migrating populations, rapid transport mechanisms and leaky borders, is that a focus to such searches is needed. The inheritance concept for isotopes means that the natural isotopic composition of waters around the world is well-known. The distribution is latitudinally-controlled, and thus it is not a case of simply measuring the isotope content of a tooth and plotting the origin on a map. Not only do broad swathes of the Earth's freshwater have similar chemistry, but the isotopic content of teeth and bone reflects many more environmental factors than simply local water. However, these complicating factors are sometimes overcome by chance, or by extensive analysis of trace elements (see above) in combination with many different isotopes, each of which will be the result of a set of environmental factors. In addition, the human isotope specialists can also use the natural chronology of bone deposits, teeth and hair growth, to define periods of isotopic influence. This may allow reconstruction of dwelling location through a rough sequence of stages, especially if the person has moved significant distances, or through varying isotopic regimes just prior to their demise.

The effect of local environment is well-known in archaeology (Evans & Tatham, 2004), and thus good proxy information exists for the interpretation of modern human populations, especially those with a strong indigenous diet.

Microfossils

The boundary between macrofossils, microfossils, micropalaeontology and microbiology in many forensic cases is blurred: each discipline has borrowed and developed techniques from the others in order to advance its science. We saw in Chapter 2 (and earlier in this chapter) how the content of ballast sand in Japanese Incendiary balloons allowed US Geological Survey military geologists to determine the origin of the sand and thus predict launch sites. Key aspects to their analysis were grain shape and mineralogy (providing a volcanic beach location) as well as the combination of macrofossils (molluscan shells), microfossils (diatoms, foraminifera) and absence of coral debris. Thus small, broken macrofossil debris was analysed in combination with genuine microfossils.

In biology and palaeontology, the definition of a fossil breaks down when dealing with recently-dead organisms: is it a fossil or not? For microfossils, we repeat our introductory discussion. A wind-blown spore, falling on a suspect's clothing and then recovered by the forensic palynologist, has not been buried, is still viable in the creation of life and is thus not a fossil (it is, of course, of massive forensic value, and frequently used alongside geoforensic techniques). A similar spore that became buried in late Cretaceous times (100 to 65 million years ago) and was recovered from a sedimentary rock is patently dead, possibly altered chemically and thus a fossil. Microbiological materials that are somewhere between these end-points are less easy to classify. A seed that blew in and fell to the floor of the Tomb of Tutankhamun, to be dried out and buried by dust, but that is still capable of propagating life, has all the attributes of a fossil. A broken piece of coral that falls to the seafloor and is buried, to be eroded out 10 years later, dead and possibly altered is a fossil but is younger than our Egyptian spore. Thus displacement, burial, preservation are all poor defining characteristics of fossil versus biological material, making discussion of forensic applications necessarily arbitrary.

Case study 1: Bluebell woods case

Kind (1982) describes this classic combined study of anthropology and diatoms (single-celled aquatic algae) centred on the discovery of scattered bones near Morpeth in Northumberland, which began in 1971 when a boy watching birds found a bone near a foxhole. Identified as a human left tibia by staff at the UK Forensic Science Laboratory, a subsequent police search of the immediate area found no more remains but more accidental discoveries occurred in the following months. A wider search recovered just over half of one skeleton and some fragmentary clothing, from which the deceased's sex, rough age and height were ascertained. These linked to an absent patient from a nearby mental hospital. Given the age and highly scattered nature of the remains, establishing a cause of death was problematic. Examination of the bone marrow allowed identification of the diatom genus *Navicula*. As described below (page 230), diatoms, caught in bone marrow are a diagnostic for drowning, especially if their type corresponds to a likely water body, suggesting that Bluebell Woods Man drowned in a river where his shoes and socks were found, was washed ashore and subject to extensive canine scavenging and dispersal of his remains.

Case study 2: Wilhelm Klaus and murder on the Danube

O'Connell (2006) uses an account of the work of Patricia Wiltshire to describe one of the first uses of fossil pollen. The story is akin to Agatha Christies' *Murder on the Nile*, with a passenger on a Danube cruise boat that periodically docked for sight-seeing suddenly vanishing. A suspect's cabin was searched and his boots noted as being muddy. Some of the mud was sent to a Dr Wilhelm Klaus (University of Vienna) who recognized extant pollen typical of a riverbank flora (alder and willow) as well as a fossil hickory pollen. Strata containing these grains were known to crop out in one location on the banks of the Danube some 20 kilometres north of Vienna. When informed of this unusual observation, the suspect confessed and led the authorities to where he had buried the body – exactly where Klaus had predicted.

Case study 3: Stockdale on the combined use of glass petrography and microbiology

In 1980 witnesses spotted a suspect arsonist at the scene of a burning primary school (Stockdale, 1982). He was apprehended and his clothing examined. Over 250 fragments of broken glass were recovered from his jeans alone. Also on his jeans was a fragment of paint that contained a blue-grey undercoat and dark blue gloss. The burnt-out scene of the school was examined and a broken fish-tank noted in the main hallway. Fire investigators observed that glass from the tank, underneath collapsed ceiling material, was not smoke-damaged, indicating breakage prior to the roof failure and most likely the fire. Among the glass were the squashed remains of the goldfish. Parts of the exterior of the building were painted with comparative paints to the fragment on the suspect. The refractive index of the fish-tank glass and that on the suspect also compared, but what of the fish? Examination of the suspect's footwear recovered similar glass as well as many fish scales, characterized using SEM. The problem now was that the debris had been cleared, no one thinking that some dead goldfish would be of any interest. Nonetheless, a search of possible comparative fish scales showed only three possible similar types (orfe, carp and goldfish) of which the latter compared best. The conjunctive use of glass optical properties, paint microstratigraphy and the microbiology of fish scales linked the suspect to the arson scene.

Case study 4: Szibor *et al.* on German(?) war graves

In a ground-breaking paper, Szibor *et al.* (1998) describe the use of 50+ year old pollen from the deceased of a mass grave. Their work began when excavations in the town of Magdeburg (Germany) revealed a burial pit containing 32 skeletons. In the absence of artefacts, documents or clothing that would establish provenance, the relatively recent state of the inhumations gave rise to two conflicting hypotheses as to their origin. One was that the bodies were of Soviet soldiers who had been involved in a revolt in the area in the summer of 1953; the other that the remains were of World War II age, possibly as a result of Gestapo murders, also known to have occurred in the area, but critically in springtime. Although the former hypothesis was deemed more likely by investigators, Szibor and colleagues conducted a clever means of obtaining more information about the victims. They injected pure water through the nasal passages of some of the skulls, releasing what were effectively fossil or subfossil pollen and spores that were typical of the kind of spores present in the area in the summer not spring. Thus 50-year-old material was compared to present-day flora and a link established. Wiltshire and Black (2006) comment that 'although scrutiny of their methods, and consequent conclusions, offer some contentious debate, it was obvious that a modified form of the technique had considerable potential for forensic investigations'. Wiltshire and Black (2006) do not continue with their critique and expand on the contentious debate: concentrating on illustrating the power of the method in the study of modern homicide victims (the realm of non-fossil material), demonstrating the blurring of biological and palaeontological examination in this instance. The combined, pioneering work of Szibor *et al.* (1998) and rigorous methodology of Wiltshire and Black (2006) demonstrates how this method will undoubtedly be used in the future for recent deaths, as well as when the taphonomy of historic and archaeological human skulls is under scrutiny.

Case study 5: Lewd behaviour on an English beach

A woman was sunbathing in a low valley incised into Palaeogene sandstone cliffs on the Dorset coast (UK) in late 2001. The beach below the cliffs is 'nourished' or made of quartz sand, dredged from the nearby English Channel (Figure 7.16). She was accosted by a male, who exhibited lewd behaviour and stood in a position whereby the woman could not leave. A passing fossil-hunter with a mobile phone caused the man to run along the base of the cliffs to the beach-exit steps. The police apprehended the man and seized his footwear. He claimed to have only been on the beach, never in the gullies/valleys in the cliffs where the crime occurred. 100 samples of the beach sand from beach and exit-steps and 100 samples of the sands found around and in the valleys of the cliffs were submitted for microscopic evaluation by simple light microscope. Broken molluscan shells were common in the quartz beach sand; the quartz cliff sand had few molluscan shells but abundant (5%) foraminifera (a calcareous protist). Furthermore, the molluscan debris of the beach

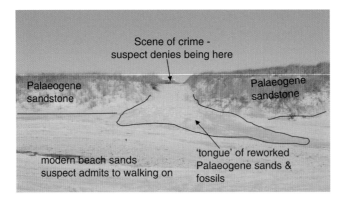

Figure 7.16 Annotated view of the scene described in case study 5. Field of view is 55 metres

sand was all of modern origin; the cliff sand ubiquitously of Palaeogene age. The suspect's footwear had abundant quartz sand, and of the biological material, less than 5 % was of modern, molluscan origin, the remaining 95 % being Palaeogene molluscan remains and foraminifera. He could not explain the presence of the cliff-sand material, which ultimately convicted him, along with footwear tread evidence found at the scene.

Diatom analysis

Living and recently-dead diatoms are really what would constitute 'bioforensic' evidence: they are included here because their analysis is sometimes carried out conjunctively with geological work and their fossil counterparts (in the form of diatomite) are sometimes associated with criminal activity. These single-celled aquatic algae have an intricate silica-wall to their tiny shells (tests) that makes them resistant to erosion and diagnostic in identification. Juggins and Cameron (1999) and Cameron (2004) summarize three known uses for diatoms. First, in the conventional compare/exclude analysis, especially where ponds, rivers, lakes and puddles are involved, other materials may be lacking, yet diatoms are abundant, producing better statistics. Dried water residues will contain diatoms (e.g., in footwear, socks and on clothes). Second, diatoms are a key diagnostic in establishing death by drowning. As diatom-bearing water enters the lungs, it may then pass to the bloodstream via the alveoli and ruptures therein. The diatoms, being siliceous, do not dissolve but become lodged in muscles and tissue, especially bone marrow. They are easily spotted because silica is not common so deep in the body. If a person is killed and then dumped in water, the diatoms do not enter and are not found. Last, diatomite, although a rare geological material, is a fire-retardant and thus used in the linings to safes and in buildings. Criminals involved in drilling or blowing-up of such inevitably get covered in the fine white dust. As Murray (2005) puts it, a safe-blowing suspect 'appeared to have a bad case of dandruff'.

Bizarrely, kieselguhr, or diatomaceous earth, is a key ingredient in the manufacture of dynamite (dynamite uses nitroglycerine and kieselguhr: the ingredients of kieselguhr, however, can be variable, gelignite uses nitroglycerine and a cellulose derivative, whose makeup can be better controlled but produces a gel-like substance, far harder to handle than a dynamite stick, hence kieselguhr remains popular in some countries), which historically was the favoured safe-blowing explosive, hence diatoms often met diatoms for a microsecond as the explosion took place! Kieselguhr is commonly used to remove dyes from materials, including illegally-sold petrol/gasoline and alcohol.

Diatom case study: Kenneth Young kidnap (Blank, 1971; Cleveland, 1973; Rogers, 2007)

Kenneth Young was the 10-year-old son of a wealthy financier when he was kidnapped from his home in Beverly Hills (California, 1967). The ransom was paid and the boy released, but not before his father noted the number-plate of the pick-up car, which was found abandoned some miles away. The car had been stolen for the job and fortuitously had been well-washed prior to the theft. The only evidence left in the vehicle was a white footprint on the back-seat. Analysis of the powder in the print revealed it to be almost entirely diatoms.

In the western USA there are two common sources of diatomite: (i) a recent/living flora from lakes and dried up lakes ('modern'); and (ii) a fossil flora from the Monterey Formation ('ancient'), an extensive range of marine deposits. Both sources are mined for diatomite – most frequently used in this area for filters in swimming pools. Analysis of the footprint diatoms by Californian Division of Mines and Geology geologists revealed both recent, non-marine and fossil marine flora, a combination rare and outcrop in nature. The supposition was that this was an unnatural mix. The question was – where would such a mix occur? Given the use of the material as a filter, the FBI visited plants that imported, stored and manufactured these items. An abandoned plant was identified; here both fossil marine and recent freshwater diatomite was imported. Sometime before the plant's closure a major rainstorm had washed the white powders all over the area, making a mix of the two sources.

Meantime, two suspects were apprehended during a bank robbery: one stated that a common place for them to hide or meet was an abandoned quarry. This led police to examine their suspect's footwear, and the imprint of one matched that in the kidnap car. Furthermore, a car belonging to the suspect had the same mix of diatoms on its exterior. Thus a series of links were established that resulted in the conviction of one Ronald Lee Miller.

Palynology

Recent pollen and spores are also part of the bioforensic world (Bock & Norris, 1997). Pollen and spores are the most frequently-used microbiological remains in forensic casework, largely because of their wide distribution and characteristic morphology. For an initial idea as to how pollen, spores and other microbiological remains may be used by investigators, Horrocks and Walsh (1998) provide an overview. They begin by outlining standard 'associative evidence' such as soil (which almost always contains pollen in sufficient numbers for such work) on a suspect's clothing being compared (or as others would say, excluded) from a scene of crime. The wide-ranging use of forensic palynology is also examined by Horrocks and Walsh, who show how the tracing of cannabis dealing, escape routes of a suspect (following a hit-and-run incident), and testing the veracity of conflicting stories following a sexual 'violation' may all be achieved using forensic palynology.

Spores, pollen and other dead or living microbiological materials generally fall under the remit of *bioforensics*: in the Introduction to this book, a plea is made for the practitioners of this science to write a review article or book in order to assist the 'geo' forensic workers. The closest work to such a review is Mildenhall *et al.* (2006), who have amassed some excellent papers on this subject. Two recent (2006) papers can justifiably be included in this text and are Case Study works reviewed below.

Pollen and spores are definite examples of that living–dead–sub-fossil–fossil problem introduced at the start to this section. Given the rarity of publications on true fossil palynology, this section could justifiably have been ignored altogether. The possible use of fossil and 'sub-fossil' material in forensic work is one reason for inclusion here. The second reason is that pollen and spores are by far the most common 'other proxy' to geological analyses when examining suspect, scene or victim samples. A casual reader of literature would not agree to such a statement – there being only a few mainstream publications where palynology and geology are used conjunctively. Nonetheless, unpublished casework in Europe, North America, Japan, Australia and New Zealand is currently combining these techniques, thus providing a second justification for minor inclusion here.

Many other organisms and their seed possess a tough exterior, allowing for extensive periods of dormancy and desiccation, blurring the line between when they can and cannot germinate or activate any longer and thus leave the possibility of living. Typical of this is the testate amoebae (Chardez, 1990), wherein these water-dwelling, soft organisms create a shell (test), and can be preserved. Testate amoebae show great potential in our allied discipline of bioforensics as an independent variable to geological analyses: they are not as common as wind-blown pollen, but are specific to microenvironments such as puddles. The spatial variability of testate amoebae versus pollen requires testing by experiment in order to establish whether the former need be championed further.

Forensic palynology uses pollen and plant/fungal spores in the same way the geologist uses mineral grains: the population statistics or unique particle concept both provide a method of comparison. Pollen is used by the plant in reproduction – millions of pollen carry the male zygote into air, onto insects (zoogamous pollen) or onto water in the chance that they will land on the female flower's anther and cause fertilization. In many cases of terrestrial soils, air-borne (anemophilous) pollen are commonly-detected. Pollen are small – too small to usually be seen by the naked eye. They are however produced in vast numbers – some individual plants producing 400 million pollen grains per year.

Case study 1: Use of forensic botany and geology in war crimes investigations in northeast Bosnia (Brown, 2006)

This paper exemplifies how the appropriate use of soil analysis contributed to a war-crimes investigation. The work was undertaken by the United Nations International Criminal Tribune for the former Yugoslavia (ICTY) during excavations from 1997 to 2002 that exhumed numerous bodies from seven mass graves in northern Bosnia. The aim of the work was to assist in the prosecution of war criminals at the International Court in The Hague. The killers and their associates exhumed the primary mass graves three months after the fall of Srebrenica and removed and re-buried the bodies elsewhere (secondary sites). The ICTY wished to link the soil found on the bodies back to the original site (often the murder site) or exclude the burial site, thus completing the story of what happened to the victims. This, for example, may allow the prosecution to show any suspect had access to the primary site more easily than the secondary. The aim of the soil/and sediment studies was to provide an environmental profile of the original site of the samples and use this to compare the relocated bodies to the original mass graves. Luckily for Brown and the ICTY, the primary sites were located in areas of contrasting geology, soil and vegetation, allowing good characterization of the sediment on the bodies and comparison to the original burial sites. The appropriate nature of the work is that optical microscopy; pollen and XRD (for mineral, especially clay mineral, identification) were used on a large number of samples, allowing a good statistical base for exclusion/comparison. The work is one of the first uses of geoforensics in an international investigation of genocide.

Case study 2: Role of forensic geoscience in wildlife crime detection by Morgan *et al.* (2006)

This work is important in that it demonstrates how not all geoforensic work is about soil on a murder/rape/robbery suspect's clothing or footwear. It is also a conjunctive approach, expounded elsewhere in this text as the soundest approach to geological materials-based investigations, depending on the case of course. Thus it neatly supports the combined 'geo' and 'bio' forensic approach advocated above. Morgan *et al.* (2006) examine three cases, two of badger 'baiting' (illegal digging and either capture for dog-fights or killing at the scene) and one of the illegal import of protected birds of prey.

In the badger-baiting cases, spades and boots (with abundant soil adhering) were seized from suspects and their vehicles following reports of digging at badger setts. Morgan *et al.* used colour, pH, chemical (AAS and Dionex), binocular and SEM microscopy and particle size analysis for their conjunctive geological approach (based presumably on operator experience, equipment availability and reliability) compared to statistical counts of pollen species.

In the bird of prey case study, Morgan *et al.* are paraphrased. A man was detained in southern England in possession of two young falcons. A subsequent search of his house revealed a bag containing climbing equipment, including a 'new' looking rope that had reddish 'dirt' adhering to one 19.9 metre end. The remainder of the rope was clean and appeared brand new, having intact paper labels along the length. The dirty rope was washed to retrieve any (probably as much as possible) soil. This was compared to four control samples taken from locations on the island of Mallorca, where this type of falcon occurs and where the suspect had been. Morgan *et al.* (2006) again employed binocular and scanning microscopy, chemical analysis (AAS and Dionex) and pollen analysis, the latter confirming a Mediterranean origin. Three of the Mallorcan samples were excluded by this combined analysis, with the third remaining comparable in all other regards, implying either that the rope had been in contact with this location or another unknown place with comparable geological makeup.

Case study 3: Manslaughter associated with attempted robbery of horses, Ireland

This unpublished work effectively mimics the above analysis of rope by *Morgan et al.* (2006), here applied to electrical cabling (extension lead, or flex) found at the scene of a suspicious death. Horses bred for racing and breeding have very high values (into millions of UK pounds) and are traditionally well-protected. Eventing horses can be of equal value, and yet in this case were known to an illegal horse-trader to be kept in less secure stables. The trader and assistant assumed the stable-owner to be away and broke into the yard with a horse-lorry with the intention of stealing one of more valuable animals. The stable-owner had in fact retired to bed early (dousing his house-lights) to prepare for a horse show the following day: he thus arrived at the stables as the theft was in progress and challenged the would-be thieves. The accomplice fled, but a struggle between the remaining two ensued, in which the trespasser attempted to restrain the stable-owner with an electricity extension lead from his horse-lorry (Figure 7.17). Thinking the owner was successfully restrained, the would-be thief drove away. The stable-owner thought he had freed himself, tripped and fell, impacting his head on the concrete with such force as to cause a fatal brain haemorrhage: he was only found seven hours later and had by then been dead for some time.

The extremely dirty nature of the scene, with mixed animal hair, blood, mud, animal food and oil made straightforward comparison challenging until it was ascertained that the electrical cable did not belong to the stable-owner and must have (a) arrived newly at the scene and (b) been associated with the incident. The cable revealed inconclusive fingerprints and DNA. A suspect's home was searched and impressions of missing rope/cable/wire in an equipment store were sampled (some 20 samples were taken by this means) with regard to depth and position. Combining binocular, petrographic, X-ray diffraction and ICP-MS (chemical) analysis, with sequentially-sampled portions of the seized cable for pollen analysis, portions of the cable showed only minor differences in makeup. Samples that compared less well were re-examined with regard to location and content, and in all cases the underlying reasons found for the minor variation, including addition of soil at scene and patches of spilt peat in the store. Such mixed materials are ideal for the application of variography (see Chapter 6), which may enable the different materials to be differentiated.

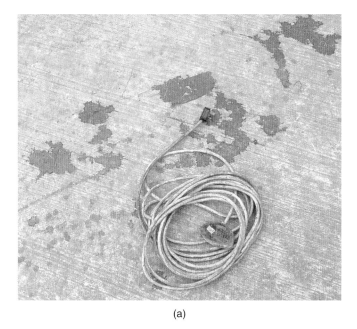

(a)

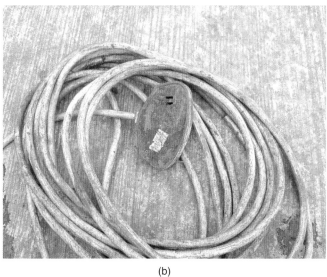

(b)

Figure 7.17 (a) View of the blood-spattered scene and electrical cable (~1 metre across): coiling from where the cable was wound around the deceased's legs still visible. (b) Close-up of the electrical cable plug, including the soil evidence

Microfossil microstratigraphy – Baden and Hennessee (1991) on the Greenwood Lake body

In April 1976, the headless, handless body of a woman was found floating in Greenwood Lake (between New York and New Jersey). The age of the woman was initially wrongly ascertained, on account of her young-looking physical appearance. The time of residence in the lake was also incorrectly assumed, the body being in a good state of preservation. Baden and Hennessee (1991) sought the analysis of biologists at the Museum of Natural History, who found two layers of algae. Furthermore, the life cycles of algae being well-known, the biologists determined that the older, lower layer (see the Law of Superposition in Chapter 2) was of a previous season's growth. Thus the assumption that the body had been in the water for only a few months was challenged, with at least two spring-seasons' worth of algal growth being present. Re-examination of the water temperatures showed that persistent coolness could indeed have re-cooled the body, providing an unusual state of good preservation. The body was presumed to be that of Katherine Howard (of New York) and her (then late) husband was suspected of murder: lack of evidence or a confession meant he was never tried and went to the grave with whatever secrets he had.

Back to the extraterrestrial: McCall on seeds in a meteorite

McCall (2004, see: *http://www.geolsoc.org.uk/template.cfm?name=HOGG09548 95486*) recounts the case of the 1864 Orgueil chondrite meteor shower (southern France), from which many samples were collected and distributed to museums around Europe. Two specimens remained (sealed in a glass jar) in Montauban. McCall recounts how some 100 years later a plant stem and seeds were noted in the surface of the specimen. The associated presence of plant gum and coal fragments proved that the seeds were not from contact between the meteorite and the ground, but a deliberate attempt to show evidence of life in extraterrestrial material. In a bizarre twist, other Montauban chondrite samples do indeed contain original, indigenous organic compounds, although whether these constitute proof of extraterrestrial life is another matter.

The next stage in multi-proxy methods: Back to prediction (Rawlins *et al.*, 2006)

In a review of the work of forensic pioneer Oscar Heinrich (see Chapter 8), we shall see how Heinrich predicted the location of a body deposition from examination of suspect-related material and comparison to a geological map of sediment and soil distributions. The prediction of a crime scene location from samples alone remains a golden chalice in the forensic geosciences and is dependent on a sufficient spatial and depth database of controls for comparison. The possible number of controls is so great that the successful comparison is one of probability. Exclusion also figures in the prediction sense, in that from the analysis of suspect or victim samples areas can be stated as unlikely points of transfer. Rawlins *et al.* (2006) use the works of Croft and Pye (2004b), Brown *et al.* (2002) and Bull *et al.* (2005) (see above) in a test of the predictive power of such techniques as scanning electron microscopy (SEM), clay mineralogy (by X-ray diffraction), palynology and the molecular structure of organic matter. They thus follow the published wisdom of using comparative geological and biological evidence from soil.

Earth-related particles found on suspects clothing, footwear, vehicles, in dwellings or on victims are most commonly used as comparative material with a scene of crime or other location (Murray, 2004; Murray & Tedrow, 1975). This comparative method has its difficulties but with multi-proxy analyses and robust statistics can be unequivocal in its outcome. Far more challenging is where earth-related particles have been recovered but the scene is not yet identified. The Earth scientist is presented with a sample of soil, a mud-spatter on clothing, a coating of dust or a fragment of vegetation and asked 'where did this come from'. The natural variability of some environments (e.g., mature woodland on sloping topography) and the homogenous nature of others (e.g., large-scale arable fields) make this question extremely challenging to answer. Without extensive (up to date) databases of the likely proxy comparators (soil mineralogy, vegetation, micro-organisms) Earth scientists are forced to use a combination of analyses to refine a predicted location.

Rawlins *et al.* (2006) attempted to show the potentials and pitfalls involved in this process, the most difficult challenge for the analytical Earth scientist: the prediction of provenance from something like a soil sample. Other published works have used some of the methods, but more commonly in the traditional, comparative type study: few attempt to do what Rawlins *et al.* did and test the predictive powers of a multi-proxy approach. Their key findings were that SEM was a highly sensitive and excellent discriminatory tool that failed in only one regard, and that was the comparison of particles from an industrial plant.

Similar uses and problems face the use of pollen, which is highly sensitive, can be widespread but so common that these problems are negated. In linking sediment to *geological* source, Rawlins *et al.* chose the XRD analysis of clay minerals as the best discriminant. Their work demonstrates that in using geological evidence in assisting the search for scenes as yet unknown, establishing the makeup of the sample is the most critical stage. This allows a key research question to be posed: 'what are you looking for?'. If the sample is dominated by mineral material, with limited organic (soil-related?) content, geology and thus the clay mineral makeup may be a useful 'lead'. If the sample is dominated by urban debris (foodstuffs, particulates, ash and burnt residues, tarmac), perhaps the SEM would prove useful.

7.5 A paradigm shift in geoforensics?

The historical legacy of Gross, Popp, Locard and Heinrich has resulted in many methods of soil and sediment analysis being tested and used in forensic comparison/exclusion and prediction, although as Croft and Pye (2004) state, there are many more methods yet to be tried. Gross, Popp and Heinrich also used biological materials, when available, but due to limitations of microscopy and preparation, the use of microbiology has been more limited to spores and pollen, diatoms and fungal/algal remains. A wealth of other microbiological material exists in soil, and soil DNA, organic biomarkers are now being used, increasing the range of forensic 'fossil' bioscience and forensic geoscience techniques. The day is not far away when even trace amounts (see next chapter) of soil and sediment can be analysed rapidly by a large number of 'geo' and 'bio' methods, compared using computationally-advanced methods and a changing robustness (for different materials) for each material created. Human judgment cannot and will not go away, it will only be strengthened.

This massive advance in the appropriate use of geological and palaeontological/biological materials may form only the first half of the new geoforensics, a holistic discipline that will develop in the future and is only introduced here. Recall that the 'geo' of our Earth comprises the geosphere (soil, rock sediment, the subject of this text) common to geoforensics, and biosphere (macro- and micro-biological materials) common to bioforensics. It also comprises the hydrosphere and atmosphere: already the macro-scale understanding of each (hydrodynamics and the washing up of bodies/contraband/shipwrecks, atmosphere and the climate/weather-crime comparison) has been used in geoforensics. The micro-scale analysis of water chemistry is likewise upon us in the location of buried remains and tracing of pollutants. For the hydrosphere then it is only the advancement of technology and problems of dilution/contamination that limit the micro-analysis of water and water residues in the sample comparison/exclusion approach.

For the cynics, an example is provided. Already we have seen how the results of geological analyses may not be truly independent, being based on essentially the same properties (mineralogy, chemistry) just measured by different means. The example of how particle size influences chemistry remains. Take then pure quartz sands, texturally similar yet of different ages in origin: minerals such as quartz contain fluid inclusions, trapped during mineral growth. Trace evidence (limited grain numbers) of apparently comparable quartz particles may be differentiated on the hydrochemistry of fluid inclusions, a use of the hydrosphere yet to be fully realized (Sugita & Marumo, 2004). Many other hydroforensics comparisons/exclusions have yet to be found, either alone (rain-, snow-soaked suspects and victims) or conjunctively (wet sediment, analysed for mineralogy/geochemistry as well as water content).

More outrageous to us at the present day would be the micro-use of atmospheric components to evidence: 'air is far too mobile to be of use'. Minerals not only contain fluid inclusions, they may also have gas or air inclusions, so enabling the same conjunctive or exclusionary analysis as above. The air or gas only has to be trapped, or leave its contact for us to detect it, analyse and discriminate it: the challenge is for us to use Locard's Principle in ways commensurate with our technology. Maybe the traditional geoforensic practitioners are by now a little convinced that the hydrosphere and atmosphere offer (not exclusively) other materials for use in our task.

The four spheres of the Earth are of course all encapsulated by one further 'sphere', the extra-terrestrial. Many readers will think the authors have 'lost the plot' at this point: however, the contact between extraterrestrial debris and spacecraft surfaces has already been established as a trace evidence – contact forensic problem.

Likewise, low- and high-atmosphere particles (aircraft jet engine glass spheres, volcanic particles [tephra]) have also been used in connecting trace material on suspects with certain locations. The volume of physical matter and energy falling on the Earth is significant, and variable over the surface of our planet: again the challenge is for Earth scientists to detect, characterize, compare and use these materials, rather than be closed-minded and claim that 'our method is best'.

Perhaps some of this speculation will be wrong and the tried and tested methods will be proven: it would be foolish not to issue the challenge and try, however, and even if partially correct, a major shift in the use of Earth materials in forensics will have occurred. Another way of looking at how the geo- of this text may integrate with the bio-, hydro- and atmo- forensics is to consider the old game of 'animal, vegetable, mineral'. The latter two we are familiar with (wrongly including fungi, algae), yet we have only touched on the use of microscopic animals and protozoa.

These predictions are but conjecture, yet they do reflect an unassailable fact: if the sophistication of the career criminal, terrorist or industrial criminal is such that we are obliged to use a large number of methods to establish innocence or guilt, the increased use of trace forensic evidence in the international courts, especially of Human Rights, will require even more analyses, of features of Earth-related evidence not yet considered, under more exacting circumstances than we 'enjoy' at the present day. Why? Because of fortune and fame: if the domestic criminal can afford the 'best' lawyers of a country in their defence, the leaders of countries will demand or can afford the 'best' available in the world. This will be reflected in how the prosecution proceed: both defence and prosecution will be looking for new analytical methods that provide supportive/confirmatory or contradictory/questioning evidence.

The dangers of continuing this cycle of competitive analysis are found throughout this book: multi-proxy analyses of the same materials do provide confirmatory evidence, but are based on re-analysing the same substance. Over-emphasis on trace geological evidence, or on highly-sophisticated methods, may be used inappropriately by prosecution or defence when other standard, bulk methods show a clear exclusion/comparison. Is this likely? The collection of papers in Robertson (2007) demonstrate how concerned many observers of the International Criminal Court are as to how impartial the process will be, and thus how fair any trials may be. Of key interest in this area are the papers by Wald (2006) and Van Den Wyngaert (2006) who examine the extent to which different national and international courts have been able to ensure that accused persons obtain a fair trial. Of importance to this book are these authors' discussions of the differences in the rules of evidence between such courts. They both argue that the varying use of the same evidence, across international borders, must be minimized in order to bring some standard to these, possibly the most important of all trials.[2] The use of geoforensics in International Criminal investigations is now established (Brown, 2006; Hunter *et al.*, 2005), making the need for cross-border discussion, exchange of methods and ideas,

---

[2]Some may take offence at this comment, countering that 'a life is a life, so every death requires as thorough investigation as is possible'. If this is your (the readers') attitude, we suggest reading Rees (2005), especially page 308: 'Yet he murdered in cold blood, standing feet away from his victims. Today, when the mass killers we read about tend to be the crazed murderers featured in the tabloid press, it is important to meet a man like Petras Zelionka who killed more than any tabloid monster and yet sat before us as composed and normal as any grandfather.' (Rees, 2005).

and eventually a shared 'best practice' of high priority. Thankfully, the international forensic science meetings that take place regularly around the world are accepting and even welcoming of Earth Science participants; likewise, established groups and learned societies such as the International Workshop on Criminal and Environmental Soil Forensics, the Geological Society of America and the Geological Society of London have also supported those working in the geoforensic arena. These meetings naturally feed into publications in academic journals, subject to peer-review: these have the ability to transcend the limitations of national courts and become standards across frontiers.

8
Trace evidence

> 'You must not fear,' said he soothingly, bending forward and patting her forearm. 'We shall soon set matters right, I have no doubt. You have come by train this morning I see.' 'You know me then?' 'No, but I observe the second half of a return ticket in the palm of your left glove. You must have started early, and yet you had a good drive in a dog-cart, along heavy roads, before you reached the station.' The lady gave a violent start and stared in bewilderment at my companion. 'There is no mystery, my dear madam,' said he, smiling. 'The left arm of your jacket is spattered with mud in no less than seven places. The marks are perfectly fresh. There is no vehicle save a dog-cart that throws up mud in that way, and then only when you sit on the left-hand side of the driver.' 'Whatever your reasons may be, you are perfectly correct,' said she.
>
> 'The Adventure of the Speckled Band' by Sir Arthur Conan Doyle.
> This version: Conan-Doyle (1988)

Many textbooks, popular works, Chapter 7 (this work) and web sites cite the 'Sign of the Four' by Sir Arthur Conan-Doyle as one of the first published works where a person is linked to a location by geological evidence. It is of note that the modern-day forensic geologist is probably faced with more situations like that from 'The Adventures of the Speckled Band' (above) than the 'here's a bag of mud, where did it come from' situation familiar to many practitioners. Why? As we shall see below, perpetrator caution and clean-up are likely major factors, caused in part by communication (inside and outside of jail, through the Internet) and by popular books and TV shows. Some may criticize this text, or indeed Murray (2004) or Martin (2007) for adding to this problem.

8.1 What is geological trace evidence?

In the illegal extraction of precious minerals or aggregate, a kilogram of removed material is considered trace; to the operator of a scanning electron microscope (SEM), the coating on a sand grain would be considered trace. The difference lies in the context of the scale of the investigation. In what many perceive as forensic geology trace would be considered as suspected of being present but either not visible to the naked eye, or barely visible. *In this sense at least, geological trace evidence is of too small a quantity to carry out normal (appropriate) destructive, statistically-meaningful analysis.* Some idea of what we consider geological trace evidence is given in Figure 8.1.

Up to the publication of Lombardi (1999), published work on the geological analysis of very small (sub-gram) quantities of material were limited, with reliable results being gained from visual observation of large samples (Murray & Tedrow, 1975) or by multiple methods (Pye & Croft, 2004). Nonetheless, work had occurred on very small samples, such as the widely-cited case (by Edmund Locard in 1912 Lyons, France) of Emile Gourbin, who denied being in contact with his strangled girlfriend after she had purchased a particular face-powder, chemical traces of which were found under his fingernails. The Gourbin case relied on the unique nature of the face powder, begging the question, what statistically-reliable analyses could be conducted on trace material with more subtle variation (soil, rock dust)? Some of the possible answers to this question are contained in this chapter.

As an addendum, many forensic geologists and chemists know of Locard's work in the Gourbin case, largely because of the sophisticated analyses he performed for the investigation. However, another case of Locard's demonstrates his mastery of the early retrieval and comparison of trace evidence. While working as an assistant for the police in Lyons, mainly making physical measurements of convicts, Locard asked to be given the clothing of three suspect coin counterfeiters. The three had avoided all links to the fake coins, and concealed any possible forger's plant. In the pockets of one of the suspects, Locard located metal shards that when analysed showed no similar chemical composition to coins of the French mint, but could be compared to the metal mix of the fake coins. This led the police to search and arrest all the suspects, as well as trust the power of Locard's detective abilities. Locard was therefore the true pioneer of using trace evidence in criminalistics.

Given the nature of such small amounts of material, it is hardly surprising that no single scientific discipline dominates the field: the makeup of trace material remains unknown until analysed. Choosing whether to deploy chemistry, mineralogy or microbiology is thus difficult: first impressions of what dusty coatings, particles or stains may be are invariably wrong: an example is given below.

A white residue found on an arrested and known drug-dealer was extensively analysed by destructive means to ascertain the nature of the drug and cutting medium to the imported 'pure' drug: the stain turned out to be semen and dried lubricant, only chance amounts of non-destroyed material allowed correct analysis (see Figure 8.1(e)).

As we shall see below, if suspects cannot see trace evidence adhering to their belongings, there is less chance that they will employ a comprehensive clean-up. Thus non-destructive methods of screening become critical, informing which destructive methods may follow. Rarely is trace evidence used in the classic manner. Murray (2004, p. 116) writes: 'It is better practice for the evidence collector to examine a questioned sample first, for colour and particle size at least, and then to search for [scene and alibi] samples with a similar appearance.' Trace geological evidence may

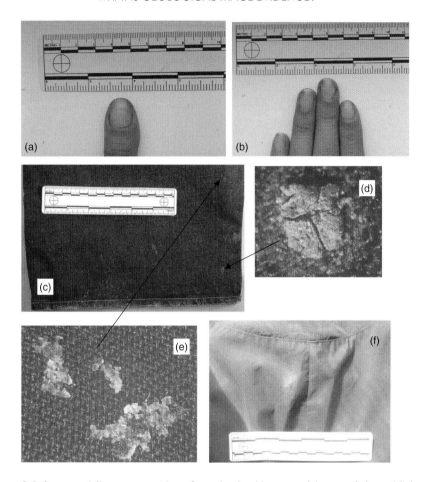

Figure 8.1 Some arbitrary examples of geological trace evidence. (a) and (b) = dirt under the fingernails of a suspect involved in the theft of horses: the classic example, as in Locard's work on the Emile Gourbin case. (c) Base of a pair of trousers, seized from a suspect in a murder. Two types of stain were identified: (d) mud-specks and (e) dried organic 'slime', unrelated to the geology of alibi or suspect locations and highlighted for biological analysis. In this case the limits of geoforensics are apparent: if it is material we do not or cannot analyse, leave this for someone else. (f) The limits of trace evidence – dust on the jacket of a rape suspect, where the scene was a dusty quarry location (see 'Agricultural Limestone' case study). Obtaining sufficient sample quantity and carrying out any corroborative (exclusionary or comparative) analyses when such small amounts of material are present is a major challenge

only be characterized by light microscopy, chemical or mineral analysis, or worst of all (for comparison) by SEM (Figure 8.2), none of which can easily be translated into simple field examination for comparable materials (Houck, 2001).

One of two courses lay open to the investigator: analyse the trace material as thoroughly as possible, so that field collection by eye and hand-lens may be informed

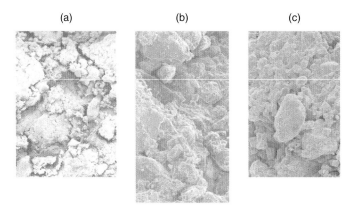

Figure 8.2 Examples of the kind of images seen under the scanning electron microscope (SEM). (a) = processed lime (calcium oxide) found at a scene of a shooting and requiring comparison to a suspect's shoes. (b) = organic matter-covered quartz grains found on a rape suspect's trousers. (c) = organic matter-covered quartz grains from the scene of the rape. Petrography, grain size analysis and palynology all showed too few grains for comparison between suspect and scene: SEM established that each has similar ingredients yet is different texturally. Field of view ∼1 millimetre

sufficiently to limit sample number; or collect sufficient scene/alibi samples, in geographically-informed locations (see Chapter 2: loose ground, points of exit/entry, located footprints/tyre-tread marks) to ensure statistical and logical representation. Morgan *et al*. (2006, p. 153) touch upon this aspect:

> There is a further difference between geological and forensic analyses which is necessitated by the relative quantities of material typically available. Geological analyses often involve quite large amounts of material whereas forensic analysis is often undertaken, particularly from anthropogenic sources, on only trace amounts of soil or sediment.

Thus the whole gambit of procedures and techniques highlighted by Palenik, as presented by Murray and Tedrow (1992) and Murray (2004: his page 195, see Figure 7.2 of this text), or the alternative schemes of Sugita and Marumo (also in Murray, 2004: his page 196, see Figure 7.3 of this text) cannot be fulfilled when such small samples are obtained. The question raised at this point is core to this chapter – what does the geoforensic practitioner do with such tiny amounts of evidence?

Two case studies from the work of Oscar Heinrich on quartz grains

Following Locard's 1912 work, the next significant examples of the use of trace evidence come in the 1920s in the United States. The first was the abduction, ransom and murder of Father Patrick Heslin, who was tricked away from his California home one summer night in 1921 (Block, 1958; Murray, 2004). Heinrich was initially employed to examine the ransom-note handwriting (his speciality), from which he amazingly predicted the writer to be a baker, on account of the concave uprights on the letters A and H and down-curves on the letter T – as used in the writing on cakes. When a baker by trade, named Hightower, approached the police with information on the beach burial location of Heslin, his reward was denied and instead he became the main suspect. Little could be found to connect him directly to the scene he took police to: coincident tarpaulins, wood and gun cartridges (found at his home and at the scene) proved non-unique. The police examined Hightower's pocket knife and pronounced it 'clean as a whistle', only for Heinrich to find shards of a white rope also found at the scene and most importantly sand grains. Block (1958) takes up what the police saw next:

> He [Heinrich] opened his portfolio and took from it a handful of photographs that looked like good-sized rocks. The police eyed them curiously. 'See these,' Heinrich went on. 'These are not stones, as you might suppose. They are grains of sand, enlargements of photo-microscopic negatives. You see, when I put the hilt of Hightower's knife under the microscope, I found a very small patch of sand within the handle at the hilt – only a few grains, three or four.'

Heinrich compared the knife grains to those collected from the wood and tarpaulins from the scene and decided on a common source. Heinrich had effectively done in 1921 what we shall see the present-day analysts of quartz-grain surface textures are doing today using the SEM, and using hundreds of grains.

Heinrich's second trace evidence case began in August 1925 with a gruesome discovery by a father and son while gathering tule stalks for decorations in a muddy marsh adjacent to San Francisco Bay. They chanced upon a human ear attached to some scalp, lying in the marshes. The police soon arrived, noted the attached earring, and searched the remaining area: all they found was a July 3rd newspaper (creased) and one other fragment of scalp on a piece of cardboard.

Opinions varied as to the origin: misappropriated remains from a medical school, someone being hit by the express train line that ran only 50 metres away, or homicide. The latter seemed likely when an extended search recovered a hatchet with blood and auburn hairs attached at a nearby ramshackle cottage. Enquiries established that a Mrs J.J. Loren was missing: relatives recounted her auburn hair. Another extensive search of the marshes recovered a few more human remains (more scalp and a jawbone, the latter not showing definitive comparison to Mrs Loren's dentist's records). Mrs Loren had led a strange life prior to her disappearance, changing her name from d'Asquith and receiving many visitors at the last hotel she stayed in. Recovery of clothes matching descriptions of hers added to the belief she was the victim of a murder: but where was she?

Heinrich was sent for, and concentrated his efforts on the dismembered ear, which he established had sand grains with a few salt crystals adhering, from which Heinrich concluded that the dismemberment location was likely a small estuary, because there was insufficient salt for a maritime location. He examined maps of nearby locations and chose one particular spot, some 20 kilometres away, where authorities searching on his instructions found the rest of the body. Heinrich used a United States Geological Survey map to limit his search, thus applying in 1925 the goal of many database-gathering projects running at the present time in the United States, United Kingdom and parts of Europe: prediction of scene location.

Interestingly, the body of Mrs Loren had been doused with lime, which Heinrich concluded was to give the remains 'age' and thus throw the authorities off the scent. 50 years after the murder, the perpetrator was still not caught: the crumpled newspaper found in the marshes contained hairs that Heinrich suggested may have come from the killer. One wonders what a 21st century investigation into the murder would add? From a geoforensic point of view, Heinrich carried out the toughest of assignments, predicting a general and then specific location of material (in this case a victim) from trace evidence alone.

In this context, we also refer the reader to Smale and Truedale (1969), who describe a murder in the Australian outback where the heavy mineral population of sandy soils from suspect and scene were compared using SEM with an attached microprobe. The study of heavy minerals can be exceptionally diagnostic, especially when used in conjunction with another method: in sands, quartz grain surface textures, thin section petrography, and QemScan are nowadays primary tools, with heavy minerals being examined thereafter. Some sands and silts have exceptionally high heavy mineral proportions ('black sands'), which are often economically valuable and thus subject to mining fraud crimes. Mining companies have developed specific methods of analysis for such materials, as part of their assaying process: these are often the best tools for comparative analysis.

Two reasons have emerged that force the geoforensic specialist to consider trace geological analysis: *perpetrator clean-up* (and avoidance) and the *urban environment*. Perpetrator clean-up occurs partly as a result of texts such as this one – where some of the methods used in evidence are outlined for the criminal to be aware of and possibly counter and partly because of more popular renditions of forensic science in the media (films and TV), which although simplified, nonetheless give an indication of what may be *avoided* or *hidden* by a criminal. Hiding evidence is a risky strategy for the criminal. If they have some geological training, can avoid other circumstantial facts (irregular activity, excessive washing, CCTV or witness information) or are just lucky, the intentional compromise of geological evidence can be successful. One the earliest known examples of such intentional compromise is recounted by Haneberg (2004):

> Knowledge of the principles of forensic geology can be used to obscure evidence or mislead investigators. Double agent Kim Philby (British, 1912–1988), who spied for the Soviet Union while at the same time working in the British intelligence service during the Cold War years, once used a small trowel to bury a camera in a wooded area near the Potomac River in Virginia. He then returned to his home and used the trowel to dig in his garden in order to obscure any soil particles that might be used to identify the location of the camera. This incident would never have been known if Philby had not described it in his autobiography. Terrorists arrested in conjunction with the Aldo Moro [Lombardi, 1999] case insisted that forensic evidence had been planted in order to steer authorities away from the true location of their activities, which might have led to the arrest of additional suspects. It appears, though, that the forensic evidence was authentic and reliable.

The rather negative reasons of avoidance by perpetrator clean-up and intentional compromise are countered by the increasing power of trace geological analysis, which allows the geologist and microbiologist/palaeontologist to operate in an *urban environment*. Traditionally, geoscience techniques of analysing trace evidence such as soil, sand and rock dust have been applied to criminalistics in environments considered 'natural' or rural. Trace geological techniques are also applicable to the urban environment, which is seen as 'cleaner' and 'harder', where abundant soft soil, sediment and rock debris for the transfer of trace evidence is limited. Traditional geoscientific methods of analysis require significant amounts of sample, found most commonly in the non-urban environment. Recent advances in analysis using non-destructive methods or very small samples have allowed the application of geoscience to the urban environment. Anyone who has seen building works excavations within cities knows that the urban environment has soil, sediment, weathered rock and stratigraphy. Likewise, the visitor to a city cannot help but notice how dirty their clothes become from pollution. In summary, the urban environment is far from 'clean' of transferable materials: the critical aspect is what these materials comprise and how we analyse them.

The preponderance of serious crime in the urban environment (gun crime in many cities, alcohol or drug-related crime) proves this point. The application of the established principles of forensic geoscience to 'mini-rural' scenes such as parks is obvious. However, serious urban crime occurs in locations other than parks or waste ground: for instance, petrol stations are common locations of armed robberies. At first it may seem that this location is nothing like a park or garden and thus the forensic geoscientist has nothing to offer: this is far from correct – the amount of transferable material at the garage, from spilt oil, water, car-washing products and foodstuffs, to the material within the garage, again foodstuffs, building materials, etc., is enough to make our microscopic examination of suspects and scenes very worthwhile. This is compounded by small or limited sample amounts also being a function of investigative procedures: should the incautious investigator choose an initial comparison/exclusion method that destroys or compromises part or all of a sample, so that none or too little remains for the type of conjunctive approach advocated in both this and other written works?

This issue is touched upon by Bull *et al.* (2005) in their published response to Cengiz *et al.*'s (2004) work on the use of SEM-EDS (energy dispersive system) analysis in soil comparison. The latter advocate their method because the standard hydration (homogenization), drying, sieving and 9 tonnes/cm^2 of pressure applied to each sample prior to automated calculation of chemical composition allows the comparison of 'like-with-like'. The result is improved reproducibility of results. Bull *et al.* (2005) accept the methods and results of Cengiz *et al.*, but voice concern at the loss of potential information in the preparation process. They imply that subtle (possibly exclusionary) differences in homogenized and standardized soils may not be recognized by the SEM-EDS process, and that furthermore the preparation method effectively precludes further analysis and the use of a conjunctive approach. The implications are obvious: the exclusionary principle may not pertain to the sample, but the possibility that a suspect is compared positively to a scene by so reducing the subtle differences in soil by extreme pressure, when in fact the soil ingredients (same bedrock, weathering, transport) were the same, but the locations different, could be enough to implicate the innocent in a crime. This, plus the fact that the preparation method severely limits other work that could test the veracity of the SEM-EDS results, could cast severe doubt in the eyes of the judge and jury.

Case study: Lombardi's (1999) work on the Red Brigade (Brigado Rosso) murder of Prime Minister Aldo Moro (1978)

The *Encyclopedia of Espionage, Intelligence, and Security* (2004) gives a very succinct account of this landmark case:

> Grains of sand and microfossils found on the body of Italian Prime Minister Aldo Moro, who was kidnapped and murdered by Red Brigade terrorists in 1978, led investigators to conclude that he had been held at least part of the time along an 11 km long stretch of beach north of Rome. The total mass of sand collected from Moro's clothing and the car in which his body was discovered was approximately 1 gram. The presence of bitumen (a tar-like substance in this case derived from oil spills dispersed by waves) and resins used in boat building further supported the beach hypothesis. Because of the high profile and political sensitivity of the case, collection of sand samples for comparison with the grains found on Moro's body occurred in secret. The geologist working on the case was accompanied by his wife, who posed as a tourist picking plants and observing the scenery while her husband surreptitiously collected sand samples.

Case study in vehicle clean-up: 'Murder and Mud in the Shenandoah' by Sever (2005)

On 26th September, 2002, a man was found shot dead along the Shenandoah River in Front Royal, Virginia. Sever describes how Erich Junger collaborated with Skip and Chris Palenik in conducting extensive soil analysis, following sampling the day after the incident. Officers carrying out the sampling were concerned at the repeated and rapid flooding of the river at this location (a low-gradient, sloping gravel and mud bank beside a concrete bridge). They concentrated their sampling on tyre-marks while the police aimed to establish a motive for the murder: links to known drug-dealers were made, but all on hearsay. The main suspect was one Lewis William Felts, who was spotted washing his red jeep seven days after the murder by a police surveillance team. Extensive searching of the vehicle failed to show any blood but did indicate chemicals consistent with shotgun cartridge discharge and mud inside the rear doorway. Much of the other mud had been removed, leaving only this trace area and some further materials on the outside of the vehicle. Unusually, the investigators only found one layer of mud: often vehicle sills, suspension and foot-wells will display many materials, sometimes layered, indicative of the environmental history of contacts between vehicle and shoes/roads, etc. The mud in this case was interpreted as the result of where mud-laden water had been flushed into the doorwell by the vehicle moving at speed through a puddle. The investigators found the minerals malachite and azurite (also found in the river sediment, washed from nearby mine workings) in both the vehicle and scene muds. As Sever (2005) states:

> After further examination, Junger felt he had several facts down: This red Jeep, which ordinarily was housed in a very clean condition in the city, far from any soil formations, clearly visited the crime scene around the time of the crime. He could not place Felts, the suspect, at the scene, but Felts' Jeep was there, he says, beyond a reasonable doubt. His work was done. He filed the paperwork and reported his findings.... Three months later, on the advice of his court-appointed defence attorney, Felts pleaded guilty to all charges against him ... basically, Felts admitted that the state had enough evidence against him that they would likely convict, but he did not admit fault.... By pleading guilty, he avoided the death penalty, and the judge sentenced him to serve life in prison.

8.2 Scanning Electron Microscope (SEM)

Following development through the 1930s, the first commercial SEM was made in 1965, since when the vast increase in uses of the machine, as well as the extra facilities the basic design now carries, is testament to the power of the device. Although the SEM can image objects at magnifications similar to standard light microscopes ($\times 15$ to $\times 30$), it can also image objects at over $\times 200,000$, allowing photography of surfaces and textures far beyond other forms of microscope (Figure 8.2). Grain textures and the material between grains in sedimentary rocks are of particular interest. The most beneficial aspect of the basic design of the SEM is the collection of back-scattered electron images, which provide the operator with information he or she can use to elucidate the chemistry/mineralogy of the subject material. Unlike many other geological analytical techniques, the SEM is well-suited to combined analysis of both inorganic (minerals) and organic phases, making it ideal for studying soil.

The SEM works by firing a focused beam of accelerated electrons in a scanning mode. Reflected or secondarily-generated electrons cause a scintillator to pulse, creating light and thus with millions of backscattered points, an image of a specimen. Eventually the microscope form or texture of a specimen is built up, with information on the chemistry and chemical variations across its surface mapped out. An additional aspect of the SEM is that X-rays are emitted when the electron beam hits the specimen surface. X-ray wavelength can be determined by spectroscopy and thus some elemental makeup determined: the most widely used system (EDS, energy dispersive system) allows accurate chemical makeup to be determined, of selected elements. Advantages of the SEM include the very small amounts of sample required, as well as preservation of the SEM sample for later analysis. Disadvantages include sample representativeness, and once prepared for SEM work, little else can be done with the specimen.

Sampling is a major issue in SEM studies – largely because the technique is imaging features that are not visible to the naked eye. Thus an area of visually-homogenous soil may contain significant variations under SEM, or be truly homogenous, forcing the collection of many samples for no reason. Contamination is a major issue for SEM work: the method is sensitive and allows detection of very small particles: geological practitioners of SEM work are very aware of this problem. As Trewin (in Tucker, 1988) states: 'Since the final sample to be used in any SEM study must be clean (i.e., uncontaminated, or representative) it is not practicable merely to collect small [rock] chips in the field. Many pitfalls exist in preparation and ample material should be collected.' This geology-based statement has obvious very serious repercussions for forensic work: often the investigator has no control over the amount of sample collected by the suspect or victim at the scene, thus making such mud splashes, adhered clumps, etc. possibly incomparable to a larger specimen extracted in the field. Careful statistics of grain numbers, or preparation/analysis whereby only like materials (e.g., quartz grains) are compared, is advisable in such circumstances because the prosecution/defence scientist working against the investigator need only quote the above Trewin statement (in Tucker, 1988) to set the seed of doubt in a juror's mind, however untested that quote is on an individual case basis.

Case study: SEM – Bull and Morgan (2006) on quartz grain surfaces

The beauty of this study is the way in which it is based on a long history of very good science, applied to the forensic arena. Quartz grain surfaces have been studied using SEM for many years – the variations were illustrated in the atlas of Krinsley and Doornkamp (1973) – although the power of the method has been diminished over the years by different authors using their own names for textures they observe. The basic idea is that characteristic textures on the surface of each grain build up in response to specific environments. Thus glacial quartz sand grains have a jagged, broken appearance compared to the smooth and polished desert grains. Indeed, some sediments have mixes of different grain origins (for instance, glacial sands that have been reworked by rivers), or some grains themselves are hybrid types. Again, the power of the method comes in statistics, experience and control samples. A few grams of sample may yield hundreds of useful quartz grains that can be classified, population charts produced and comparisons made.

As well as terminology, other problems exist in using quartz grains. First is the matter of operator subjectivity – different observers may view the same grain texture in a different manner, based on experience and the highlighting of different features to their eye. Second is the problem of hybrid textures on multiple grains, or of processes in different environments producing similar surface textures. These are minor issues, however: quartz grains are so common in soil, rock, aggregate and other samples as to be nearly ubiquitous. Bull and Morgan (2006) demonstrate the usefulness of the quartz grain texture classification by opting to show how their database of samples from England and Wales may achieve one of the ultimate goals of forensic geology – the location of a suspect sample, as well as the classic use of any soil or sediment, in linking or excluding suspect materials to and from a scene of crime.

Extraterrestrial applications (spacecraft surfaces)

Graham *et al.* (2004) consider the problems caused by cosmic dust grains when impacting spacecraft surfaces. These they describe specifically as they are so abundant, and thus the chance of contact between craft and dust is high, as with Locard's Principle of contact (applied to materials such as soil). The abundance of such dust has been measured with capture media as well as sensitive film that give an impact rating. Analysis of the impact allows an estimate of how fast such grains can travel, which may be compared to models of their gravitational attraction. An impact at a vulnerable point on a spacecraft, or repeated impacts, could have disastrous effects. In addition to space dust there is the added hazard of 'space junk' or discarded anthropogenic materials causing the damage. Thus the two types of problem are analogous to two terrestrial scenarios: (i) impact from natural materials such as lava bombs, waves or sand storms; and (ii) impact between two human materials such as car crashes, hit and run or aircraft collision. In both cases, what caused the impact is of great concern and Locard's Principle is paramount. When every contact leaves a trace, this may not be a residue, it may be a characteristic mark. Thus visual, photographic and SEM analysis of surfaces and textures, as with ballistics analysis, is carried out. The recovery of trace physical evidence is then subject to chemical and mineralogical analysis, as in hit-and-run incidents with car paint.

Graham *et al.* (2004) show how control measurements of captured cosmic dust particles may be compared to fragments recovered at impact sites on spacecraft surfaces in order to determine whether they are natural or not. Graham *et al.* (2004, p. 145) suggest that such comparison of impactor to human-made origin (space junk) may never be used for litigation. We rather think that the veracity of this statement will depend on the future of space exploration. If 'littering' of Earth orbits becomes an international offence, and a spacecraft is critically damaged through the impact of another country's negligent astronauts, then when the provenance and age is proven, litigation could occur.

QemScan – the work of Duncan Pirrie and Matthew Power (Camborne School of Mines, UK) and Alan Butcher (CSIRO, Australia)

Until the invention of this automated, combined SEM and energy-dispersive X-ray system, investigators were obliged to visually scan polished sections of rock or sediment/soil (embedded in a solid medium) and use their own judgment as to which grains should be subjected to chemical analysis (see Pye, 2004b). The QemScan method is an advanced SEM technique and was developed by government scientists in Australia for the scanning of old mine waste for precious mineral phases that had been overlooked by previous processing techniques (McVicar & Graves, 1997). Duncan Pirrie, Matthew Power and Alan Butcher have championed the application of the method to forensic geology (Pirrie *et al.*, 2003, 2004).

QemScan is an automated SEM system, providing rapid quantitative mineral analyses using a LEO 440 scanning electron microscope fitted with three energy dispersive X-ray spectrometers. Soil, sediment or other mineral samples are embedded within resin that is allowed to harden and then the blocks are polished. Analysis is based upon the location of particles present within the polished block using the contrast in backscatter coefficient between the resin and soil/sediment particles. A variety of resolutions (pixel spacing) at which the X-ray spectrum is measured are available, with each compared to a database of known spectra (the 'species identification programme'), which was developed by CSIRO Minerals (Australia). A mineral or phase name is assigned to each pixel every 10 milliseconds, resulting in over 100,000 pixels being measured in an hour. The system can thus quantify and map the composition of approximately 1000, 1–10 μm sized particles in one hour. The system can image and analyse non-geological materials such as industrial products (lime, plasterboard, smelting waste), crystalline drugs and cartridge discharge residue, but it cannot be used on organic materials, which it simply does not 'see'. This negates the use of organic solvents, albeit that a soil sample may look quite large, but actually contain very few measurable particles. The system provides data on particle size (area), particle shape and particle density. Particle mineral analysis systematically maps the composition of each discrete particle, or a pre-defined number of particles, within the prepared sample. This provides quantitative data on mineralogy or phase composition along with particle grain size and shape. Once the operator selects the pixel spacing and mode of operation, data collection is entirely automated. The same groupings are used for all the samples measured allowing data output to be comparable between different samples – all samples are examined under the same set of operator-defined rules during analysis. Because the phase analysis is based upon the acquisition of X-ray energy spectra, different phases with the same chemical composition cannot be separately classified.

For example, the different forms of silicon oxide, sodium chloride or calcium sulphate cannot be differentiated. This is a minor weakness rectified by the conjunctive use of QemScan with a mineralogical technique such as X-ray diffraction. The automated nature of the QemScan system makes it ideal for quantitative work in which exclusion or comparison are to be achieved. The method works best on materials with a reasonable number of grains, including some variation. This is because the output is part visual (mineral maps of grains); the convincing forensic aspect is the summary statistics of grain numbers, their chemistry, and thus inferred mineralogy. Thus sediment in very small quantity, or of a pure nature (quartz sands for instance), or worst of all both, would provide a less-varied output than many soils. The system does not detect mineralogy, and thus different minerals with identical chemical compositions may also prove to be difficult to differentiate, especially when of clay-grade. The other problem of automated systems is their attractiveness to the non-specialist and the exclusion they create for human-based observation. The non-specialist may be tempted to throw sediment into an automated system such as the QemScan and not bother trying to understand the output. This is analogous to both early work on comparative DNA as well the use of the soil density column through the 1950s and 1960s. The visual comparison of the black and white stripes produced in both methods made for tempting and easy comparison, without questioning the natural variability, origin or methods behind the technique. The human eye and brain is still better at spotting the unusual or repetitive pattern or shape. Thus the sound reliance on statistics generated by the QemScan system may preclude discovery of the unique particle or unusual grain surface texture (the grains having been cut).

Pye and Croft's (2004) proposition of a multi-proxy methodology in forensic soil/sediment analyses did not use the newly-arriving method (at the time) of QemScan, yet this method is patently suited to their approach, in conjunction with light microscope, petrography/mineralogy (using XRD if need be), SEM and microbiological analysis.

8.3 Laser Raman spectroscopy

Certain light scattering is termed the Raman effect, named after one of its discoverers, Sir C.V. Raman (1928, together with K.S. Krishnan and independently by Grigory Landsberg and Leonid Mandelstam). Raman won the Nobel Prize in Physics in 1930 for his discovery. The Raman effect occurs when light impinges upon a symmetric molecule and interacts with the electron bonds. The deformation of the electron cloud reflects any polarization within the molecule, which determines the intensity and frequency of the Raman shift. Directing a laser (near infrared is commonly used) at a target and observing the Raman shift in the light provides a method of molecular diagnosis, using the inferred information about how the material is bonded.

The advantage of Raman spectroscopy is that samples need not be cut or impregnated in resin for analysis, preserving their crystal or surface form (see the descriptions of quartz grain surfaces, above), making this a good complementary analytical method, especially for very small particles (<1 μm in diameter). Raman spectroscopy can also be used on human-made products as well as minerals, whether hydrated or not. Often both direct and hyperspectral (thousands of Raman spectra) are used conjunctively in mineral analysis. Specialist methods of enhancing the Raman shift by gold coating, or by measuring resonance, provide specific advantages for certain material. Geoforensic applications on Earth materials have so far been limited, yet the wider use on materials such as human tooth dentine, bone and archaeological materials shows great potential.

8.4 Inductively-coupled plasma spectroscopy

This method was introduced in Chapter 7. Plasma comprises ionized atoms and molecules in a gas form, created by passing a gas through a quartz glass torch at super-high (10,000 kelvin) temperatures, then through a nebulizer for measurement by optical emission of atomic spectra (ICP-AES and ICP-OES) or mass spectrometry (ICP-MS). These methods have different advantages in terms of which elements are best analysed, etc. For trace quantities of material, ICP-AES needs only 1–10 mg of material. ICP-MS can be used to measure the different quantities of isotopes of the same element in minerals, bone and other material. Isotope ratios provide enormously useful information on mineral provenance, genesis, material similarity and organic matter growth and origin. For this reason, the uses of isotope measurements are here developed more fully.

8.5 Isotope analysis

The mass spectrometer accurately measures the differences in abundance of certain isotopes compared relative to one another, as well as to laboratory standards. The sample is reduced to gas, which is ionized by an electrical field: ions accelerate in this field and can be focused or collimated into an ion beam. These ions can then be separated by atomic mass by passage through a magnetic field. Each mass can be collected as an energy discharge, measured electronically and summarized as a ratio to another isotope of the same element. Commonly measured geological isotopes of sulphur, nitrogen, hydrogen, oxygen and carbon all require highly-specific preparation methods, which when complete provide accurate, reproducible information on compared samples, the origin, and possible changes to occur in materials. The uses of isotope geochemistry are so varied that complete texts have been written on the subject, far beyond the remit of the current text.

8.6 X-ray diffraction and trace evidence

In situ X-ray diffraction: work of Kugler (Krill, 2003; Kugler, 2003) and 'X-ray diffraction analysis in forensic science: the last resort in many criminal cases'

In this important work, Kugler considers how 'Smears, minute contact traces, small sample quantities or tiny sample areas can be successfully analysed as well as large quantities of materials' (Krill, 2003; Kugler 2003), through description of the work of the Forensic Science Laboratory of Stuttgart and how it evaluates material evidence. The need for 'reliable, definite and accurate results' has enforced the importance of scanning such trace materials with an X-ray diffractometer. Materials analysed include 'paints from automobiles, buildings and tools, building materials, minerals, ceramics, asbestos, metals, alloys, explosives, gemstones, soils, abrasives and drug impurities and extenders'. Kugler relates how the X-ray diffraction method creates a unique pattern that is open to visual or statistical comparison. He uses the analysis of iron oxide pigments in paint as a simple example – normally the comparative analysis of paint is a complex procedure, but here Kugler shows how non-destructive X-ray diffraction screening of samples can be used both to include or preclude samples, as well as add X-ray diffraction data to a multi-proxy database, with chemical data, for improved statistical comparison. Such analysis can prove scientifically cost- and time-effective in the analysis of soil, rock dust and other crystalline materials adhered to fabric and other materials associated with crime suspects as a rapid screening method for further forensic-based scientific analysis. The reasoning is that trace (sub-gram) soil, rock and other crystalline debris evidence has to be analysed qualitatively, non-destructively and rapidly to inform both searches by crime investigators and which subsequent analyses should take place. The X-ray diffraction analysis of *in situ* materials can achieve this but has yet to gain widespread use or be fully tested.

This problem needs to be placed in context (see the following list):

1. In Chapter 7 we examined the comprehensive flow charts of best practice from Palenik and from Sugita and Marumo for the sequence and range of analyses that can and should be undertaken when comparing crystalline materials for forensic comparison (Figures 7.2 and 7.3).

2. This best practice is tried and tested, follows other scientific protocols and should be adhered to where possible.

3. Recent cases worldwide, however, demonstrate why this best practice could not be followed, forcing a review of methods that was untested in a full scientific analysis such as this. These were:

    — where only very small amounts of possible clay-rich soil and rock debris (>1 mg) were recovered as mud spatter on clothing, footwear and vehicle interiors;
    — where large numbers (hundreds) of seized clothing and footwear items, each with the small spatter marks described above, had to be screened for more expensive, time-consuming but ultimately more accurate palynological, isotopic, QemScan or DNA analysis. In each case, there was insufficient sample material to be sequentially analysed, as in standard practice. Similarly, it was desirable for screening to be as non-destructive as possible, so that subsequent analysis could be performed.

4. These scenarios will become common for two reasons.

    — As criminals become forensically aware, their clean-up operations become more efficient, leaving only the trace evidence described above. This is especially relevant to organized gangs be they terrorist or not, who have the finances and facilities for efficient clean-up operations.
    — The need for sophisticated forensic analysis requiring pristine samples will increase as defence or prosecution use whatever techniques are not available to their opposition to establish suspect to scene comparisons. The maintenance of sample integrity by non-destructive analysis will be of major benefit in this latter case.

5. Two problems arise with Murray and Tedrow's (1975) best practice protocol when applied to the above scenarios:

    — Although such small amounts of material as described could all be picked from fabric, examined, split by grain size with coarse material mounted for optical microscopy and fine material analysed by SEM and XRD, this sequence takes time that could be critical in the continued prosecution of the case (Murray & Solebello, 2002).

— In the studies embarked upon by the authors, it was found that the reason these materials were mud specks in the first place was due to their fine-grained nature, thus the traditional, rigorous sequence of analyses provided only limited information on the materials.

6. Recent work by Werner Kugler (Krill, 2003; Kugler, 2003) has demonstrated the use of X-ray diffraction analysis on fine-grained dust-covered or mud-spattered sections of the original suspect material (clothing, footwear) as a qualitative mineral determinant for comparative purposes. The quality of X-ray diffraction analysis data is dependent on substrate type, sample preparation and analytical software. Otherwise, the technique is well-established and trustworthy. Thus, Kugler's work establishes the efficiency of the technique.

In addition to the above problems of trace geological evidence, Kugler considers other applications. These include the comparison of suspect and scene trace materials from burnt residue from campfires; hoax anthrax threat materials sent following the first World Trade Centre attack; *in situ* analysis of paint on pliers, car trim and water faucets; and the analysis of fake gems (embedded in a bible).

Cold case – murder of teenager, West Belfast work (Keaney *et al.* (2006))

In a conference abstract, Keaney (2006) describes this case of rape and murder in 2004 in West Belfast. The victim was seen at one location before being discovered murdered at a second. At the first, she obtained some fast food with friends in a shopping area, where the suspect was also seen. She then went missing for about 10 hours (overnight). The suspect claimed to be 'worried' by her disappearance, and took two friends (witnesses) on a drive to look for her. They (he, because he was driving and thus dictated the route) 'chanced' upon her battered body near an abandoned quarry: the suspect moved from the car, contacting the scene, and touching the victim. Thus all chance of denying an alibi had gone.

The murder shocked the local community and witnesses from a temporary accommodation site came forward to say that they had seen a vehicle similar to that owned by the accused, at the bottom of a lane near their site. A search of the lane revealed items of the deceased's clothing at the top of the lane. Footwear tread and DNA analysis failed to link the suspect to the lane or items. Visual examination of his (mainly clean) clothing did reveal mud-specks along the rear of trousers and a jacket. Each of these specks was subjected to irradiation by X-ray diffraction, and although visually-indistinguishable, two types of mineral content were observed: 22 specks were of red clay with dolomite and calcite (calcium carbonates), while another 30 were of red clay with zeolite and feldspar. Samples from the body deposition site were analysed by standard X-ray diffraction, but also specks on the clothing of investigators were examined: these all had zeolite and feldspar. Mud from puddles on the lane were also analysed: these all contained dolomite and calcite. The suspect denied having been at the lane. Further analysis recovered DNA and other materials that linked the suspect to the lane, denying his alibi. He later pleaded guilty to the murder.

Cold case opportunity? – Scott Petersen accused of murdering his wife (Lee & Labriola, 2006)

In this famous case (Lee & Labriola, 2006), the abundance of cement dust in the accused's workplace warehouse and the presence of concrete weights in his boat, with no associated rope for usage, caused police to be suspicious. One possible theory to explain the concrete was that weights had been made and used to weigh down the body of Scott's murdered wife (her remains, as well as those of the near-term baby she was pregnant with, were found some months after the initial searches). As Lee and Labriola (2006) state:

> Detective Hendee was curious about a possible match between the color of the gray powder (at the garage) and the color of several small chunks of gray matter in Scott's pickup truck.... At the conclusion of the warehouse search, Hendee returned to the truck and, indeed, verified by looking that the colors were the same. However, this finding had no forensic value since almost every type of cement is gray in color.

Comparing the two, in our opinion (given what information we have on the case) would only have provided the possibility of denying an alibi reason Scott may have provided for the grey 'matter' in his truck. Had he denied that it was concrete dust from the warehouse, a textural, petrographic and most conclusively, isotopic analysis of each may have verified or denied his statement.

Unique particles – time since offence

In the introduction to this chapter, the problems and opportunities that perpetrator clean-up and the urban environment present for trace evidence analysis were introduced. A subsidiary problem to both is that of the passage of time since the offence. There are many examples of how crimes have been solved (actually, or theoretically) in the literature. Because they all involve some unusual circumstances, materials and situations, little generic advice can be given on how to approach such problems. With this in mind, to learn by example is probably a good way of approaching any future problem, hence the following case studies.

Case study: Trace evidence – JonBenet Ramsay, Colorado, 1996; Sheridan (2006), Davis and Thomas (2000)

This very famous case has at its conclusion the combined retrieval of stored historic evidence and DNA exclusion of a suspect. For those not familiar with the case, or wanting some perspectives other than those provided here, Sheridan (2006) and Davis and Thomas (2000) both give accounts. In brief, JonBenet Ramsay was the six-year old pageant beauty queen daughter of a wealthy Boulder couple. She was murdered, most likely by strangulation, and left in the cluttered and apparently sealed basement to the family home, on 26th December 1996. A two-page ransom note was left in the kitchen, using pen and paper apparently from within the house. An initial police search did not discover her and kidnap was assumed until JonBenet's father discovered her corpse. The crime scene was gradually established and evidence retrieved. The widespread belief by media and many locals, that either one or other of the parents were suspects, initially gained momentum, latterly waning when no evidence of a murder link or motive appeared. In addition, many complications arose, such as unidentified DNA, fibres on JonBenet and footprints and fibres (the latter on a baseball bat) outside the residence. The bizarre and unexpected twist came 10 years later, when a known offender (John Mark Karr) then on the run abroad following a charge of possessing child pornography, came forward to claim responsibility. Luckily, DNA evidence, as well as the critical trace material that had been held by geologists who specialize in the microscopic identification of such material, had been stored by Skip Palenik (Microtrace Inc.). This allowed the 'suspect' to be have his claims denied (Houde, 2006).

Trace element and isotope content of ammonium nitrate and potash

During the period 1970 to 1995 the Provisional IRA conducted a series of large-scale bombings of military and civilian targets, concentrated in Northern Ireland and England. Some of these included the largest explosive terrestrial detonations on UK soil since World War II. Between 2 and 3,000 pounds of explosive were used to destroy the Forensic Service of Northern Ireland laboratories in 1992. The large amounts of explosive required either time-consuming collection of urea, black powder, or ammonium nitrate, the favoured method of most terrorist groups. ANFO stands for ammonium nitrate/fuel oil (most often diesel fuel, sometimes kerosene, molasses or caster sugar). This is mixed in certain proportions and detonated with a commercial or home-made detonator, primary charge, which causes explosion of the ANFO mix. The key ingredient is ammonium nitrate or fertilizer, which is commonly ground down to increase surface area and thus explosive capacity. In the period under question, large quantities of all fertilizer were ordered legally and illegally, and consignments often went missing and bags split, causing mixing.

An unsuccessful detonation by a rogue branch of the IRA in 1995 was ascribed to the ammonium nitrate having been mixed at some point with potassium chloride (potash), another common form of fertilizer. At that time, potash was being imported either from Yorkshire or Germany: in each location it is mined at depth, amazingly enough from strata that are equivalent in age and that occur extensively beneath the North Sea. Nonetheless, it is possible to distinguish the origin of the potash by isotopic and trace element content: Permian potash has a reddish tinge, ammonium nitrate being milky white, making removal of potash crystals for individual or collective analysis possible. Once analysed, the rogue bomb was determined to have Yorkshire potash in it, cutting intelligence efforts on the ingredient origins by almost half, and freeing up time for other enquiries.

Unique particle concept – work on Japanese soils by Sugita and Marumo (2004)

Sugita and Marumo (2004) open their article with a statement that could be direct from Hans Gross, who indicated that toilsome investigations could be made easier by the examination of soil on a suspect's shoe. They imply that when unusual particles are discovered in both suspect/victim materials and at a potential scene, the standard methods of colour, particle size, and mineralogical analysis are strengthened and the need for extensive control samples becomes diminished. This latter point of course depends on the rarity of the 'unique' particle, leading some to use the phrase 'unique identifier'. Sugita and Marumo (2004) show how fortunate they are in having some unusual features in many Japanese soils, including aeolian dust from China, spatially-restricted areas of volcanic glass and shards (tephra), peculiar secondary volcanic minerals (zeolites) and adapted plants, spores and pollen. The ultrastructure of a fragment of camphor leaf found in soil from a suspect's shoe linked the perpetrator to a location with just such a tree. They also comment on the existence of plant opals, a silicate found in cells and having characteristic morphology in different plants. Such plant opals allowed positive comparison between mud found on a screwdriver head and holes drilled in a mud wall in order to carry out an arson attack.

8.7 Manufactured or processed materials that geoscience techniques can analyse

Many geological materials are processed and mixed to form commercial products. If involved in some form of criminal activity, or humanitarian problem, then the techniques developed for geoscientific reasons are best combined with engineering, chemical or biological analysis in order to establish provenance or a comparison between suspect and scene or victim. The range of materials that may be subject to such analysis is as great as their use in crime. Here we give some case studies in order to demonstrate what may be achieved: many more examples exist.

Bentonite links to a murder suspect: Raymond Murray (2005) on the Dodson murder (Colorado, 1995)

Murray (2005) recounts this case as one where geological materials were one of the most important pieces of evidence in a murder case. On the 15th October 1995, John Dodson was found dead from two gunshot wounds while on a hunting trip with his wife Janice in the Uncompahgre Mountains (Colorado). An off-duty policeman heard Janice shouting from his nearby camp, rushed to the scene and presumed a hunting accident had occurred. If only one shot had been fired, this would likely have been believed. Critically, Murray recounts: 'Prior to calling for help, Janice had returned to her camp and removed her hunting coveralls, which were covered with mud from the knees down. She later told investigators that she had stepped into a mud bog along the fence near camp.' Also critical to the story was the bizarre fact that Janice's ex-husband was also on a hunting trip, only a short distance from the scene, and had reported his gun and ammunition stolen. The route from the scene to the ex-husband's camp went through a pond that had been lined with the water-retarding clay, bentonite. Janice denied having obtained the gun from her ex-husband, yet the mud on her discarded clothing was a bentonite. Janice claimed the mud to be from a nearby bog, which contained no bentonite, denying her alibi reason for the soiled clothing. In fact extensive sampling showed *none* of the surrounding bogs to have bentonite in them. She was convicted in Colorado of murder and is currently in jail.

Glass and plastics in soil

This case is used as an example of where it was presumed that geological/microbiological analysis of soil would be useful, but was not necessary because of the observation and caution that showed a far easier solution to the problem. Two men (armed covertly with guns) hired a private taxi in a country town in Northern Ireland. The men were part of a gang that controlled property; the taxi was controlled by a rival organization. As they arrived at their bogus *cul de sac* destination, an accomplice drove in behind, blocking the taxi's escape route. The gunmen shot the taxi-driver in the back and escaped. The taxi driver reversed, ramming the accomplice car violently. The taxi-man, having sustained numerous gunshot wounds still managed to run off. The gunmen and accomplice drove away.

A similar car to that rammed by the taxi was found damaged and partly burnt some distance away, but close to where suspects lived. The problem was testing whether this was the suspect vehicle or not. Impact specialists were undecided on whether the taxi was the cause of the impacts seen on the suspect car, it being so badly damaged. Lumps of mud found at the scene were compared to material retrieved from the wheel-arches of the burnt-out car. The results were surprisingly simple to analyse, both containing visually-similar soil (not a sound comparison method) with chips of tar-coated dolerite (common roadstone in the area), 3 millimetre-diameter 'balls' of polystyrene, yellow polystyrene (Figure 8.3), white rock fragments and foodstuffs, and the clincher, both had yellow indicator glass that was sub-exhibited for specialist analysis. Following this, all non-rock materials were considered for specialist, non-geology-based analysis.

This study goes to show that the geoforensic specialist should be open-minded, flexible, observant and not so dogmatic as to presume that only geoscience can assist in the analysis of mud. Why make life difficult when an easier solution presents itself?

Case study: Establishing the location of a murder scene from victim clothing

During a loyalist feud in the late 1980s in Northern Ireland, a bizarre manner of one group showing its power was the random murder of members of the Catholic population.[1] These were often shop-workers or taxi-drivers. In this case, a well known community figure in North Belfast (Northern Ireland) was left to close up the village centre following a women's group meeting. After she closed the gates and walked down the lane to the main road (Figure 8.4(a)), a gang of armed men appeared. What happened next is not known: witnesses claim to have heard shots at this location, but no unequivocal evidence of weapons discharge was

---

[1]The details of this case have been intentionally obscured, on advice from the Police Service of Northern Ireland.

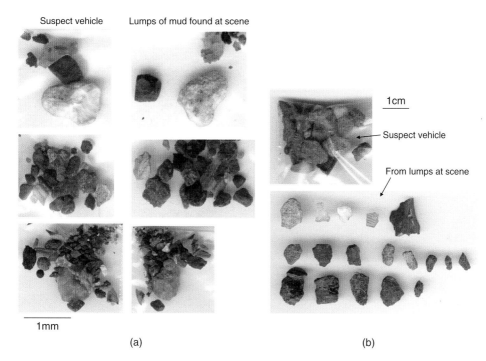

Figure 8.3 Extracted granule-grade particles from mud found adhered to a vehicle and at the scene of an attempted murder. Few of the granules that appeared visually similar to geologists working on the case turned out to be diagnostic rock fragments, the bulk of the more visually-striking material being of human origin and thus recommended for appropriate analysis. (b) Visual comparison of the critical suite of fragments from suspect and scene. Visual comparison alone would indicate that the samples do appear similar. Is one polystyrene fragment the same as another? How are they analysed? Does one piece of what looks like fried food compare to another? How are they analysed?

found. Footprints and other evidence showed the narrow part of the lane to be the abduction site. All that is known is that about half an hour later, the victim's car was found burning around 2 to 3 kilometres away, with the victim dead beside it (Figure 8.4(b)).

The investigation into the murder bore little information and the case was closed, to be re-opened 10 years later by a cold case review team and following community pressure. At this time, very little evidence still existed, bar the scenes, the victim's mud-stained trousers (Figures 8.4(c) and 8.4(g)) and some items seized from a possible suspect (Figures 8.4(d), (e) and (f)).

The question raised by this cold case review was where was the victim shot? Did her trousers and jacket contact the ground at the community centre or the body deposition site? The various stains on the victim's trousers and jacket as well as suspect clothing (e.g., Figure 8.4(e) and (f)) were mapped out, and any microstratigraphy noted, to be separately sampled, as was each of the two scenes.

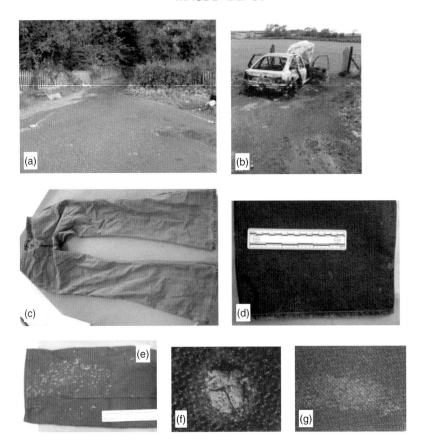

Figure 8.4 Supposed abduction site, the lane to a community centre the victim had closed for the night when abducted. (b) Body deposition site and victim's burnt car. (c) Victim's mud-stained trousers (from police photograph, taken at the Police Scientific Laboratory). (d) and (e) Suspect's mud-stained trousers. (f) Detail of mud-speck on the suspect's trousers (speck/lump is ∼1 millimetre in diameter). (g) Presumed mud-stain on victim's jacket sleeve. Field of view ∼10 centimetres. Context of location not shown for legal reasons

Every area of loose (possibly transferable) ground was examined, related back to the original photographs (apparently, not much had changed) and analysed. Two methods of trace evidence analysis proved useful: mineralogical and microbiological. Mineralogy was determined by X-ray diffraction of the mud scraped from each patch on the trousers and from each scene puddle or patch of ground (Figure 8.5).

Samples were also examined for testate amoebae (discussed in Chapter 7). The two sets of analyses showed that all the community centre samples could be excluded (Figure 8.6), and that comparison was good between the trousers and the body deposition site, implying that the victim was not likely to have been shot at the first scene (possibly subdued), but likely murdered at the second, body deposition

MANUFACTURED OR PROCESSED MATERIALS 263

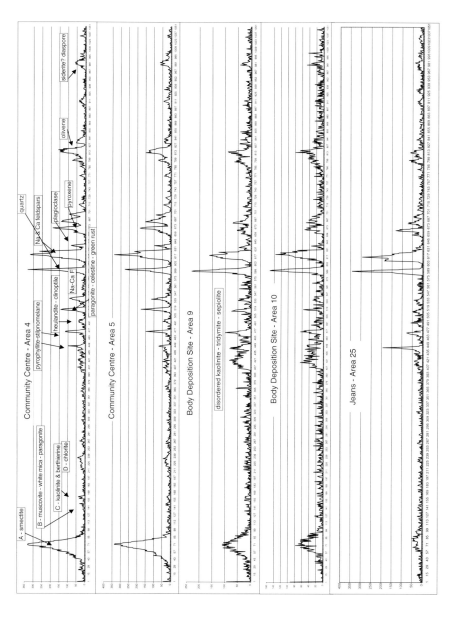

Figure 8.5 Selected X-ray diffractograms from the two possible murder scenes and the trousers

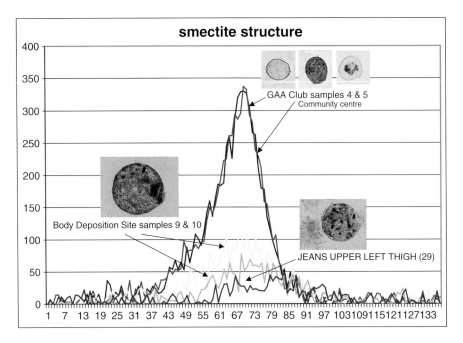

Figure 8.6 Expanded view of the smectite (a clay mineral) part of the X-ray diffractogram and key testate amoebae species from each sample. Testate amoebae analysis courtesy of Graeme Swindles

site. These data were later used to assist a call for public assistance on a television programme.

The work raises serious questions about the timing of sampling of scenes: in this case, because of the lack of use or renovation of the community centre or body deposition site, no visually-obvious environmental changes were seen. Had either area been renovated, this work would have proven impossible. The apparent lack of environmental change could also be incorrect: what microscopic mineralogical or biological changes had taken place? Many environmental scientists will attest to the effect 10 years of climate change has on such locations. A series of time-sequence analyses is required of the mineralogy, chemistry and biology of loose ground, puddles and other micro-environments where material may be transferred.

Matches – the multi-proxy work of Farmer *et al.* (2007) in a case of suspicious burning and arson following a crime

Sound advice for anyone asked to provide specialist advice on legal matters is not to stray from one's specialism for there will no doubt be an expert for the opposition who knows more than you. Only luck will get the non-specialist out of such trouble. However, when confronted with a problem for which no established analytical procedure has yet been developed, a safe first attempt is the cautious approach adopted by Pye and Croft (2004) for soil analyses, and that is the multi-proxy use of comparative yet independent analyses.

The background to Farmer *et al.*'s (2007) study is somewhat sensitive, but it is stated by the authors that un-burnt matches (but not match-box) were recovered from the scene of a failed attempt to burn illicit materials (clothing, etc.) following a crime, as well as from the suspect's home. Isotope ratio mass spectrometry (IRMS) was at the time a very novel method of excluding and comparing organic materials (e.g., drug-doping in sport, environmental pollution, the characterization of explosives and food additives). Thus it could be used to examine whether the loose matches found at the scene could be compared to either control matches (bought at various locations locally and globally) or to those found in boxes at the suspect's home. As this was the first published use of IRMS for matchstick comparison, the results were compared independently to two other methods: X-ray diffraction (XRD) analysis of the mineralogy contained within the match heads and thin-section comparison of the wooden match-sticks. XRD[2] was useful in characterizing both the 'geological' content of the match heads (quartz, mica) as well as the explosive content (potassium chlorate). However, due to the common makeup of ingredients, many control matches were indistinguishable by their mineralogy. Thin-sections of the wood successfully discriminated the matches found at the scene of crime from those seized from the suspect's house but provided no differentiation between the various aspen woods in the control matches. Thus IRMS was not only tested by two other proxies, the level of discrimination among the aspen woods was also shown to be great. In this case, the unusual nature of the matches found at the suspect's house allowed these to be excluded by thin section analysis. However, had these matches been of standard manufacture, IRMS would have provided a discriminant where XRD and thin sectioning did not. What is not known from this study is whether other, sophisticated geological techniques, such as quartz-grain surface texture or QemScan (see Chapters 6 and 7 for descriptions of QemScan) on the match-heads or their burnt residue, would have bettered IRMS. One may think that the applications of this kind of work are somewhat esoteric, yet examination of the literature and personal correspondence (Marianne Stam, January 2007) indicates that burnt and unburnt wooden matches are commonly encountered at many scenes of crime such as arson or self-immolation: cardboard matches and their booklets have been the subject of far more scrutiny than wooden matches.

Case study: Agricultural limestone

A fatal shooting during a gangland feud in the Greater Dublin area in 2001 occurred when one gang led the leader of another to believe they had a common enemy that their joined forces could defeat. The trickery went as far as the gang taking the rival leader to a secret weapons cache and discussing the division of arms and tactics. Thus when the following night the gang returned, neither their rival nor his family had much cause for concern. The rival was led to the cache location, which was an abandoned Carboniferous limestone quarry weighbridge. An awaiting

---

[2]The images of the prepared match-heads are shown in Figures 10.4(a) and (b).

gunman appeared and shot the rival a number of times. The gang extended a nearby excavation, placed their victim in it, scattered some agricultural 'lime' (stored nearby) on top and crudely covered the victim with sandstone slabs, rubble and soil. The 'lime' was crushed limestone ($CaCo_3$), a slow-release acid soil improver as opposed to the rapid action of CaO or oxidized limestone. It is presumed the gang undertook extensive clean-up measures and established good alibis; their rival was not reported missing for some days, but almost as soon as he was, a man walking his dog near the abandoned quarry body deposition site noted the disturbed ground as well as some patches of clothing protruding through the earth and rubble of the grave.

Murder having been established, various suspects were arrested, with little to connect them to the crime other than known association, some witness reports and inconsistent alibi accounts. An expensive make of jacket was recovered at the gang leader's home, which had light-coloured 'smears' on it indicative of the surface being wiped and then dried (see Figure 8.1(f) on page 243). Detailed *in situ* examination indicated this light coating to be calcium carbonate with no microbiological materials visible. Sections of the dust-coated jacket were cut away and placed in an X-ray diffractometer. This established the mineralogy of the 'dust' as predominantly calcite but also indicated the presence of the mineral cookeite (a rare lithium-aluminium-hydroxy-silicate). Scanning electron microscopy confirmed this result, and in addition established the presence of Carboniferous microfossils and a 'crushed' or jagged-edge to the calcite crystals. The combination of unique minerals (cookeite), microfossils from a limited part of the Carboniferous succession, and the crushed nature of the calcite made it difficult to find a source for the material on the jacket other than the quarry. The defendant pleaded to a lesser charge, admitting his presence in the quarry. He had been reluctant to destroy his jacket and had his wife wipe the clothing with a wet cloth: this merely made the light coating wet, and thus darker, only for it to reappear when dry!

Fraud

One of the greatest frauds ever known involved the analysis of geological materials. Henri (Hans) Van Meegeran was one of the best art fakers known. As a celebrated student, he became less and less successful, demeaned into painting portraits and designing Christmas cards until he came upon the idea of faking an old master. He spent over five years copying the work of Vermeer, until an elaborate series of 'discoveries' allowed his fake onto the market, having been 'discovered' in an old farmhouse and sold through Van Meegeran's corrupt solicitor. His extreme patience and caution came to light, partly through his use of ageing techniques, but also by analysis of his blue paint, which he used lapis lazuli to create, in order to avoid detection of modern paint.

8.8 Some conjecture on the future of trace evidence

> The end has not been attained simply by the elucidation of the affair in an ordinary way. It is very convenient to say, 'It is impossible to go further.' But if

one says continually, 'another step forward must be taken,' one finally advances several steps.

Hans Gross (1891)

Pye (2007) makes an excellent and realistic comment regarding Locard's Principle, suggesting that although it is correct, all contacts do leave a trace, in his (Pye's) practical experience of collecting trace geological evidence, this is sometimes just not practical. Pye's comments are based on experience, and serve as good advice for those collecting evidence. We agree, but take the more open-minded approach that Locard's Principle pertains: if every contact leaves a trace, yet the perpetrator takes extreme measures to avoid that contact, to remove or confuse it, so solutions to the problem are sought. Locard is correct, even if the only contact a person has with what appears an inert environment is walking on a surface and breathing the air, both are contacts: the footwear on the floor and the airways with the air. Pye is also right: in the present-day, establishing such a contact, with our early 21st century technology would be problematic and most likely insufficient for the purposes of the courts of law. Hans Gross provides the best perspective on this: seen in 100 years time, our modern-day methods will appear crude and insensitive. Already we are detecting trace evidence that would not have been possible 100 years ago, so who knows whether the 'inert' environment of today will provide abundant contact evidence in the future?

Among other possible ways around the problem, two come to mind that rely on sophisticated analysis. First is the non-destructive, cost-effective and rapid use of multi-proxy screening methods. Research here is needed in order to establish the limits of detection and hence (hopefully? maybe?) differentiate the environmental history of the host material (e.g., fabric) from any suspicious contact. Second is to examine what the suspect cannot avoid contact with. We have already discussed how popular films, books and television shows, as well as prison-contact, educate the criminal in forensic avoidance.

Hence a second-guessing game begins: if the dust, sediment and soil of the geosphere are known to be 'problem' materials to the criminal, do rain and water, the microbiology and air/particulates of the hydrosphere, biosphere and atmosphere register? Experienced soil- and rock-based forensic analysers will at this point be both scared that their art is under threat by such sophisticated methods and cynical of the opportunities. On the former point, we should be scared, challenged and determined: very sophisticated forensic precautions and clean-up are already being practised by the educated killer and the well-funded terrorist. On the latter point, it would be irresponsible as scientists to throw our hands in the air and state that 'our methods won't work in the face of such precaution'. Or 'this is all crazy conjecture'. Quite the converse, if our methods can detect compromise in a sample handled by an accredited forensic laboratory, surely we can do the same with the criminal? We should be determined not just for the sake of forensic science, justice and the advance of criminalistics: improved methods can enhance environmental, engineering and medical sciences (among them all) in the pursuit of accuracy and result replication.

To try and both reassure and get the cynics on board, let us take the conservative approach: soil and sediment are established sources of earth materials we can analyse in comparing/excluding suspects, victims and scenes. Particulates in precipitation (one thinks of rain, but dew, fog and snow are as effective or better transfer vehicles) are no different, only the quantity changes, and as we have seen, trace evidence is already a growing branch of materials-based geoforensics. The particulate and chemical content (e.g., chlorine isotopes) of rainwater is know from GIS-based spatial analysis to change over time and distance, especially with regard to the passage of air-masses over various environments (rural, urban, freshwater, sea). Climate change, urbanization and population growth may be of some advantage: climate scientists tell us to expect increased seasonality, with dry, stable periods interspersed with increased storminess and heavy precipitation. Irritating it may be to set out without one's coat or umbrella, but imagine how the clear, still night selected by criminals for some nefarious activity turns to a deluge, with the associated confusion, transfer of wet materials, sloppiness in clean-up, etc.: a geoforensic opportunity where there was none before!

When considering any use of precipitation-based trace evidence, it is thus more critical to have samples of the control precipitation-borne particles than those from suspect/victim/alibi. Of the other 'geo'-based transfer-contact media, workers using the materials of the biosphere, especially the microbiosphere, are already well-advanced in their methods and applications. The trace evidence use of water and air/gases alone remains the greatest challenge, and until the advent of modern isotope discrimination (especially of water) would have been seen as impossible: residence time on and in the body, dilution, compromise, etc. are all problems for using these 'materials' in comparing/excluding suspects/victims from scenes, etc. Now we see this very difficult discrimination in the same light as soil and sediment compromise, dilution and discrimination: it precludes the lazy and provides the tenacious with possibilities. Geological fluids such as fluid inclusions (Sugita & Marumo, 2004) and hot waters related to crimes (Fushinuki et al., 2001) are already being analysed. If the fluid is a contact, and it leaves a trace, advances in recovery and analysis of minute quantities of material are all that stand in the way of establishing whether or not (exclusion/comparison) the contact took place. The tenacious will be initially rewarded with problems of misinterpretation but latterly with advancement of science. The challengers to established science will be forgotten if proven wrong, hailed as heroes if proven correct! Readers interested in this idea should read Davis (1926) on his support for the 'outrageous geological hypothesis'.

For Earth-related evidence (soil, sediment, non-biogenic dust, manufactured mineral-type particles), the ability to analyse small fragments will undoubtedly feature in the future analysis of trace evidence. SEM, XRD of small particles, Qem-Scan and micropalaeontology will likely feature as the conjunctive (or multi-proxy) methods that will be used, with non-destructive methods being deployed first, and to inform which destructive methods should follow. The challenge will be statistical significance: we may find out what a material is from only a tiny amount, but can we reliably compare it to another sample? The problem of handling, splitting (for conjunctive analysis) and contamination of trace geological (and biological) evidence becomes more acute where such small amounts as to be barely visible to the naked eye are concerned. Thus the future of research into trace materials and particles, for whatever reason (valuable, rare, forensic), will face all these handling, splitting and contamination issues.

9
The search for buried materials

> We buried him darkly at dead of night,
> The sods with our bayonets turning.
>
> Charles Wolfe (1791–1823) 'The Burial of Sir John Moore at Corunna' (1817)

9.1 Introduction

The search for buried materials can use many of the techniques outlined in all previous chapters, especially using geomorphology, geophysics, remote sensing, mineralogical and geochemical sampling. Search methods are dependent on the nature of the target and the terrain: however, some basic concepts may assist those involved[1] in the search for buried, hidden, sunk or remotely-accessed materials. Initially, as much background information as can be gathered will be useful. The key here is not to get bogged down in chasing what promises to be useful information: the searcher will never obtain all the information he or she needs. A preliminary data source can be worked on and interpreted while other work continues, which with management and timetabling may provide results when needed by another stage of the investigation.

Although subsurface terrestrial and sub-aqua searches are commonly made for pollution sources, weapons, drugs, vehicles and hidden or discarded contraband (jewels, computer parts, toxins), the most written-about searches are definitely made for missing people. The latter tend to fall into three groups:

- high-profile individuals;
- victims of mass murder, genocide or serial killer(s);
- disaster victims (landslides, building collapse, floods).

---

[1] for example, environmental agencies looking for pollution sources, humanitarian workers looking for disaster victims, search and rescue teams, and people looking for lost, buried items (archaeologists, discarded stolen goods).

Geoforensics Alastair Ruffell and Jennifer McKinley
© 2008 John Wiley & Sons, Ltd

Because of their background, some missing persons are not presumed murdered; others have almost certainly been killed. Both types of person(s) are initially searched for using similar intelligence and technology. Lord Lucan (Mack *et al.*, 1986) is presumed to have been responsible for his disappearance, yet until his remains are found it is not proven whether he was assisted in his vanishing, if he committed suicide, or was murdered. Popular belief would have the latter as unlikely. Jimmy Hoffa (Brandt, 2004), because of his union/Mafia connections, is presumed to have been murdered: like Lucan, however, until his remains are recovered and scientific analysis conducted, this is not proven, making it possible he took his own life, has 'disappeared' himself or befallen an accident.[2] Again, all unlikely.

The appearance that high-profile people vanish more effectively than others is an illusion, borne of publicity. It is a frightening fact that many people vanish each year without a trace: we do not hear about them because they are not famous. Indeed, it is useful to see how effectively a famous person like Hoffa can disappear, and how many conflicting theories abound as to the manner of their removal from society, in order to see how with time, resources and luck others can equally vanish without trace. The background of individuals can equally affect what is presumed to have happened in their disappearance: few would look for children missing from Bosnia since the mid-1980s in the ranks of the Middle East slave trade, just as the many 'vanished' people (especially young women) from the Far East would not be expected to turn up in mass graves in Mediterranean eastern Europe: nonetheless strange occurrences like this are recorded. The global network of people trafficking has resulted in indigenous peoples of one region turning up (dead or alive) as individuals or in groups, in locations on the other side of the world. It is often forensic geoscience that established both their origin (isotopes from bone/hair), last and recently-lived in locations (pollen from nasal passages, microbiology of associated materials) as well as movement of containers, vehicles and ships (mineralogy/geochemistry of external surfaces).

---

[2] The disappearance of Union boss, Teamster, racketeer and Mafia-associate Jimmy Hoffa is a classic in the intelligence–search cycle. Convicted in 1967 of attempted bribery of a grand juror, Hoffa was sentenced to 15 years in prison. In 1971, President Richard Nixon commuted his sentence on condition he desist from union activities. While fighting this restriction through the courts, Hoffa strove to reassert his power and met two prominent Mafia leaders from Detroit and New Jersey in Michigan on 30th July 1975. He was never seen by any public witness alive again, although a succession of high- and low-value witnesses have successively come forward. Two witnesses in particular have caused excavations and the establishment of a subsurface search methodology. In the first, DNA evidence examined in 2001 indicated Hoffa had been in the car of a Teamster associate, Charles O'Brien, despite O'Brien's protestations that this was not the case. Convicted killer Richard Powell told the authorities that the syringe used to administer drugs and subdue Hoffa, was hidden in a case and buried in a Michigan backyard, resulting in a fruitless excavation. The second witness (unknown, but named by some as Donovan Wells who owned a Detroit-based trucking company and was convicted of moving marijuana from Texas) indicated that the Hidden Dreams Farm, Michigan, was the location of Hoffa's burial site. A massive excavation of the farm followed, with no remains recovered. Hoffa's disappearance remains a mystery.

Multidisciplinary approach to the selection of clandestine graves by France et al. (1992)

France et al. (1992) make an important preliminary observation in their paper, stating 'only a few studies listed in the literature concentrate on multidisciplinary methods directed toward the location of buried human remains'. Of the five papers they consider, Boyd's (1979) 'buried body cases' comprises advice for the investigator, and includes a possible list of experts to be deployed. Imaizumi (1974), while considering landscape features as well as distances/access from suspect's known locations, was the first description of the probe (see Chapter 2). Bass and Birkby (1978) are like Boyd's technical advisory work, with the latter concentrating on the work of the anthropologist/archaeologist in body/remains recovery. In essence then, Boyd and France et al. (1992) were the first to consider integrating scientists of widely-varying disciplines in order to maximize information recovery at a body recovery site.

France et al. (1992) describe their project as 'Pigs in Ground', an experimental site in Colorado, where, prior to burial, control aerial photographs, geological and geophysical measurements were collected. The same, plus additional data, were collected at intervals following the burial of pig cadavers, in order to assess clear indications of burial. Aerial photography included black and white, colour and infrared: terrestrial thermal imaging was also carried out. Geological descriptions concentrated on the nature of the site and on the stratigraphy and alteration to geology following excavation. Botanical assessment was made by comparing grid-counts of control flora to that which grew at the burial sites following internment. Entomology likewise involved insect traps laid at regular intervals before and after burial. Geophysical methods used included magnetics, electromagnetics and ground-penetrating radar (GPR), surveyed over the pre-burial ground and then following internment. Soil gas measurements, scent dog reactions and scavenging patterns were also noted.

France et al. (1992) then give the advantages and disadvantages of all the above methods, as well as some further considerations. Two important messages emerge. First, that geographic location, time of year, season and day and available funding all play an important role in how the investigator chooses to deploy these techniques and their operators. Second, that because of these limitations, there is no single body-locating apparatus, making the multidisciplinary approach advantageous. A subtext to this is apparent from the authors' own experience: result quality is nearly always directly correlated with effort expended. Even accidental discoveries are associated with someone expending energy (walking, climbing, digging). A second message emerges for those investigators with limited access to equipment or expertise: lack of one or more method of examination should not be an excuse to give up the search. The investigator may fear the criticism of the courts or peers for not using a certain technique. Instead, there is every reason to argue for the total energy expended approach. The multidisciplinary method may involve 50 people working for 2 days, equal to a single person spending 100 days searching using a map and a probe or dog. The multidisciplinary approach is of course recommended, but must be balanced against time, cost and effort in deploying techniques with limited availability, especially in dangerous, poverty-stricken, poorly-connected areas.

Example of the selective use of burial-location techniques (Ellwood *et al.* 1994)

In their 'Search for the Grave of the Hanged Texas Gunfighter, William Preston Longley', Ellwood *et al.* (1994) provide the background to their investigation. This centred on the conflicting reports of what happened to this notorious and dangerous man around the time of his hanging on 11th October, 1878, in Texas. Some said the hanging was rigged, and in fact Longley escaped to Nicaragua or Mexico, or died in less noteworthy circumstances. Ellwood and others report how a search of the Texas graveyard he was supposedly buried in was made. The background to the search is fascinating, with good documentation of how the original Longley grave marker was moved and stolen, leaving the searchers with various tangible indications such as old photographs and newspaper cuttings describing the gravesite. Historical background is essential in such cases and demonstrates the need for specialized searches of archives, where appropriate.

Extant markers were used to indicate possible original locations, compared to later modifications. Ellwood *et al.*'s (1994) resultant map is analogous to the mixed geomorphological/geographical maps of crime locations first shown by Gross (1891). With this framework, Ellwood *et al.* deployed magnetometer and resistivity surveys. Their experience of the former has been borne out in subsequent examinations of the use of magnetometry: the background noise in a location such as a graveyard makes interpretation very difficult, if not impossible. Nonetheless, in less 'noisy' locations the outline of some graves was recorded. Resistivity showed similar promises and problems. Interestingly, in such historic cases, the use of the probe and auger was not questioned, with the authors bringing coffin wood and fragments of human skull to the surface! Such destructive analysis in a recent burial would be considered a last resort, to be undertaken only with full prior discussion as to the technical merits of such an operation, and probably not to be used at all. Ellwood *et al.* use their auger results to define a lens-shaped gravel bed in one location, which they suggest would impair GPR or EM surveys, without testing these devices on the ground. Compared to France *et al.* (published 1992, but working four years after Ellwood *et al.*), the techniques used by Ellwood *et al.* are limited, the analogue being made above; where resources are unavailable, then what techniques are available should be used. In this case, Ellwood *et al.* were vindicated, in that although Longley's grave was not located, the likely locations were without evidence of his remains. The work was recently re-published and updated with DNA evidence of living relatives by Owsley, Ellwood and Melton (2006: *http://www.sha.org/publications/ha-sha/2006.htm*).

Selective use of techniques at the 'Body Farm' – Rodriguez and Bass (1985)

This work makes a second useful comparison to both France *et al.* (1992) and Ellwood *et al.* (1994), in that the former is a very comprehensive trial using pig cadavers, the latter is more selective but uses historical and archival information in the search for human remains. Rodriguez and Bass had the advantage of using real (donated) human cadavers, where they used air, soil, cadaver temperature, soil pH and texture, insect activity and vegetational changes as proxies for the discovery and dating of inhumations. The cadavers of six adult white males were placed in trenches (graves) of varying depths at different times of the year, to be periodically exhumed using the authors' skills as forensic anthropologists. Their data thus comprise mainly time-series temperature and pH data, backed up with observations on the timing and level of collapse above the burial site, carrion insect activity and changes to topsoil colour and texture. Again, the difference between this study and France *et al.* (1992) is notable, with no aerial photography, thermal imaging, geophysics, gas detectors or dog indications. This work is one of the first to describe the work undertaken at the 'Body Farm' (University of Tennessee, 1982–1984) and is essentially an anthropological study of decomposition rates in buried, unembalmed human cadavers. As such, much of the data presented are of incidental concern to the geoscientist. One aspect is discussed that concerns geoscience, and one feature of the work has subsequently become important (removed remains).

Rodriguez and Bass (1985) are clear to stress that their work not only establishes some baseline data for decay rates, and thus some suppositions as to time since death (in the geoscience [anthropological] sense of weeks, months and years, as opposed to the pathological sense of minutes and hours), but also using the decomposition of buried human bodies as the best secondary material for their detection. This is clear from the title of their work ('Decomposition of buried bodies and methods that may aid in their detection'), yet in their second paragraph, 'Methods that might aid in the location of a buried corpse were also considered in this study', almost relegates search and recovery to second best. At each burial location, pH, surface conditions, entomology, plant growth and soil temperature were noted, these being the important (bio) geographical and morphological indicators. At each location, pH measurements showed significant increases in alkalinity at ~1 centimetre above and immediately below the decomposing cadaver. Deeper burials created deeper depressions in the ground surface with time, often showing a primary and secondary collapse of the soil due to settlement and compaction followed by decay of the torso and thence later collapse. Plant growth (re-growth) was found to be variable above cadavers, suspected by the authors to be due to differences in soil disturbance. This has major implications for the detection of disturbed ground by remote sensing – as absolute rules cannot (apparently) be applied. The slower decomposition rates noted by Rodriguez and Bass for buried versus above-ground human remains is of note to the geoscientist involved in established ('cold case') investigations of murder or genocide. The insulating effect of soil that impedes decay results in the geoscientist being relevant and then possibly essential to an investigation long after surface searches and public appeals are over. Rodriguez and Bass's work has implications for the search operation: deploy pH and temperature (probes or infrared) sensors over vegetation; maintain repeated examination for collapse (primary, secondary, shallow, deep); use entomology as much as possible, remembering that deep burials will have limited insect (and other) carrion activity; and remain sceptical regarding the interpretation of vegetational changes (deploy a botanist).

9.2 Possible methodologies for non-urban underground searches

As we have stressed with the collection of bulk and trace evidence, as well as the recovery of suspect materials and human remains, no two scenes are the same, requiring on-site flexibility in strategy. On the one hand, search methodology is more straightforward, wherein a macro- to micro-scale approach is best, but on the other hand more complicated in that a complete suite of analytical tools is rarely available. In addition, the search specialist must never be afraid to examine all the data, select target areas for search, and if unsuccessful, return to re-examine the data and gain a second opinion.

The reader is referred to Killam (2004), who provides the most comprehensive published list on search planning (in his Appendices). This listing gives advantages and disadvantages of the various methods: possible uses are considered in the text and the order in which Killam lists the data and the Appendices gives some idea of his reasoning. To plan a search you have to know what you are looking for (in Killam's case, human remains, but this applies equally to any remote or buried target, especially size, chemistry and degradation/longevity in/on the ground).

Boyd on 'Buried Body Cases'

Boyd (1979) considers the specific search for buried bodies (our approach is to widen context to all covertly-buried materials). His summary diagram is particularly useful as a way of conveying what the investigator should be considering (Figure 9.1). Many people concentrate on the surface expression of a burial – naturally because they are fixating on finding the victim, etc. As Boyd shows, the succession of activity prior to, during, and after the burial are also critical; as events that alter the landscape and give rise to the surface expression one is looking for, but also to reconstruct what has happened (see the notes on stratigraphy in Chapter 2).

Initially, there is the choice of a burial location: this is not considered by Boyd, but features heavily here, because landscape plays a huge part in the behaviour of offenders. At the would-be burial location, there is likely to be soil or sediment stratigraphy, probably vegetation. These are successively disturbed and have to be stored or removed (the latter is unlikely but does happen). Figure 9.1 demonstrates this. The body (or other object[s]) is placed in the burial location, commonly elongate but sometimes foetal, depending on constraints imposed by burial pit geometry (space available, tree roots, etc.) or by *rigor mortis* in the victim. The grave is infilled with what is now foreign material: the excavated vegetation, soil, sediment and rock layers have now all been homogenized to make this effectively 'new' material. In a case in Northern Ireland some five years ago, for some reason the protagonists removed the excavated soil and brought turf in from elsewhere: likewise concrete, aggregate and rubbish (see Owlsey, 1995, on the burials at the Branch Davidian Compound at Waco, Texas) may all be placed in the grave. Attempts to conceal the surface of the grave are rare in rural locations, common in urban; the former being often 'left to nature', while in the latter, building materials, flower-beds, etc. are common, whereas bare earth is not, requiring concealment. Boyd (1979)

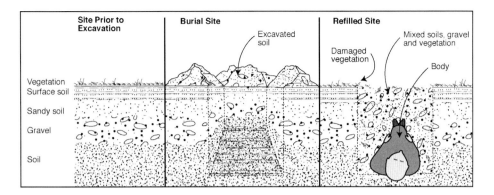

Figure 9.1 The succession of events at a burial site, from layered stratigraphy in soil and sediment (may not always be present), to removal and stratigraphy disruption, vegetation damage, body deposition, infill of homogenized and foreign material, establishment of an odd local geochemical environment. Re-drawn, with modifications, after Boyd (1979)

summarizes the offenders as having left remnants of the excavated material around the edge of the grave, disrupted stratigraphy, damaged vegetation, foreign materials introduced (infill, human bodies, weapons, etc.) and an unusual local geochemical environment. These are the features we now use to locate the grave.

Boyd (1979) was considering the local effects of burial caused by one body. Two influences have been key in changing how Boyd's work would now be considered. First, the wider influence of the surrounding landscape is now seen as important in determining where a burial is located (offender activity) and how it is located. This is largely the result of the improved resolution of satellites and aerial photography, the creation of digital terrain models and thus the possibility of understanding more about the landscape (rural or urban, or other) in which the burial has occurred. Second, the publicized discovery of large burial sites (mass graves, illegal dumps, weapons and associated facilities) has increased since 1979 when Boyd wrote his article. Thus we see anthropologists now concerned with remote sensing, geophysics and geomorphology when 20 years ago these were separate specialisms. Even if large-area data (maps, satellite photos) are not required in order to narrow down a search, they are often requested (i) in case the first search is unsuccessful (why have to re-trace your steps?) and (ii) to understand the landscape in context of other factors – offender behaviour, access, visibility, what may have happened in the area since burial (landslides, bog slides, flooding, dune migration). Hence many searches still begin with all the large-scale data available on topography (maps, Digital Terrain Models [DTM]), vegetation (maps, remotely-sensed data), soils and geology (maps, remotely-sensed data), drainage (Figure 9.2). These can be combined in the GIS environment for comparison and possible secondary remotely-sensed data acquisition, or for onsite, non-destructive surveys by geophysics, search experts, dogs and proxy geochemistry of gas, soil and water. Collectively, these data will produce a range or hierarchy of targets, again for possible investigation prior to excavation by surveying, laser scanning, fingertip search and monitoring of probe holes, gas

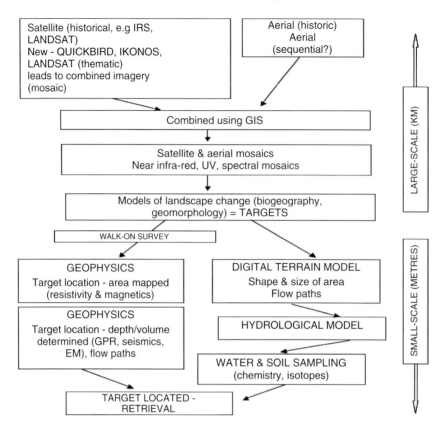

Figure 9.2 Possible search methodologies for buried materials, presumed to be of significantly different chemistry to their surroundings (human, animal cadavers, explosives, drugs, toxic materials)

emission and water chemistry. Finally a decision is made to dig or not dig, but it will have been based on solid evidence, and if unsuccessful, can be used to excavate secondary targets and understand why they have not been successful. Large-area maps (topographic, hydrology), DTMs and landscape interpretation are also needed when planning excavation activity for health and safety assessment of natural and induced hazards (flooding, landslip, subsidence).

9.3 Underwater searches and scene mapping (remote sensing, geophysics)

In Chapter 2 we saw the importance of water to the work of anyone conducting searches and indeed collecting evidence. Water covers over two thirds of the planet, is largely impenetrable to regular human activity and is thus both an ideal location for the covert hiding of objects or movement, as well as an unpredictable and dangerous environment for the suspect, victim and search personnel. Water-cover on the planet can be thought of at a range of scales, from the micro (ponds, ditches, streams, wells) through to the macro (rivers, lakes, seas, glaciers, oceans). A number of attributes dictate the search and mapping methodology of water: target makeup (organic, metal, radioactive, unusual chemistry), area, depth, chemistry, physical properties (frozen, suspended solids/vegetation) and access, among others. Thus the method of searching and mapping or sampling water is always dictated by 'fit for purpose'. Here we examine a range of aqueous search locations and consider different objects or events in each. These are nowhere near comprehensive (as for any criminal investigation in this book), but serve as examples. Often, the combination of location and target will not have been searched for or examined before, requiring consultation with experts in specific fields, brought together under one operation.

Small water bodies

These are generally in non-marine locations, the only small marine searches being in shoreline pools, embayments and harbours where visual/manual examination and draining are possible. However, not all non-marine small (typically, a few metres in one dimension) water bodies are necessarily fresh, as organic content (peat, slurry), pollutants and salts (marine ingress, salt-prone soils and bedrock) can occur. Thus when a pond, ditch, well, water-hole, stream requires searching, mapping and sampling, the size of the water body must be compared to what search methods may be deployed.

Prior to any diving or wading, which may dislodge evidence, the available remote-sensing and geophysical tools that can operate in the space available ought to be considered. These may define a range of targets, hopefully with a hierarchy of positive attributes, as in the remote-sensing case study on large burials (page 110 in Chapter 4). In clear waters, remotely-controlled cameras are invaluable. Metal targets may be scanned for by boat-towed magnetometers or a waterproof metal detector, with locations marked by small buoys or differential GPS. Organic targets may be established by UV irradiation (in darkness) or analysis by UV filter from sunlight, by scent dogs deployed from shore or boat, by geochemical sampling for organic compounds or diagnostic isotopes of the air immediately above the water and by the water itself. The latter generally require laboratory analysis and thus a time delay in establishing targets: a solution is the deployment of gas detection systems such as the laser spectroscope arrays discussed in Chapter 4, with the disadvantage of atmospheric and non-diagnostic gas generation. Many commercially-available GPR systems are small enough to be deployed from walkways, over the target or from small flotation devices (polystyrene boards, small rubber or fibreglass boats). The automated function of GPR may not require a person to be adjacent (Figure 9.3).

Conversely, commercial SONAR systems are designed for deployment from larger boats searching larger areas: depth echo sounders are smaller but do not achieve the penetration of sediment that CHIRPS and similar sub-bottom sounders or GPR systems obtain. SONAR and CHIRPS are advantageous over GPR when searching organic-rich water (or other waters with high conductivities) for organic targets, but rarely, if ever adapted for use in small areas. There are two schools of thought when

Figure 9.3 The use of remotely-controlled GPR for non-destructive and non-compromise (divers or waders disturbing sediment) search of small water-bodies (see text for explanation)

such problematic areas are searched: first, that science can overcome the problem and effort expended in obtaining a sampling or search device, or second that if no method is found after an established period of effort, one gives up and moves to manual search by diver or drainage, with attendant compromise of evidence or environmental damage.

Intermediate-size water bodies (rivers, lakes, reservoirs, estuaries)

These are common search locations as they are not as easily compromised by discovery as smaller water bodies, which may drain, the water clarifies, or be observed by witnesses. Unlike the open ocean, they allow easier, less hazardous access. Smaller and intermediate water bodies may also prove attractive to the criminal mindset where control is desired: either by viewing the location of a victim, or by return to a hidden store of weapons, explosives, drugs or valuables. Although the open sea and oceans are far more challenging for the search team, these places are dangerous, unpredictable (creating superstition in the suspect) and cannot be controlled or returned to without GPS or landmark and specialized equipment that may prove incriminating.

Remote sensing and geophysics are essential in the search of intermediate-scale water bodies, as physical contact (diving) is time-consuming and possibly dangerous. Again, chemistry, size and target dominate the choice of methods used: metal objects in freshwater and marine locations may be mapped using boat-borne magnetometers, where location is critical because their sensitivity, especially in urban environments, may provide hundreds of targets. Organic and non-metal objects in all navigable locations may be imaged by SONAR or a high-resolution seismic method, CHIRPS (Chapter 3). Small craft may use SONAR for sediment surface objects combined with GPR, which can also be used alone if SONAR is unavailable.

One of the most extensive uses of SONAR and CHIRPS sub-bottom imaging to date was the search of San Francisco Bay around Berkeley Marina (Lee & Labriola, 2006). This was stimulated by the disappearance of the 8-month pregnant Laci Peterson and the coincidental decision of her husband to go fishing alone, with inadequate tackle, not long after he confessed to having an affair. Trawling, diving, boat-borne cadaver dogs and the above geophysical devices failed to locate Laci's body, which was washed ashore, badly decomposed a few months later.

Case studies: GPR used in searches of small, intermediate and large water bodies

Small water body. Local authorities in the Irish Republic carried out a combined investigation into illegal dumping on land at the border between counties Fermanagh (Northern Ireland) and Cavan (Irish Republic). Animal care charities were later contacted because there was evidence of animal abuse. Most seriously, there were suggestions that diseased animals had been illicitly disposed of in a ditch (on the accused's land) and adjacent flooded quarry. This accusation needed to be verified before the police were asked to take control of the case. The possible deposition areas could not be easily drained for environmental and logistical reasons, nor could divers be used because both contained very high volumes of suspended solids: some areas of the quarry were close to 50 % water, 50 % sediment. Most specifically, intelligence from a local itinerant farm worker suggested that a calf cadaver whose suspected death was from foot and mouth disease, was in the ditch, and that numerous sheep that may have died from scrapie, were in the flooded quarry. The ditch was searched by placing 200 MHz (see Chapter 3) GPR antennae in a small rubber boat, which was towed along the ditch and in the flooded quarry (Figure 9.3). The only problem encountered was the occasional snagging of tow-ropes on overhanging vegetation. A number of small, hyperbolic targets were identified as discrete objects that could be rocks, implements, rubbish or bones, but only one significant anomaly was observed (Figure 9.4).

Although the shape of this anomaly was apparently inconsistent with a calf, the dimensions were correct, and as the only anomaly it was decided to dam the ditch either side, drain as much as possible into secure containers (for recycling) and examine the anomaly. This was confirmed as a calf. Later intelligence indicated that the witness had been accurate in his locating the calf, but mistaken in when deposition had occurred. Thus the animal was highly degraded, with part of the

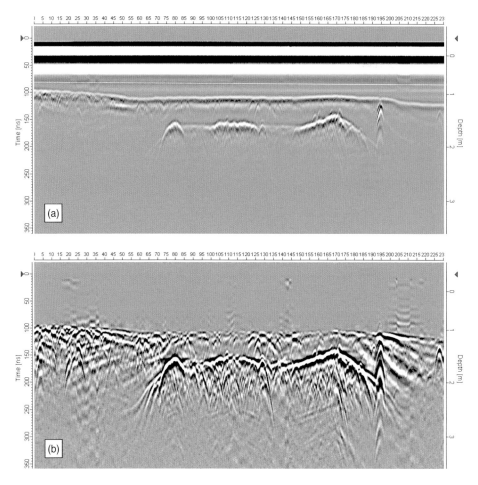

Figure 9.4 Unprocessed (a) and processed (b) 200 MHz GPR data from the ditch (Figure 9.3) where the anomaly shown was observed. This was later recovered as a diseased calf, considered a serious offence due to the threat to human and animal life. Profile is 2.3 metres long

skull smashed and the body cavity collapsed, causing the unusual shape on the key GPR profile. Flooded areas of the adjacent quarry were also searched by the same means, with less success. The target areas were filled with liquefied mud. This area could not be walked on nor have a boat effectively floated on. The boat, with both 100 and 200 MHz antennae, was dragged across the target area: the resultant GPR record was dominated by the excessive ringing and laterally-inconsistent depth penetration typical of highly conductive materials, such as seawater (Figure 9.5). No discrete anomalies could be identified for recovery and the GPR operation was abandoned. Unlike publications on terrestrial GPR and burial location, there are no published case studies of a human cadaver alone being detected by water-borne GPR. M. Harrison (personal communication, October 2004) has, nonetheless, used

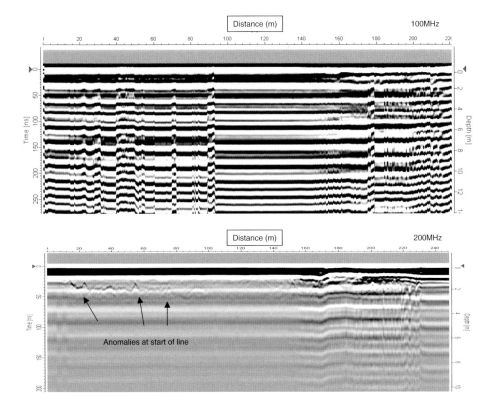

Figure 9.5 100 and 200 MHz data collected remotely (using a small rubber boat and Bluetooth control unit) along the length of the infilled quarry ditch (Figure 9.3). No water analyses were taken in order to establish where the radar data is so poor, but the similarity between the 'ringing' on these profiles and that recorded on salt water is identical, suggesting a high level of electrolytes in this water. It is important to show poor data and where methods do not work!

the method many times and successfully imaged a drowning victim in a lake in North America.

Intermediate water body. Ruffell (2007) reports on the GPR-based search for a submerged, partly-buried (by silt) jet ski involved in a collision with a speedboat in a shallow lake. The location of both was crucial in establishing events at the time of the accident. An evening collision between a jet-skier (adjacent to the walkway) and shallow-draft speedboat caused severe damage to both vessels, submerging the speedboat, sinking the jet ski and severing the jet-skier's leg at the knee. GPR was considered a useful method by which targets could be identified in shallow water (sometimes 50 centimetres deep), with the ability to image the sediment subsurface, because of silt deposition. The small width of the embayment and shallow water made use of boat-borne sonar or seismic sub-bottom profiling impractical (see above). Previous work suggested that a major problem in GPR surveying on (or in) freshwater is the 'ringing' caused by impedance mismatches between the antennae and the water or from bouncing off terrestrial objects. Freshwater GPR is generally successful (see above) if a range of antennae and survey methods are deployed.

In this case, flotation of GPR equipment by a variety of means allowed collection of a grid of 100, 200 and 400 MHz data. In waters over 6 metres deep, GPR data showed the form of the lake floor, obscured by ringing in the data. In waters less than 6 metres deep, this ringing diminished on both 100 and 200 MHz data, the latter displaying the best trade-off between depth penetration and horizontal object resolution. Surface objects such as a wooden walkway caused interference on 200 and 400 MHz data when antennae were oriented both normal and parallel to survey direction; this may be a function of the low attenuation of radar waves in freshwater, allowing excellent lateral and vertical radar wave movement. Data were viewed in the field as raw output and with limited processing: background removal eliminated 'ringing' on the records from deeper water but caused a problem as this processing step removed flat areas of the lakebed itself. Problems were encountered included high winds moving the survey dinghy off line and by the lack of manoeuvrability in the dinghy near obstacles, which had to be dragged over reeds or shallow areas.

100 MHz antennae, with no preferred orientation on this open lake with no isolated upstanding structures, produced the best images of the shallow subsurface of the lake floor. In areas separate to the scene of the accident, few lake-floor or shallow-buried anomalies could be seen (Figure 9.6).

The identified anomalies comprised 1–3-metre-long objects of 10–20 centimetres height. Closer to the likely scene, numerous high-relief anomalies were observed (Figure 9.7).

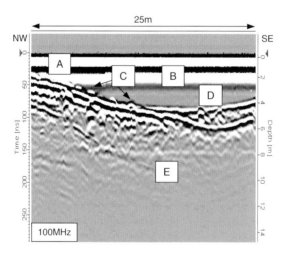

Figure 9.6 100 MHz GPR data obtained by placing the antennae inside a rubber dinghy and rowing over the search area (modified after Ruffell, 2007). A = 'ringing' in the data; B = water; C = isolated objects on lake floor; D = objects in lake sediment; E = lake sediment – bedrock boundary

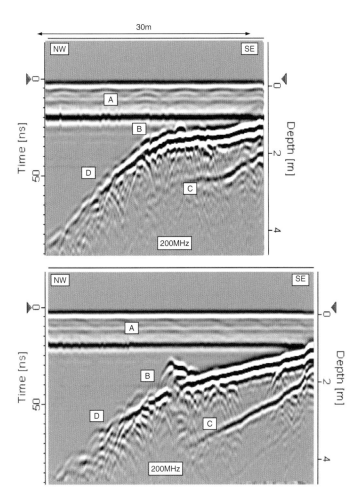

Figure 9.7 Profiles adjacent to the wooden walkway. The anomalies observed are debris from the submerged jet ski, (modified after Ruffell, 2007). A = interference from walkway; B = targets on lake floor; C = lake sediment – bedrock boundary; D = base of lake

These comprised features less than 1 metre across, often creating hyperbolae at the sediment–water interface or just below. The main feature of interest comprised an asymmetric feature about 30–40 centimetres in height, with the shallow-side to the up-dip, southeast. This and other anomalies appear to 'float' in the water column: physical investigation by divers found such objects close to, but not directly under the survey line, suggesting they are out-of-plane features on and in the lake-bed.

Large area: murder victim in reservoir. The last-known location and associates of a missing person in Northern Ireland prompted a wide-scale search of large areas of farmland, quarries, ditches, building sites and the sea. These failed to produce significant information, and so a public appeal with details of the last-known sightings and clothing of the missing person were circulated. This generated one particular piece of information. A large (2–3-kilometre long, 1–2 kilometre across) freshwater lake was known at the limits of the then designated search area, used for water abstraction and treatment for potable water, and by local duck-hunters and freshwater fishermen. The keeper of the hunting boats noticed one boat had been used at about the time of the supposed abduction, and thought this may be poachers. However, a second witness noticed activity at the remote end of this lake at the same time. Thus the entire lake became subject to searching, with some focus at one end. Scent dogs were deployed on boats that likewise showed more interest in certain areas of the lake. The problem remained of how to focus any diver-based search in waters of 4 metres depth with unknown thickness of silt below. Boat-borne GPR was considered and the following positive and negative attributes determined.

The lake had low conductivities and a neutral to very slightly acid pH; access was easy and safe from the boating jetty. Conversely, the area to be surveyed was very large and in a windy, upland location where storm wave height could be up to 50 centimetres, making surveying hazardous. The lake had jetties for water abstraction, buoys for fishing vessels and at the periphery, semi-submerged trees. These were all likely to create anomalies on the GPR profiles. Determining boat location in such open water, without differential GPS was to be a problem. The size of the survey area was overcome by use of low-frequency (50 MHz) unshielded antennae: these have deep penetration and large footprint, effectively 'seeing' more of the water and sediment that the focused nature of the shielded antennae. When anomalies (Figure 9.8) were located, a weighted buoy was dropped: in other circumstances differential GPS would have been preferred. A problem with these 50 MHz antennae was interference from out-of-plane objects (trees, buoys, jetties) that obscured submerged features. All targets were then re-surveyed using lower-frequency, 100 and 200 MHz antennae, in order to better locate any anomalies. Although the missing person was not found, one anomaly that was repeatedly imaged (Figure 9.8) was located by divers and discovered to be a semi-submerged tree stump of 35–55 centimetres diameter and 1.3 metre length, proving the efficacy of the technique. Marine inlets, estuaries and seas have similar problems to this survey area, where low-resolution, wide footprint devices such as SONAR and CHIRPS may be deployed in collecting grids of data that allow target definition for divers.

Case study: Sampling assists searches 'Tiger' Kidnap in northwest Ireland

The 'Tiger' kidnap method has a long history. The web site *http://www.doubletongued.org/index.php/citations/tiger_kidnapping_3/* states that 'More than 80 per cent of "tiger" kidnappings involve ethnic minorities. There has been a particular problem with Chinese nationals who have entered the country illegally. They are often snatched by gangs linked to the Triads and then their families in China are pressured into paying a ransom.'

This phenomenon has now permeated many societies where intimidation is routine, including those where gang-led control of territory, drug-dealing, the sex

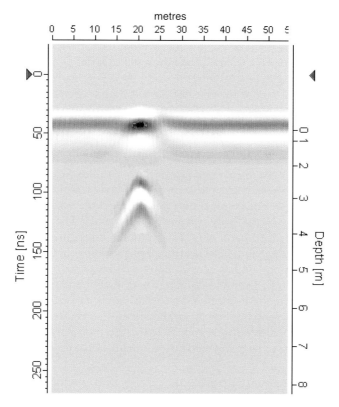

Figure 9.8 50 MHz GPR data obtained on a large, harsh environment (0.5 metres high standing waves, 45 mph winds, 2 °C above freezing) inland reservoir in Northern Ireland, during the search for a missing person. The 50 MHz cable antennae (see Chapter 3, Figure 3.7) was encased in water-proofing and towed behind a motorized boat with DGPS. The electric engine of the boat had no effect on the radar. The only anomaly proved to be a buried tree stump, proving that the method worked (the police and army divers had identified no targets until this survey was run). The victim in this case is still missing, but not presumed buried in this location

trade, etc. occur. The web site (*http://www.ibp-group.com/news/news_archive/2005/20_sep_05.htm*) goes on to state 'The term "Tiger Kidnap" is national police shorthand for a particular type of kidnap where a manager, or similar person who is in a position of trust, or a relative of that person, is kidnapped to enforce their cooperation in the theft of money or goods under their control.'

In the present case study, the chief finance manager of a manufacturing company in northwest Ireland was abducted and instructed to bring the end-of-month wages to a rear door of the plant some minutes prior to the factory closing for the Christmas (December) national holiday in 1999. At the same time as abducting the manager, other members of the gang took his wife to a safe house location, and let the manager know they had her and she was alive. The gang had been extremely astute in their timing and forensic planning: the pre-Christmas break meant that the wages would contain any bonuses due. In addition, the large number of temporary staff at this time of year, plus recently-arrived Eastern European workers, meant that nearly all the wages were paid in cash. The gang wore forensic suits, took the manager's wife to the safe house in her own car, lined the house with plastic flooring and made her sit in one chair during the operation. Following this, the victim was taken in her own car again to an abandoned quarry, her car burnt and she was left to walk some way to raise the alarm.

Her clothing, shoes and other effects were all examined and control materials taken from her home and work. Few fragments of trace evidence on her clothing showed any indication of being from the hostage location. The only geological clue was provided by chance. As she stepped from the 'clean' hostage location, into her washed shoes, she lost her balance and leaned heavily against the lime-washed wall of the porch area. The side of her coat and hair brushed the wall heavily, a contact not noticed by the gang. A two-way transfer had occurred. The historic lime-wash was found to contain distinctive minerals (epidote) associated, from geological surveys, with a known nearby beach location. Thus to find a lime, beach sand, and epidote-bearing mineralogy on her coat in hair-brushings was excellent. However, many houses in the vicinity have old buildings with lime made from the calcareous beach sand. These were surveyed, and one location in particular was examined. Here, on one side of the porch, fibres from the victim's jacket were discovered. The geological analysis had directed a search that led to confirmatory evidence being found. This trial has yet to be completed, and so the details are necessarily vague.

Underground scenes (mines, caves, sub-ice)

It may seem of limited application, but mines, caves and specifically excavated underground locations (bunkers, shooting galleries) are often the scenes of covert activity as well as industrial accidents that the forensic scientist or investigating officer has to consider and help investigate. The principles of spatial awareness (Chapter 2) pertain but measuring and sampling become difficult. As with normal communication devices, GPS will not work underground, so location is made by traditional tape and laser range-finder, or total station, referenced to some area or object that can be preserved against excavation or rock-fall.

The advantage of underground scenes of crime is entrance and exit routes are limited, and thus sampling points have a forced location. Mines are constructed in order to extract valuable materials, which may have extensive previous analytical work on them and have a less widespread distribution than surface soils and sediment. This makes initial comparisons of suspects who deny being at the mine to such a scene more straightforward. Similarly caves are also filled with characteristic deposits not regularly found elsewhere, although in extensive cave systems, obtaining sufficient samples may prove challenging, especially if a suspect admits to being in *a* cave, if not *the* cave.

Geoforensic practitioners may well find themselves assisting search and rescue operations in land- and mudslides, building collapses and post-flood burials following earthquakes, volcanic eruptions and tsunamis. This is because the materials covering victims, vehicles, dwellings, etc. are almost always geological or sedimentological in nature, and the geoscientist is used to the behaviour of such dynamic materials and the methods needed to assess them. The behaviour of materials is often of great importance in this regard: in the Boxing Day Tsunami a number of people went missing on a Thailand beach. They were presumed buried by sand, as the witness reports stated seawater covered the beach and adjacent seawall and road at the height of the inundation. A coastal geomorphologist was consulted, who realized that the beach had only been smoothed by the wave, with no significant deposition. Instead the collapsed sea wall was noted, and compared to postcards of the location before the disaster. This showed how beach visitors liked to sit with their backs to the seawall: lifting of the concrete slabs retaining this wall revealed the majority of the victims. A similar problem occurred following the massive Indian earthquake centred on Gujarat province (26th January, 2001; approx 20,000 killed, 166,000 injured, 600,000 made homeless). Here, when victim recovery dogs arrived, the number of buried dead was so great that a means of searching for the living was required. Thermal imagers appeared not to function well, until they were deployed early in the morning, when contrasts between cool debris and warm humans was greatest, allowing the focused use of excavation and dogs during the day.

Ice may seem an unusual medium for searches, and yet materials and people hidden below ice present particular problems for the search and recovery team. Although such items will inevitably become unfrozen and exposed at some time (summer, or due to long-term melting), the capture of materials in ice is different to many other burial media. For instance, in the rush to recover an object, the evidential value of the ice itself is often overlooked – it being dug out or melted away with warm water, when it may capture materials from the suspect and victim or cache from the time of deposition that without care may be washed away or left out of context. Ice may form a stratigraphic layering, like soil and sediment, and only a glaciologist will be able to spot how snow and ice has been excavated and infilled.

Two examples (large scale and medium scale) are here summarized to show the challenges associated with the search for and recovery of evidence from ice. At the large scale, the story of the 'Glacier Girl' airplane shows how ice may be searched. In July 1942, a squadron of six P-38 fighter aircraft and two B-17 flying fortresses left their air base in Maine to join the build-up of US forces in Britain. Flying the short-circle route, the squadron hit a snow-storm and made a forced landing on an ice floe in Greenland. All 25 squadron crew were rescued, abandoning their aircraft on the ice. The aircraft were inundated and buried, later to be termed 'The Lost Squadron'. Vintage aircraft enthusiasts used a magnetometer to locate one of the B-17s in 1990, at a depth of around 150 metres. Continued use of the magnetometer was abandoned because the effects of polar magnetism compromised data interpretation. The remaining aircraft were located using a GPR system, and one ('Glacier Girl') was selected for retrieval. This P-38 was recovered, restored and flew for the first time in 2002 (Fildes & Williams, 2001).

At the medium scale, the problems of body recovery associated with the 'Man in the Ice' (Spindler, 1993) demonstrate how poor positioning of the remains (no GPS was available) and the inability to preserve such an inhospitable and isolated scene led to argument as to which country the body was recovered in.

9.4 Gas monitoring, organic remains and the decomposition of bodies

Boyd (1979) provided one the first descriptions of the use of soil temperature and gas probes in defining covert burial sites. He gives well-founded advice on the deployment of such probes, being obviously (if not overtly) aware of spatial and stratigraphic sampling issues. The investigator cannot hope to randomly-place a probe in a large search area and produce a definitive result. The large search area must either be subject to some form of target-defining method (thermal imagery, geophysics, subsident or soft ground, offender access/behaviour) at which the probe is inserted, or a suite of probes deployed by informed opinion about soil permeability, gas movement and age/decay of inhumation. Exceptions do occur, however. Boyd himself recounts the use of the probe in a hole drilled into concrete, where a burial was suspected. The concrete acted as a form of cap, trapping the gas: if the probe was inserted at the greatest elevation in the concrete, it would capture a wider area of gas escape than otherwise expected. Boyd also envisages stratigraphic control on gas probe results, with probes placed too far or too deep not recording any escape gases (Figure 9.9).

The spatial distribution of gas monitors and the effect of soil gas permeability were not dealt with explicitly by Boyd, and the detection of organic remains by remote sensing or dogs was not re-visited in the literature until Ruffell (2002), who examined the application of soil gas probes and considered intervening advances in spatial sampling, especially by laser monitoring (see Chapter 3 – this aspect is more

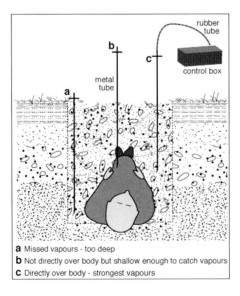

Figure 9.9 The emplacement of gas-capture probes in and around a buried body. These devices may be interfered with by animals and humans, but can be easily recovered. Redrawn after Boyd (1979)

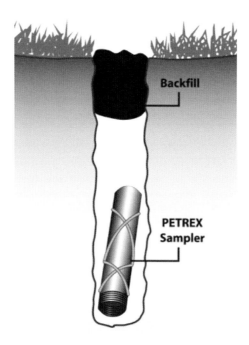

Figure 9.10 The buried gas sampler usually has a permeable base and a deactivated charcoal (or similar) capture medium. May be hidden covertly and thus not compromised by animals or humans, but without DGPS or a mild radioactive source, can be hard to re-locate. As in Figure 9.9, incautious deployment of such probes may damage evidence. Modified after Ruffell (2002)

akin to remote sensing). The advantage of using the scent of gaseous emissions in the search for buried bodies, illicit animal disposal sites and toxic waste pits is the rapid and wide spread of the scent, as well as the rapid walk-over. The disadvantage is the subjective nature of what is being detected and the specific location (wind direction, rate and amount of gas generation over time and season, precipitation). Ultimately, gases such as methane are valuable indicators, not to be relied on alone, and thus if detected by search dogs, *in situ* probes or laser interferometry, all provide 'targets' for subsequent exploration. The gas probe may be completely buried in a pit to avoid compromise (Figure 9.10).

Nothing is new – Heinrich did most of the above 80 years ago

At the beginning of this chapter, the search for large objects (mass graves, bodies, explosives, weapons, contraband and drugs) was outlined and then developed. The other reason for a search is to recover evidence, and it thus falls also into the realm of sample-taking and recovery. Areas to be searched usually comprise the scene of crime(s), suspect alibi locations, victim locations and control locations. Scientifically, all require the same methodology and attention to detail. However, forensically the scene of crime is concentrated upon, it being likely to yield the most valuable clues concerning the last point of contact between suspect, scene and victim or objects (drugs, explosives, contraband, pollutants). The geomorphological-based approaches to searching an area were pioneered by Gross (1891), who recommended that investigators visit and familiarize themselves with possible covert locations for activities (forests, brothels, hidden areas) or storage/dumping (ponds, wells, streams, forests) that could be connected then (or later) to criminal activities.

The modern methodology, of beginning a search at the scene of crime, and radiating out, was probably known to Gross, but elucidated by Heinrich. In an investigation of a massive road traffic accident in August 1930 to the south of San Francisco, Heinrich was asked to reconstruct events previous to and during the collision(s). He deployed a classic methodology, having the highway closed and *beginning at the known point of impact (in effect, starting at the last-known scene of crime), radiating out*, mapping all the features he noted and giving them a hierarchy. A fresh impact into the asphalt was of particular interest, because when the suspect vehicle was examined, a bolt on the front underside matched the imprint in the tarmac perfectly. Heinrich did not stop with this simple link: he widened his search to encompass the larger part of the highway and recovered yet more evidence of what had happened. His methodology, while pertaining to a motor vehicle accident, set a standard for all searches. Geoscientists, when asked to assist in the search for materials, are wise to deploy one such methodology: they can do more though, and see the whole environment in its wider context, bringing to the investigation what they see in terms of landscape change, peculiar materials and soil disturbance, what the scenes of crime or environmental or litigation-inspecting officer(s) may not observe.

9.5 Weird and wonderful burial locations

It seems very dark humour, but there are a number of strange burial locations, hoaxes and mistakes that deserve mention here (all illustrated in Figure 9.11).

Figure 9.11(a) shows how the dead never give up their location. In this case, the grave of a person involved in the 1798 rebellion in Ireland remained undetected for 200 years in this upland location until this hollow appeared in 2003. Presumably the cadaver had remained intact long enough to keep the weight of soil and rock from falling in all this time.

Figure 9.11(b) shows a gravel-covering to a driveway where a major toxic oil-spill occurred. The gravel remained wet, no matter how dry the weather, as oil does not evaporate! Thus a dark-staining remained as evidence of the oil-spill.

A drug dealer in Scotland had indications that he may be under surveillance by the police and hid his store in a variety of locations. The disturbed earth was more permeable than that surrounding, and thus in dry weather caused the grass to die: the locations are shown in Figures 9.11(c) and (d).

Figure 9.11(e) is the supposed location of well-known missing person in North America. The tarmac forecourt to the garage continues to subside in a roughly-grave shaped manner, adding to speculation that the person is buried here.

Figure 9.11 Various strange burial locations, hoaxes and mistakes – see text for full details

In Figure 9.11(f) a coffin- or body-shaped depression can be seen in a scrub woodland location near industrial units and a young offender's centre in Northern Ireland. The trench is also only some 10 miles from the last-known location of a no-body murder victim. The subsident area was noted by a man walking his dog (they always are!), and the police informed late on a Friday afternoon. By Sunday evening, two visits by GPR experts, scent dogs and a full archaeological excavation revealed this to be a test-trench for the monitoring of the local water-table, a common type of excavation used by engineers to avoid drilling of boreholes.

10
Circuit complete

At the start of this text in Chapter 1, we suggested that to give a definition of our understanding of 'geoforensics' would require a book as long as this one. If you have stayed on board and read even a fraction of this work, hopefully you will now see what we were driving at: the book is itself the definition. The broad range of the scope of geoforensics shows that far from being a text on soil and people's shoes, amazing and sometimes weird applications of Earth science to litigious problems occur.

In *Chapter 1*, the case for each sub-discipline learning from the others, and not trying to out-compete one another, was made. Just as geoforensic workers usually have a former discipline (geophysics, micropalaeontology, mineralogy), which they apply to new problems, so environmental, humanitarian, criminalistic, (or blend of all three) specialists can greatly assist fellow forensic practitioners by sharing their experiences. People love stories and they like to hear about personalities: as demonstrated in Chapter 1, strange and important stories abound in the geoforensic literature. As for personalities, we can only use the literature of the authors or their biographers to conjecture on what Gross, Popp, Locard and Heinrich were like. This is a young science, with practitioners alive (and still quite young!) who worked with Heinrich and Walter McCrone, yet in many ways the reality of their characters becomes wonderfully blended with their remarkable achievements. Gross (1891) wrote a book that contains more common sense than any current text. In 100 years' time, perhaps today's forensic textbooks will appear over-burdened with bureaucratic attention to detail. Maybe by then much outdoor and laboratory-based analysis will be automated, allowing us to return to Conan-Doyle's intuitive, thinking detective. Today we think nothing of taking any evidence that links a suspect to a scene to court: it is our challenge to explain these materials and their makeup to judge and jury. Yet imagine Georg Popp bringing soil and other evidence to the police and justice system for the first time. We touched upon the equivalent today in Chapter 8 ('Trace Evidence') where what appears intangible or out of control to us today (contact with water and air) will perhaps in 100 years' time be viewed as Popp was.

The present-day media vogue for interest in the 'forensic' has its unfortunate repercussions, such as misrepresenting the time and effort required to carry out the

Geoforensics   Alastair Ruffell and Jennifer McKinley
© 2008 John Wiley & Sons, Ltd

work of the search coordinator and trace evidence analyst. However, perhaps this publicity may embed itself in some thinking, questioning and truly inspirational people who can do what Locard did: establish so simple a principle, yet as applicable in the early 20th century as it is today, that a real challenge for the geoforensic practitioner is set, and instead of us playing catch-up with the criminal, we find ourselves once again ahead of the game. With Heinrich and McCrone, our conjecture becomes based on more facts: Heinrich 'The Wizard of Berkeley' and McCrone 'Debunker of Legends' (Lewis, 2002) were both legendary in their use of microscopes and other analytical techniques: unlike Popp, Gross or Locard, their legacy remains but has yet to be realized.

In *Chapter 2*, those who know geoforensics as the application of soil, sediment and rock analysis in exclusion/comparison of suspects/victims to scenes are challenged. The 'geo' in the title applies equally to all aspects of physical geography as to geology, with meteorology (Japanese incendiary balloons and Kaplan's (1960) 'weather and climate' (see Figure 10.1)), geomorphology (assisting searches for buried materials such as pollutants) and hydrodynamics (see Figure 10.2) playing important roles in establishing what happened, when, and why it occurred.

Chapter 3 demonstrates how perhaps a popular concept of a scientific method can be so wrong. When the search for dinosaur bones, the mapping of toxic waste, the definition of mass graves or the search for buried murder victims all fall under the same 'geophysics' chapter heading, then what appears as a rather specialized set of tools for ground investigation becomes an essential part of subsurface investigations. The same effect has been achieved by the popular UK television programme 'Time

Figure 10.1 The weather and crime. In this case, a small fire was started to burn waste materials associated with the processing of illegal fuel, so as to avoid detection. In trying to avoid detection, a second crime was committed: strong winds fanned the flames, causing a far larger fire than anticipated, and creating such toxic smoke that two down-wind people with respiratory complaints were hospitalized, one fatally

Figure 10.2 Oil spill in Northern France. Hydrodynamics can be used for a variety of purposes in investigating where materials will end up. The one everyone thinks of, of course, is the dead body, but far more common is the massive environmental crime of oil spills. Without the oceanographer, disaster management (directing resources) as well as back-locating the spill to a likely origin, and thus maybe the offending ship, would not be possible

Team', where the historian, landscape interpreter and geophysicist work initially to provide the excavators targets for their dig. Even a chapter on geophysics has the same problem as the whole book: rarely is one geophysical technique used in the assessment of buried materials. Commonly, landform, geology, remote sensing and some limited form of intrusive survey go hand-in-hand with geophysics, which is not magic and will only be as good as the effort put into the whole survey, including associated investigations.

In *Chapter 4*, the casual reader will initially be intimidated by the terminology and physics/chemistry behind remote sensing, when actually the types of data are very familiar to us through GoogleEarth™, cinema films (*Enemy of the State*, *The Conversation*, *Mission Impossible*) with a military or surveillance aspect to them, and from news reports of war and mass disaster around the world.

Remote sensing sometimes provides the solution to the forensic problem (for instance, the actual burial location); more often than not, remotely-sensed data does much the same as geomorphology or geophysics, helping us to understand the landscape, to look for patterns or anomalies, and thus to focus resources. This may be true of the satellite image, or of the hand-held spectroscope: both are providing proxy data for actual ground conditions, that if necessary, will have to be dug up, scraped off, drilled or removed for analysis. The 'Star Trek' day when we point a device at the ground and learn all we need to about it is still a long way off. Chapter 2 described the use of very traditional techniques in a new context, Chapters 3 and 4 introduced technology that is rapidly changing, so it is logical that *Chapter 5* be on the most recent advance in geoscience, the improvement in spatial location that is borne of the Global Positioning System(s) analysed using some form of GIS software (e.g., ArcGIS™). Like the familiarity we have with geophysics and remote sensing from popular media, so the importance of GPS is easily understood by reference to the modern motor vehicle, which even middle-price range models now possess: 'built-in GPS for route finding'. Geography and allied disciplines have always been about maps, and the GPS is just a better, interactive way of displaying and describing locations in map form.

Popular familiarity with these technologies begins to erode, however, when discussing the means of manipulating and displaying such data. When purchasing Ordnance Survey maps, a digital output is commonly the first stage in assessing customer needs, yet a flat map on paper is the ultimate output. The Digital Terrain Model is not commonly seen, nor needed, being a specialist device for the visualization of the land. Yet the DTM is incredibly important for the geomorphologist, who can assess much of the area to be investigated before even visiting the location.

Chapter 6 likewise covers topics rarely considered in forensic geology or even in general forensic science articles: geostatistics. The soil scientists have discussed widely the issue of sampling; geoforensic practitioners have likewise touched upon issues of sample representativeness and statistical tests of such for display and testing in court (Figure 10.3).

Like GIS, the power of modern computers to provide even the most simple of statistical or graphical tests of data comparability has resulted in increasing use of computerized data handling in forensic cases. As seen in following chapters, the increase in automated sample analysis and the need for multi-proxy data sources necessitates computer power to store and display these multivariate data.

Figure 10.3 Spatial sampling of shallow water for total organic carbon and isotopic characterization: in this no body missing person (murder) enquiry, behavioural profiling, police intelligence and scent-dog indications led to this location, where this innovative method of sampling and GIS-based data handling were used to better locate a location to excavate

Once taken, a sample requires analysis: *Chapter 7* displays the mix of established methods of rock, soil and sediment analysis and newly-developed techniques that now make up the battery of techniques available to the physical evidence analyst. Tried-and-tested methods are of course favoured by scientists and by the courts: the need for multi-proxy analyses allows the increasing integration of both traditional and newly-arrived methods, providing both internal control and an assessment of new and old. The problem created by the repeated analysis of essentially the same material by many different methods shows, of course, that results will be comparable. Hence the power of integrating a non-geological method of analysis: this has traditionally been microbiology (spores, pollen, diatoms and more rare organisms). Therefore, elements of the solid geosphere and biosphere are used. There is evidence of an increasing use of material from the hydrosphere (water residues, fluid inclusions) and even atmosphere (gas inclusions, trapped air, particulates, pollutants) in material forensic analysis. The challenge provided by analysis of gas, air and water should not dissuade us from the power another independent proxy will provide in our exclusion/comparison of materials in criminal cases. The same is happening to our allied forensic disciplines: in Chapters 7 and 8 we also examine how geoscientific methods may be applied to unusual materials such as builder's products (concrete, lime, bentonite), wooden safety matches (Figure 10.4; see the case study by Farmer *et al.*, 2007, in Chapter 8) and explosives.

Chapter 8 follows in the same manner as previous chapters – from the traditional to the innovative, from the simple to the complex. This does not denigrate traditional geological analysis but more reflects the problem inherent in this book: criminal concealment, cover-up and clean-up. One way of looking at this problem is that it is only a matter of education: at the start of Chapter 8 we heard how Kim Philby, the British spy, knew about the cover-up of trace evidence by deliberately compromising a trowel used to bury incriminating evidence. He was an intelligent person. Popular media and offender communication merely bring such ideas into the criminal fraternity: even so, an understanding of why people take such precautions is necessary, otherwise a mistake will be made.

Two examples of this come to mind. In the first, an associate of a convicted murderer was asked why they put lime (tiny amounts of which were found on his clothes and at the scene) onto their victim before dumping his body in a ditch: 'to make the body dissolve' was the answer! Of course lime, being hydroscopic dries out tissue and can assist in desiccation/mummification. When crushed to dust, it is also highly-mobile and blows onto adjacent surfaces such as the murderer's clothing. In the second case, a suspect walked through broken plaster on escaping from a gun attack. Just as with the case study on 'Agricultural limestone' Chapter 8, the person assiduously wiped the plaster from his/her black leather shoes – which like any pebble from the ocean looked so shiny and clean, only to turn dull on drying, in this case revealing all the smears of plaster dust left from the wiping. The person had failed to remember childhood detention – and being given the chore of cleaning the blackboards that when wet were black and shiny, when dry, all covered with grey smears. Common manufactured blackboard chalk is of course not chalk at all: it is the far softer gypsum, the same as is used in plasterboard.

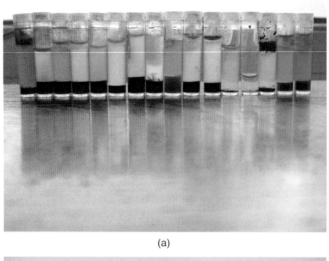

(a)

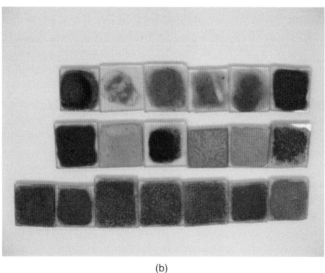

(b)

Figure 10.4 Dissolved heads from wooden safety matches are settled (a) and then dried on glass tiles (b) for irradiation by X-ray diffraction; from work carried out by Farmer *et al.* (2007). Each tile is 2.2 centimetres in diameter

This mistake suggests two ways (among others) that allow the geoforensic scientist to stay one step ahead of the criminal: (i) by using their mistakes or (ii) by bringing ever-more sophisticated analytical methods to the problem of using small amounts of Earth material (Figure 10.5) in solving crime, establishing guilt or innocence and assisting justice. In Chapter 8 we saw how geological trace geological evidence is currently commonly characterized by light microscopy, chemical or mineral analysis, or worst of all (for comparison) scanning electron microscope

Figure 10.5 Mud left on a gate found on an escape route used by two gunmen involved in a drug-gang murder in Northern Ireland. Gate bars are 7 centimetres in diameter

(Figure 8.2), none of which can easily be translated into simple field examination for comparable materials, leaving the investigator two possible avenues of sample collection. First, use what non- or semi-destructive methods as are possible to give a sample characterization, and seek similar materials in the field, from descriptions. Second, carry out an extensive sampling regime, taking samples of all variant materials, as well as numerous samples of visually-homogenous materials, in case variations are not obvious to the handlens and naked eye in the field. Proxy measurements of variation may be employed (remote sensing). A combination of these methods is ideal.

The ultimate challenge for the geoscientist is explored in Chapters 7 and 8 and completed in *Chapter 9*: traditional uses of Earth-based physical evidence have been in comparing suspect to scenes. Yet even in the 1920s Heinrich was using such material in a *predictive* capacity: this still remains a challenge and depends on comparison to databases, geological maps and vegetational distributions. Ultimately, the search for organic materials (human remains, some drugs, pollutants, biological agents) in an essentially-organic based medium such as the soil that covers much of the Earth, is trying almost to differentiate like from like. This requires integration of many (if not all) the methods of the preceding chapters. Hence the modern search method relies on all this, plus more, including the one thing rarely discussed here, *operator/interpreter experience*, combined with open-mindedness and intuition. In the *Appendices*, some ideas on and links to advice for best practice are considered, albeit lightly. At the outset, this text was not written as a manual on geoscience methods, but as a consideration of how geoscience philosophy can be applied to domestic, international, terrorist and environmental crimes, and to humanitarian ventures such as emergency response, disaster management and genocide.

Appendix 1
Search methods

Although this book has advocated the use of geography, remote sensing, GIS, geophysics and geology in criminal and humanitarian investigations, these are but a few tools available to the investigator and they are rarely used in isolation. This is the case for the search, which must be intelligence-led. Given no background and a problem of a crime or other incident occurring in country X, there is little chance that the combined power of all the methods advocated here would discover any associated location. The nature of the crime or other problem may have a strong geographical bias. Large volumes of covert materials (weapons, bodies, drugs, toxic waste) require large areas for burial such as waste ground, quarries and industrial building sites. Type of incident must then be combined with a likely suspect and time of activity to provide a geographic profile of what happened, where geography can be combined with offender profiling. An investigator will have priority location(s) established, at which point a number of maps are desirable: recall (Chapter 5) that all maps are cognitive to some degree and based on scale and training method. Likewise, aerial photographs are also scale-dependent, subject to lighting, platform, film/pixel resolution and camera type, etc. Maps need to show roads, paths and areas with easy (flat, dry ground, vegetation windbreaks) versus limited access (trees, thick vegetation, water courses that are considered permanent). The same or different maps need to provide topographic information (height of ground, vegetation and buildings) that dictates views of and from the scene. These may be generated by visits, marking views on the ground, or by creation of a Digital Terrain Model (DTM) that allows a fly-through visualization of topography. Terrain maps need to have more detailed information on vegetation type and age, soil type and distribution, especially inferred thickness. Variations in soil density may be quickly ascertained by probing: probe insertions can be marked for later soil, water or gas sampling, or visits by scent dogs.

These combined attributes will allow a number of smaller areas to be prioritized, again dictated by target size, whether whole vehicles, water-courses or large numbers of buried people, down to small bags, infants, drug caches, weapons and small valuables. Target type dictates search method. Shallow metal objects may be found with metal detectors, deeper metal targets will require a magnetometer. Human

---

Geoforensics Alastair Ruffell and Jennifer McKinley
© 2008 John Wiley & Sons, Ltd

or animal remains, other organic debris and some drugs are detectable using gas probes, cadaver dogs and ground-penetrating radar. Negative returns from each proxy method can be used to create or change the hierarchy of each small search area, the end result of which may be that the entire location is demoted and a new set of targets established some distance away.

Killam (2004) provides some guidance on possible search strategies, from initial coverage of terrain, through to the appropriate ('fit for purpose') use of technology or other assets. His method is a blend of search and rescue (especially mountain environments), emergency response and some no-body, suspected homicide based methods. Although this blend is compatible with the ethos of this book (*geoforensics is not just about helping the police*), when advice is being given that may save a person's life, we must be very cautious.

Once secured, the below-ground scene requires management by trained professionals. Geologists and geographers are not well-trained in the methods that need to be deployed: they don't need to be, because they dig holes for other purposes. The geoscientist, especially an appropriate one, can however be essential at a scene managed by an experienced person and many geographers and geologists have gained this knowledge after their standard training by attending archaeological and criminal investigation-related digs. Recent papers have shown the importance of stratigraphic information (Hanson, 2004): thus a suspected burial in sand-dunes would best be attended by a specialist in this area; in glacial till, fluvial outwash, deep soils, etc., likewise, the specialist will understand the natural stratigraphy, note un-natural features and, with a scientific reputation, can/may make a good expert witness in any subsequent court of law.

Bass and Birkby (1978) published one of the first papers dealing comprehensively with the excavation and recovery of buried human remains, largely from an archaeological/anthropological point of view. At two points in their advisory article they do delve into areas familiar to the geoscientist: on page 10 mentioning the use of gas-sensing probes to determine burial likelihood/location and throughout the article, and especially on page 11, the observations of soil discolouration and disturbance. Bass and Birkby (1978) touch upon the geomorphological aspects of search (Chapter 2) in their penultimate paragraph, mentioning site visits in the morning and evening in order to observe slight ground depressions.

Boyd (1979) gives some excellent advice (not significantly improved upon) in one of the first published articles on how the burial scene may be dealt with. He urges planning and the need a range of specialists such as: pathologist, archaeologist, anthropologist, odontologist, toxicologist, psychiatrist, entomologist and botanist. Boyd goes on to point out just how frequent accidental discovery is (hunter, construction worker, dog-walker), even of collapsed ground. Following this, there is rarely any urgency, as long as the scene is controlled and that acquisition of evidence is not going to assist in tracking down a serial-killer and thus prevent further murders.

There are instances where urgency is required: personal experience includes a body in a river-bank that was likely to flood, a body on a demolition site where unstable structures were liable to fall onto the scene, a body in a coastal refuse tip that the next tide was likely to inundate and a body in soil above a tunnel that delay in removing was likely to bring a large part of a major city to a standstill. In the latter case, the science that directs the work had to balanced against a billion-pound economy, and as the body was not proven to be a murder victim, finance won.

Boyd suggests a photographer be one of the first people to enter the scene (and gives guidance on scene of crime photography), and then implies that the archaeologist, pathologist be next in. The fragility of botanical evidence has accelerated the importance of this part of the investigation in recent years, yet is balanced against the possibility that the disturbed ground or suspect grave may not be a grave at all – in which case specialists are brought in for no good reason. Again finances need to be balanced against intelligence. In a recent excavation in Northern Ireland, a suspect grave was reported, yet with no missing people reported, and the public-access site of the collapse, a compromise was taken: the only specialists brought in were geophysicists (to evaluate the 'grave' contents) and archaeologists (to determine what was there). Both entered and exited on defined paths, took full forensic precaution and minimized compromise of the scene. The burial turned out to be a trial pit for groundwater evaluation, justifying the use of minimal resources. Had the pit contained human remains, compromise of the scene from botanical, geological and possibly entomological aspects would have been likely: basically the scene investigator cannot win the budget versus science argument!

Appendix 2
Soil sampling

A.1 Sampling protocol suggestions

Scientists, and especially geoscientists are employed in many fields where sampling, sample representativeness, storage, transport, documentation and testing are critical to the work being undertaken. The type of material being sampled, and what it is being sampled for, dictates much of what happens. However, geoscience protocols can be informed by other disciplines, just as we have developed methods that will enable our cognate sciences. These notes cover soils, sediment (rivers, beaches), rock dust (quarries) and building sites. They should not be stand-alone but used with mappers, reconstruction and other samplers' consent and discussion. Each scene is unique; each soil may be unique so ask others their opinion and form your own judgement: ultimately you will be the person who defends what you did and why. *Of over-riding concern is continuity of evidence – the chain of custody, which must be maintained at all times and is often alien to the scientist not used to the forensic arena. With increasing amounts of litigation in environmental, humanitarian, human rights enquires, industrial tribunals and insurance claims, it is an unfortunate and inevitable fact that all scientific procedure will be discoverable, questioned in court, and thus subject to the most rigorous procedures of sample/data handling available.*

Some basic knowledge of how to describe soils and sediments is useful (INTERPOL, 2001). It is assumed that the strict forensic protocols provided by accredited courses, training and mainstream textbooks are followed.

The conflict between personal training (from a particular course, experience in certain countries) and the methods of co-investigators (possibly with different training, in different countries) is noted but not further developed here. Needless to say that parts of this book have been written in the USA, Canada, Spain, Italy and Greece, where such conflicts arose: we acknowledge that ours is a United Kingdom-based perspective, possibly better and worse that other countries standard methodology, depending on necessity. The requirement is to learn a 'fit for purpose' method to suit terrain, conditions and population. The key is to maintain scientific protocol and use this rigour to report in manner acceptable to the authorities in charge. The opposite, carrying out science to the specification of others, is not acceptable.

Grain size, sorting, colour, organic content and soil stratigraphy are useful starting points, covered in Murray and Tedrow's excellent *Forensic Geology* (1975) but also found in Saferstein (2001) and similar texts. Find out which investigating officer has a geology, physical geography, biology or archaeology qualification to assist (*not replace!*), or contact an expert. The reason? Once the desired sampling programme (described below) has been undertaken, it is wise to stand back and reconsider the scene. Is there any material that looks (grain-size, colour, texture, location) different to the main soils that occur around the scene? If the answer is 'yes' then consider whether any of these 'odd' materials may have come into contact with a suspect. A good guide is to ask whether the footwear, clothes and car tyres of all the visitors to the scene have passed through or over something that has not been sampled? The ultimate reason? If it looks different, it probably will be different, and if unique, a very small amount of a unique material (soil, manure, building waste, vegetation, rock material or dust) may be enough to link a suspect to a scene or exonerate the innocent, compared to a large amount of non-unique material. Likewise if it looks different or out of place, it maybe non-representative, possibly fallen from footwear or a vehicle: the eye is drawn to such materials, which is good, they should be sampled (see above), but *not* at the expense of the bland, the common and thus the possibly representative. Be aware of the aim of the investigation. Simply applying the disciplines of one's own specialism, however rigorous, may not be appropriate. Time and financial constraints will limit the possibilities: the view of the law courts inevitably dictates what is required. For environmental crime, some of the procedures developed for other crimes suffice, yet there are also distinct differences: Suggs *et al.* (2002) provide excellent guidelines in this context.

Soil samples may link suspects to the scene of crime or lead police to a scene from a suspect's clothing, vehicle or other material. Visual match, microscopic analysis, mineralogical determination, geochemistry, palynology or other microbiology may achieve this aim. Samples have to be taken with all the above in mind.

Amount of sample

Current practice favours a number of analyses being undertaken, so enough sample for splits should be taken – 100 grams (enough to fit in the palm of your hand, although of course the sample won't be in your hand! – see contamination below!) is sufficient, 200 grams sensible.

What to sample

Too many samples are never enough. However, common sense allows certain constraints. If the questioned (or suspect/alibi/victim) material and its descriptions are being taken to the scene, then comparable materials should all be collected, some of which (with sophisticated analysis) will be excluded and become controls. With or without the questioned sample, likely points of contact with footwear, clothing, hands and vehicles will constrain the locations from which soil can be obtained and thus the total amount available. Thereafter, in the case of foot, hand, body and tyre prints, the whole area should be sampled. In the case of a larger area, such as an inhumation, laneway, garden or field, a *grid* of sample points should be set up with likely points of entry, exit and activity (diagnosed by thermal imaging, logistics, dog scent, geophysics, etc.) sampled completely, the rest with unique X Y coordinate markers.

Number of samples

Too many are not enough. Once again, the amount needed constrains the number of sample points available. If only the edge of a footprint or tyre is observed in mud on hard ground, and this is represented by about 100 grams of sample, then only this can be taken. Points of entry and exit by people, vehicles and areas of disturbance may again only allow one or a few samples. Areas more than the width of the likely contact (footprint, tyre) should be sampled in transect or gridded, with a sample taken at each grid centre and uniquely numbered. Sample unusual materials as well as the common soils: if a patch of dust or soil looks odd to you, chances are it has got something rare in it to make it look odd, thus a chance contact with a suspect is more likely to provide a link than a series of homogenous soils: these 'oddities' are quickly discounted in forensic geology examinations; it may also look odd because it is a foreign material that needs to be excluded. A classic mistake is similar to this example: in a hit and run vehicle incident in England, officers who had been examining possible suspect vehicles found mud at the scene. The mud appeared at the scene after the accident – it fell from the police officers' vehicle and thus wrongly connected one (innocent) suspect location to the scene when neither they not their vehicle had been near the locus.

Depth

Caution is good. Small-scale (millimetre, but visible to the naked eye) variations in the soil to be sampled should be considered. Soil often possesses micro-stratigraphy, with a top layer of live vegetation, animal faeces or dust, a layer of dead vegetation or humus, faeces, and layer of mixed soil and humus. A captured sample in footwear or tyre tread may only represent one layer: thus a soil sample may not match a suspect sample, even though they came from the same scene of crime. Common sense once again prevails: examination of the soil will show its small scale variation or homogenous nature: likewise, careful extraction of embedded soil in footwear and tyres will show whether microstratigraphy has been preserved. A handlens is useful in both cases. Sampling that fails to account for microstratigraphy may still allow comparison/exclusion of the whole homogenized scene and questioned sample: it will make historical reconstruction very difficult.

Contamination

Soil, by its very nature, is already contaminated! Most methods of dealing with soil are insensitive enough to cope with the minor introduction of a rare grain of sand: it is the contamination between scene of crime and suspect that is dangerous. Palynology is far more sensitive than mineralogy, and thus contamination here is possible. In the likely event of palynology being employed, use pristine sampling tools. In cases where a scene may be further contaminated or destroyed (heavy rain, traffic, fire) it is better to get the sample than not at all, so a clean knife and pristine bag can be used. Contrary to the view of many textbooks, soil samples can of course be taken after plaster casting of prints: any mineralogist worth their salt should be able to spot anhydrite–gypsum (plaster) in a sample: the only problem comes if the scene also has plaster in it! Do not attempt to dry a sample: water-borne organisms may be useful in linking victim to scene of crime to suspect. Dry the samples only after microbiologists have been consulted.

Recording

Sketches and photographs, located by the scene mapper, are essential. Any photographs should have an accompanying labelled sketch, with any samples taken annotated. Do not be afraid to sub-sample and record likewise. This may seem cumbersome, but it is essential. You may consider a system of numbers where you record your case number, the general location number, the location within the grid and then any variation (such as depth). An example might be:

$$1135/01/A6/1$$

where 1135 is your case number, 01 a location at the scene, A6 the grid coordinates of the sample at 01 and 1 is layer 1 of a number of sampled layers.

The legal process

The geoscientists' involvement in most investigations (criminal, environmental, terrorist, military) will inevitably end up being examined and tested by the legal system. This may be the courts, or a political or military enquiry. Before geoscientists are even contacted by an investigator, they must be aware of this likely outcome, and at the first discussion of the problem, the ultimate presentation of information and witnesses in court is to be anticipated and incorporated. Scientific methods ought to be rigorous enough to stand any examination under legal constraints. The reality of other scientist's information being used and compared, the high financial and humanitarian stakes, and media interest can cause geoforensic practitioners' methods to be significantly altered. Presentation is one such element. Morgan and Bull (2007, p. 2) state:

> It would be difficult to imagine a discipline within geomorphology or geology which based its published research upon personal opinions. These personal interpretations are not peer-reviewed but rather are judged on general public understanding and acceptance (by judge and jury) in a court of law. This indeed is the case in forensic geoscience....

Data recording is a second element. As soon as the geoscientist is involved in a case, all materials become discoverable and subject to intense scrutiny by the opposition. Inconsistencies are seized upon. The expert will know of no other similar experience: peer review is a lengthy process, given to reflection and consideration over time. To account for one's activities and results in a matter of seconds to minutes when under legal obligation requires contemporaneous note-taking, excellent data recording, storage and recall, accredited methods, repeatable results and confidence in one's discipline.

This text is mostly about philosophy, the reasons why and the meaning behind geoforensics. A whole text could be (and ought to be) written on the geoscientific legal aspects of data recording, sample taking and handling, and continuity of evidence. Most of these are country-specific and are to be found, and conform with, other forensic sample and method handling. The reader who is likely to face the legal system is urged to do a number of things. Read up extensively on general and specific forensic method textbooks; attend specialist expert witness training, including going to a real court of law (if possible) and what is expected of a witness statement; ask colleagues who have been in court, etc. what to expect, and then expect this and more. Then, when the time comes, presume from the outset that everything you do will be examined under oath, from notes, to samples, to data and opinions. Never presume anything of the courts: of the 50 or so criminal, terrorist and environmental investigations undertaken personally, it is a consistent feature that the case one expects to flow smoothly ends with a problem and vice versa.

The media

Like it or not, trial by media is sometimes more significant than trial in the courts in the telecommunications and Internet-driven world of today, complete with 'sentences' such as being shunned by colleagues, fired from work and harassed by journalists (BBC News, 2005: *http://news.bbc.co.uk/1/hi/uk/4317926.stm*). For this reason, many humanitarian, police and environmental investigators (and to a more limited extent, the military) now work with the media, providing them what they can, when they can, or actively use the media for information. This has been common practice for many years in problems of missing persons/mass disasters: we have shown how similar such cases are to criminal investigations, in terms of science, so why not use of the media? The forensic geoscientist is well-advised to cooperate with the media in the same way as the legal process: do not attempt anything without advice, training, gradual experience or legal cover.

The fate of the forensic geoscientist is unfortunately like all other scientists when the media are involved: we are portrayed as mad, lab-coat wearing, obsessed individuals, out of touch with reality. All major scientific societies have attempted to deal with this problem that is only increased where criminal investigations are concerned. There is no easy solution: the one offered by this book's title and scope is of fairness, the immutability of physical evidence (it cannot lie or be corrupted) and the correct use of geoscience in the whole legal system. The issues involved when dealing with the media are not restricted to criminal cases: as much attention is lavished upon environmental, equal opportunities, fraud and other enquiries. Thus anyone, reporting anything about work colleagues, neighbours and law-enforcers need to be wary of the media attention they may face.

Other investigators

Luckily, soils, rocks and similar materials such as glass or builder's materials do not have the capacity for contamination that many other materials have. The general precautions of scene of crime are generally better than needed for bulk soil forensic analysis. Nonetheless, issues remain. First, visits to the scene and then contact with suspects, their clothing, vehicles, materials or houses do have the capacity for secondary material transfer. Likewise, samples taken for soil analysis will very likely be the subject of other analyses (insects, pollen, CDR, DNA) and thus these, more sensitive techniques *must* take precedence in terms of contamination. Other investigators at scenes will have different strategies and objectives to those listed above (the entomologist will sample locations where insects will reside, the anthropologist/archaeologist will sample around inhumations and materials of human origin).

In all cases, remember that the ultimate aim of all this work is a robust defence in court, best achieved by consistent and conjunctive means. This requires communication between samplers and often sample 'splits' of the same location, for later comparison. This may even come to splits of four or five from one sample point. These can be made onsite, with (for example) an area of soil measuring $4 \times 4 \times 1$ centimetre, being bagged and then divided into 4 or 5 smaller bags. These can all be placed bag in the master bag with continuity label attached and a note to the effect that a number of samples have been taken for different analyses: this may speed up the analytical process.

In the field, the forensic geoscientist should *consult* with a range of other specialists, and may then *collaborate* with the same. Each case is of course different, and the availability of specialists will also vary. Below are some examples of these people and what they may offer.

Crime scene reconstructors. Probably the most frequent collaborators, these specialists may themselves be, or already use, any one of the 'ologists' listed below. However, crime scene reconstructors have specific guidelines for what they do, all of which have been developed through years of investigation, but yet still worthy of challenge by forensic geoscientists, as we have different, sometimes equally-valid, ways of reconstructing what happened, when and where. Chisum and Turvey (2007) recently published an excellent guide to crime scene reconstruction that is very compatible with many geoforensic methods.

Vegetation specialists and ecologists. The geologist or physical geographer is not expected to know all the plants of the forest, but can work with (for a successful investigation and presentation at court) and learn from botanists, ecologists and agriculturalists. These scientists can spot windbreaks in banks of vegetation that may confirm or contradict what the physical geographer or behaviour specialist sees;

they may contribute to information on how the land has been altered (cut/broken vegetation, application of fertilizer/pesticide) that again the geoscientist would not be expected to know. Insect activity is related to all the features of geoscience, from climate and weather (temperature, water availability), soil type, slope and aspect, and pollutants.

Remote sensors and geophysicists. These specialists have wide knowledge of data acquisition and interpretation from geological, physiographic, biological (vegetation) and human perspectives. They can benefit from input of ideas from all these specialists, just as the geologist who may mis-interpret (for instance) a vegetational feature can benefit from the remote sensor. Likewise, remote sensing specialists or geophysicists will know the limits of their method when the geoscientist may be prone to over-enthusiasm or misplaced cynicism.

Dog handlers. The rescue and cadaver dog is especially suited to being observed by a geoscientist, because their work is so concerned with buried (rock, soil, sediment, water, snow) people and cadavers. Thus the dog's reaction to changing topography and ground conditions is informative because the dog does not care where he or she is – dogs are concerned with scent and their handler's instructions. An indication from the dog at a geologically or geographically sensible location (low ground, loose ground, disturbed ground, water, windbreaks) confirms suppositions and limits the search area. An indication from the dog at what the geoscientist sees as an absurd location is not crazy to the dog – it has to be explained by us (wind direction, the ground not being what it looks like [loose rocks, made ground]). The same applies to the conjunctive work of the dog and the remote sensor, and especially the geophysicist (the latter work at the same time and at similar scales). The dog indicating at a geophysical anomaly is an example of a dual-proxy approach, raising the importance of this location in the target hierarchy. If the dog indication coincides with a geographic, remotely-sensed, behavioural and geophysical target, so it goes even further up the target hierarchy until only a few places remain where digging should occur.

Search and recovery specialists. Anthropologists looking for human remains, disaster recovery specialists and police officers looking for trace evidence all get into intimate contact with the ground, and thus the inquisitive cannot help but spot changes. In the authors' experience, sitting (or kneeling) and staring at ground, rock, sediment or water causes the curious human brain to ask exactly what the Earth detective asks: 'how did this get here?' and 'when did it get here'. The geoforensic worker must be open-minded to the lack of specialist language in such co-workers and instead use their observations. The two most critical disciplines in this respect are the *archaeologists* and *anthropologists* (Rinehart, 2003).

Footprint and tyre impression specialists. These workers are examining trace fossils – an indication of past life in rocks (especially sediment and soil). Each vehicle or shoe is akin to an organism, with its own scars, gait, imprint, etc. Tracks cross one another and become eroded, providing a relative age for movement, and tyres and footwear preserve a microstratigraphy that can be interpreted using the laws of superposition.

Behavioural psychologists. The obvious collaboration here is the plotting in space and time of criminal activity and its links to GIS-based interpretation. However, as we see above, features of the landscape are also important factors with which the geographer can assist, and be instructed in, by the behaviourist: points of access, viewpoints and covert locations are all controlled by the environment but may be cause for repetitive offender behaviour and thus prediction.

Glossary

aspect	the compass direction of the maximum rate of change, measured in degrees with respect to a compass bearing
cadaver	the dead body, especially of a human
CAT or CT scanning	computed axial tomography uses a rotating circular X-ray source and electronic detector mechanism to provide a 3-D image via a series of slices through the object of interest
coefficient of determination	the squared correlation coefficient
conductivity	a measure of a material's ability to conduct an electric current
convexity	the rate of change in slope, and can be divided into plan convexity and concavity (that is, negative convexity)
correlation coefficient	a measure of how well a curve (or line) 'fits' a set of data
delinquency	an illegal act, committed by a juvenile
deposition	the accumulation of these same particles and objects
deviance	behaviour that is considered to violate social norms
diatom	unicellular algae with a silica cell wall called a frustrule; if found in bone marrow, it is a diagnostic of drowning
diatomite	chalk-like in appearance, but a more abrasive material (fossilized diatoms) used to filter out solid waste in wastewater treatment plants; also used as an active ingredient in some powdered pesticides and explosives
digital terrain model	a data model that attempts to provide a 3-D representation of a continuous surface, usually of the land
Dionex	a chromatographic and extraction system used to separate, isolate and identify the components of a chemical mixture
Doppler shift	the apparent change in the frequency of a signal caused by the relative motion of a transmitter and receiver
drift geology	soils, alluvium, glacial deposits above solid geology
electrode	an electrical conductor used to make contact with a nonmetallic part of a circuit such as a semiconductor, an electrolyte or vacuum

electro-magnetism	the study of the interaction between magnetic and electric fields
epidote	a silicate of alumina, carbonate and oxide of iron, or manganese, commonly of a yellowish green (pistachio) colour
erosion (e.g., fluvial)	the removal of material from the base and sides of the channel
fluvial	pertaining to rivers
foram(inifera)	unicellular animals, mostly microscopic and marine, that secrete hard coverings composed of calcium carbonates or build them of cemented sedimentary grains
fossil	from the Latin fossus, literally 'having been dug up' the mineralized or otherwise preserved remains or traces (such as footprints) of animals, plants, and other organisms
gamma-rays	a penetrating electromagnetic wave (ray) emitted by some radioactive isotopes during decay, similar to X-rays
geomorphology	the surface form of the Earth, the general configuration of its surface, and the changes that take place in the evolution of landforms
geophysics	the study of the physics of the Earth
GPS (global positioning system)	a worldwide navigation system based on coordinated satellite signals
gravimeter	a precision instrument that measures minute variations in the Earth's gravitational attraction
histogram	where each vertical bar is proportional in area to the frequency in each class or group
hydrodynamics	the study of fluid motion, here, specifically the movement of river, lake, seaway and oceanic waters
hydrology	the behaviour of water as it occurs in the atmosphere, on the surface of the ground, and underground
hydrolysis	addition of water to create hydrous molecules
laser	a device that concentrates light into an intense, narrow beam
median	divides the normal frequency curve into two equal parts (when the individual values are arranged in increasing order, the value in the middle of the list is the median)
mode	the most commonly occurring value – highest point of a frequency curve or steepest part of cumulative frequency curve
Mulberry Harbour	a floating harbour used in the D-day invasions of northern France
organized crime	organized criminal or criminal organizations that are groups or operations run by criminals, most commonly for the purpose of generating a monetary profit

oxidation	addition of oxygen, an example being rusting of iron
pedology	soil science and the study of soil-forming processes
placer	a sediment containing valuable dense minerals, often fluvial, or shallow marine
precipitation	the quantity of water falling to Earth at a specific place within a specified period of time as rain, dew, fog, hail, snow, etc.
radiometrics	A measure of the natural radiation in the Earth's surface
raster	A 2-D array of black and white cells, called pixels or picture elements, which when displayed on a screen or paper, form an image
regression	the procedure for finding the line of best fit
remote sensing	the science of deriving information about the Earth's land and water areas from images acquired at a distance
resistivity	A measure of how strongly a material opposes the flow of electric current
seismic	relating to waves of elastic energy, such as that transmitted by P-waves and S-waves; seismic energy is studied by scientists to interpret the composition, fluid content, extent and geometry of rocks in the subsurface
seismogram	a record of an Earth tremor using a seismograph
SEM (scanning electron microscope)	an electron-beam-based microscope used to examine, in a 3-D screen image, the surface structure of prepared specimens
slope	composed of gradient (maximum rate of change of altitude), usually measured as per cent, degrees or radians
solid geology	bedrock underlying soils and drift geology
spectra (l)	relating to a spectrum, such as spectral colours or spectral analysis
spline	a curve in 3-D space defined by control points; splines can be cut or joined at their edit points
strike (and dip)	relates to the attitude (slope) and altitude of geological planes, strike being 90° to dip
subsidence	a gradual sinking of land with respect to its previous level
tailings	the residue of raw material or waste separated out during the processing of mineral ore
tectonic	the geological processes by which rocks are deformed and that produces features of the Earth's crust
transportation	the movement, by suspension or saltation of particles and objects
vandalism	willful or malicious destruction, injury, or disfigurement of any public or private property
viewshed	a polygon map resulting from a visibility analysis showing all the locations visible from a specified viewpoint

X-ray diffraction	analytical technique used to determine the structures of crystalline solids: a monochromatic beam of X-rays (usually Cu-Kα) is diffracted off repeating planes of atoms in crystalline samples to produce a diffraction pattern
X-rays	a form of electromagnetic radiation with a wavelength in the range of 10 to 0.01 nanometres, primarily used for medical imaging and crystallography

References

Abbott, D.M. 2005. Investigating mining fraud. *Geotimes*, January 2005 Issue, 30–32.

Aitchison, A. 1999. Logratios and natural laws in compositional data analysis. *Mathematical Geology*, **31**(5), 563–580.

Aitchison, J. 1986. *The Statistical Analysis of Compositional data*. Monographs on Statistics and Applied Probability. Chapman & Hall Ltd, London. Reprinted (2003) with additional material by The Blackburn Press, Caldwell, NJ.

Aitken, C.G.G. 1995. *Statistics and the Evaluation of Evidence for Forensic Scientists*. John Wiley & Sons, Ltd, Chichester.

Aitken, C.G.G. & Stoney, D.A. 1991. *The Use of Statistics in Forensic Science* (Ellis Horwood Series in Forensic Science). Taylor & Francis, London.

Atkinson, P.M. 2004. Resolution Manipulation and Sub-pixel Mapping, In: De Jong, S.M. & van der Meer, F. (eds), *Remote Sensing Image Analysis: including the Spatial Domain*, Kluwer Academic Publishers Group, Netherlands, 51–70.

Atkinson, P.M. & Lloyd, C.D. 2007. Geostatistics and GIS, *Manual of GIS* (ed. M. Madden) (in press).

Bass, W.M. & Birkby, W.H. 1978. Exhumation: the method could make the difference. *FBI Law Enforcement Bulletin*. **47**, 6–11.

Baden, M.M. & Hennessee, J.A. 1993. *Unnatural Death*. Time Warner Books, London.

Bergslien, E., Bush, P. & Bush, M. 2006. Application of Field Portable X-ray Fluorescence (FPXRF) Spectrometry in Forensic and Environmental Geology. Conference Abstract. Geoscientists at Crime Scenes. Geological Society of London, 20th December, 2006.

Bevan, B. 1991. The search for graves. *Geophysics*, **56**, 1310–1319.

Bischoff, M.A. 1966. Prof. August Bruning, Ph.D. and LL.D. h.c. (1877–1965). *Deutsch Zeitschreift Gesamte Gerichtl Medikal*, **58**, 1–2.

Blank, J.P. 1971. The almost perfect kidnapping. *Reader's Digest*, **98**, 140–144.

Bleakley, H. 1905. The love philtre: the case of Mary Blandy, 1751–2. In: *Some Distinguished Victims of the Scaffold*. Kegan Paul, Trench, Trubner and Co. Ltd., London.

Block, E.B. 1958. *The Wizard of Berkeley*. Coward-McCann, New York.

Blott, S.J., Croft, D.J., Pye, K., Saye, S. & Wilson, H.E. 2004. Particle size analysis by laser diffraction. In: Pye, K. & Croft, D.J. (eds), *Forensic Geoscience: Principles, Techniques and Applications*. Special Publication of the Geological Society of London, 232, 63–73.

Bock, J.H. & Norris, D.O. 1997. Forensic botany: An under-utilized resource, *Journal of Forensic Sciences*, **42**, 364–367.

Bommarito, C.R., Sturdevant, A.B. & Szymanski, D.W. 2007. Analysis of forensic soil samples via high-performance liquid chromatography and ion chromatography. *Journal of Forensic Sciences*, **52**, 24–30.

Boyd, R.M. 1979. Buried body cases. *FBI Law Enforcement Bulletin*, **48**(2), 1–7.

Brandt, C. 2004. *I Heard you Paint Houses: Frank 'the Irishman' Sheeran and the Inside Story of the Mafia, the Teamsters and the Last Rites of Jimmy Hoffa*. Steerforth Press, Hanover, New Hampshire.

Briener, S. 1981. Magnetometers for geophysical applications. In: Weinstock, H. & Overton, W.C. (eds), *Applications to Geophysics*. Society of Exploration Geophysicists, Tulsa, 3–12.

Brilis, G.M., Gerlach, C.L. & van Waasbergen, R.J. 2000a. Remote sensing tools assist in environmental forensics: Part I – digital tools – traditional methods. *Environmental Forensics*, **1**, 63–67.

Brilis, G.M., van Waasbergen, R.J., Stokely, P.M. & Gerlach, C.L. 2000b. Remote sensing tools assist in environmental forensics: Part II – digital tools. *Environmental Forensics*, **1**, 1–7.

Brown, A.G. 2006. The use of forensic botany and geology in war crimes investigations in NE Bosnia. *Forensic Science International, Forensic Palynology Special Issue*, **163**, 161–248.

Brown, A.G., Smith, A. & Elmhurst, O. 2002. The combined use of pollen and soil analyses in a search and subsequent murder investigation. *Journal of Forensic Sciences*, **47**, 614–618.

Buccianti, A., Mateu-Figueras, G. & Pawlowsky-Glahn, V. 2006 (eds). *Compositional Data Analysis in the Geosciences : From theory to practice*. Geological Society, London, Special Publications 264.

Buck, S.C. 2003. Searching for graves using geophysical technology: Field tests with ground penetrating radar, magnetometry, and electrical resistivity. *Journal of Forensic Sciences*, **48(1)**, 5–11.

Bull, P.A. & Morgan, R.M. 2005. Always be careful of a book of matches. Review of *Forensic Geoscience: Principles, Techniques and Applications*, by Pye, K. & Croft, D.J. (eds). *Science & Justice*, **46**, 107–24.

Bull, P.A. & Morgan, R.M. 2006. Sediment fingerprints: A forensic technique using quartz sand grains. *Science and Justice*, **46(2)**, 64–8.

Bull, P.A., Morgan, R.M. & Dunkerley, S. 2005. 'SEM-EDS analysis and discrimination of forensic soil' by Cengiz et al., A comment. *Forensic Science International*, **155**, 222–224.

Bull, P.A., Morgan, R.M., Wilson, H.E. & Dunkerely, S. 2004. 'Multi-technique comparison of source and primary transfer soil samples: an experimental investigation' by Croft, D. J. & Pye, K., A comment. *Science & Justice*, **44**, 173–176.

Bull, P.A., Parker, A. & Morgan, R.M. 2006. The forensic analysis of soils and sediment taken from the cast of a footprint. *Forensic Science International*, **162**, 6–12.

Burrough, P.A. & McDonnell, R.A. 1998. *Principles of Geographical Information Systems*. Oxford University Press, Oxford.

Butler, J.C. 1979. The effects of closure on the moments of a distribution. *Mathematical Geology*, **11(1)**, 75–84.

Cameron, N.G. 2004. The use of diatom analysis in forensic geoscience. In: Pye, K. & Croft, D.J. (eds), *Forensic Geoscience: Principles, Techniques and Applications*. Special Publication of the Geological Society of London, 232, 277–280.

Canter, D. 2003. *Mapping Murder*. Virgin Books, London.

Canter, D., Coffey, T., Huntley, M. & Missen, C. 2000. Predicting serial killers home base using a decision support system. *Journal of Quantitative Criminology*, **16**, 4.

Cave, M.R. & Wragg, J. 1997. Measurement of trace element distributions in soils and sediments using sequential leach data and a non-specific extraction system with chemometric data processing, *Analyst*, **122**, 1211–1221.

Cengiz, S., Cengiz Karaca, A., Cakir, I., Bulent Uner, H. & Sevendik, A. 2004. SEM-EDS analysis and discrimination of forensic soil. *Forensic Science International*, **141**, 33–37.

Chainey, S. & Ratcliffe, J. 2005. *GIS and Crime Mapping*. John Wiley & Sons, Ltd, Chichester.

Chambliss, W.J. 1978. *On the Take: from Petty Crooks to Presidents*. Indiana University Press, Bloomington.

Chang, K. 2002. *Introduction to Geographic Information Systems*, McGraw-Hill, Boston.

Chaperlin, K. & Howarth, P.S. 1983. Soil comparison by the density gradient method – a review and evaluation. *Forensic Science International*, **23**, 161–177.

Chardez, D. 1990. 'Thecamoebologie et expertises juridiques'. Travaille Laboratoire Unitie Zoologique Géneral Applice Facultie Science Agricole, Gembloux 22.

Chayes, F. 1971. *Ratio Correlation*. University of Chicago Press, Chicago.

Chilès, J.P. & Delfiner, P. 1999. *Geostatistics: Modeling Uncertainty*. John Wiley & Sons, Inc., New York.

Chiozzi, P., Pasquale, V. & Verdoye, M. 1998. Ground radiometric survey of U, Th and K on the Lipari Island, Italy. *Journal of Applied Geophysics*, **38**, 209–217.

Chisum, W. & Turvey, B. 2000. Evidence dynamics: Locard's exchange principle and crime reconstruction. *Journal of Behavioral Profiling*, **1**, 1.

Chisum, W. & Turvey, B. 2007. *Crime Reconstruction*. Elsevier, Burlington, San Diego, and London.

Cleveland, G.B. 1973. CDMG helps find kidnapper. *California Geology*, (no issue number), 240–241. *www.consrv.ca.gov/CGS/information/p.*

Coates, J. 2000. Interpretation of infrared spectra, a practical approach. In: Meyers, R.A., (ed.), *Encyclopedia of Analytical Chemistry*. John Wiley & Sons, Ltd, Chichester, 1–23.

Conan-Doyle, A. 1988. *The Extraordinary Cases of Sherlock Holmes*. Puffin Books, London.

Cooke, R.U., & Doornkamp, J.C., 1990. *Geomorphology in environmental management*. Clarendon Press, Oxford.

Corre, F. 1968. Profile of Edmond Locard. *The Criminologist*, **3**(10), 119–128.

Cox, N.J. 1978. Hillslope profile analysis (comment). *Area*, **10**, 131–133.

Cox, N.J. 1990. Hillslope profiles. In: Goudie, A., (ed.), *Geomorphological Techniques*, 2nd edition, Unwin Hyman, London, 92–96.

Cox, R.J., Peterson, H.L., Young, J., Cusik, C. & Espinoza, E.O. 2000. The forensic analysis of soil organic by FTIR. *Forensic Science International*, **108**, 107–116.

Cressie, N.A.C. 1985. Fitting variogram models by weighted least squares. *Mathematical Geology*, **17**, 563–586.

Croft, D.J. 2002. The use of CF-IRMS as a tool in forensic soil analysis. In: *Forensic Isotope Ratio Mass Spectrometry*, Conference Proceedings 2002. *http://www.forensic-isotopes.rdg.ac.uk/conf/conf2002.htm.*

Croft, D.J. & Pye, K. 2004a. Colour theory and the evaluation of an instrumental method of measurement using geological samples for forensic applications. In: Pye, K. & Croft, D.J. (eds), *Forensic Geoscience: Principles, Techniques and Applications*. Geological Society, London, Special Publication, 232, 49–62.

Croft, D.J. & Pye, K. 2004b. Multi-technique comparison of source and primary transfer soil samples: an experimental investigation. *Science & Justice*, **44**, 21–28.

Crofts, R.S., 1974. Detailed geomorphological mapping and land evaluation in Highland Scotland. In: Brown, E.H., Waters, R.S. (eds), *Progress in Geomorphology: Papers in Honour of David L. Linton*, Institute of British Geographers Special Publication No 7, London, pp. 231–249.

Cummins, B. 2003. *Missing: Missing Without a Trace in Ireland*. Gill & McMillan Ltd, Dublin.

Cummins, B. 2007. *Unsolved*. Gill & McMillan, London.

Darnley, A.G. & Ford, K.L. 1987. Regional airborne gamma-ray surveys: a review. In: *Exploration '87: Third Decennial International Conference on Geophysical and Geochemical Exploration for Minerals and Groundwater*. Proceedings Geophysical Methods, Advancements in the State of the Art. Toronto, Canada, 30–34.

Davenport, G.C. 2001a. Remote sensing applications in forensic investigations. *Historical Archaeology*, **35**(1), 87–100.

Davenport, G.C. 2001b. *http://www.technos-inc.com/Surface.html#122*

Davenport, G.C., Griffin, T.J., Lindemann. J.W. & Heimmer, D. 1990. Geoscientists and law enforcement professionals work together in Colorado. *Geotimes*, July, 13–15.

Davis, J.L., Heginbottom, J.A., Annan, A.P., Daniels, R.S., Berdal, B.P *et al.* 2000. Ground penetrating radar surveys to locate 1918 Spanish Flu victims in Permafrost. *Journal of Forensic Science*, **45**(1), 68–76.

Davis, J.C. 1986. *Statistics and Data Analysis in Geology*. John Wiley & Sons, Inc., New York.

Davis, S. & Thomas, D. 2000. *JonBenet: Inside the Ramsay Murder Investigation from a Leading Detective on the Case*. St. Martin's Press.

Davis, W.M. 1926. The value of outrageous geological hypotheses. *Science*, **63**, 463–468.

Dawson, L.A., Towers, W., Mayes, R.W., Craig, J., Vaisanen, R.K. & Waterhouse, E.C. 2004. The use of plant hydrocarbon signatures in characterising soil organic matter. In: Pye, K. & Croft, D.J. (eds) *Forensic Geoscience: Principles, Techniques and Applications*. Geological Society, London, Special Publication, 232, 269–276.

Deutsch, C.V. 2002. *Geostatistical Reservoir Modelling*. Oxford University Press, New York.

Deutsch, C.V. & Journel, A.G., 1998. *GSLIB: Geostatistical Software Library and User's Guide*, second edition. Oxford University Press, New York.

Dillon, M. 1990. *The Shankhill Butchers: A Case Study of Mass Murder*. Arrow Books Ltd., London.

Doyle, P. & Bennett, M.R. 1997. Military geography: Terrain evaluation and the British Western Front 1914–1918. *The Geographical Journal*, **163**(1), 1–24.

Dungan, J.L., 1999. Conditional simulation. In: Stein, A., van der Meer. F. & Gorte, B. (eds), *Spatial Statistics for Remote Sensing*. Kluwer Academic Publishers, Dordrecht, 135–152.

Eggar, S.A. 1999. Psychological profiling. *Journal of Contemporary Criminal Justice*, **15**(3), 242–261.

Ellwood, B.B., Owsley, D.W., Ellwood, S.H. & Mercado-Allinger, P.A. 1994. Search for the grave of the hanged Texas gunfighter, William Preston Longley. *Historical Archaeology*, **28**, 94–112.

Encyclopedia of Espionage, Intelligence, and Security. Lerner, K.L., Wilmoth, B. & Lerner, J.B. (eds) 2004. The Gale Group (Thomson Gale).

Equitas. 2006. Innovative Contributions to the Documentation of Missing Persons and Clandestine Cemeteries in Colombia's Conflict. Equitas.

Evans, I.S. & Cox, N.J. 1999. Relations between land surface properties: Altitude, slope and curvature. In: Hergarten, S. & Neugebauer, H.J. (eds) *Process modelling and landform evolution*, Springer, Berlin, 13–45.

Evans, J.A. & Tatham, S. 2004. Defining 'local signature' in terms of Sr isotope composition using a tenth- to twelfth-century Anglo-Saxon population living on a Jurassic clay-carbonate terrain, Rutland, UK. In: Pye, K. & Croft, D.J. (eds) *Forensic Geoscience: Principles, Techniques and Applications*. Special Publication of the Geological Society of London, 232, 237–248.

Farmer, N.L., Ruffell, A., Meier-Augenstein, W., Meneely, J. & Kalin, R.M. 2007. Forensic analysis of wooden safety matches. *Science & Justice*, **46**(3) 88–98.

Fennessy, M.J. 2006. The application of forensic oceanography to cases of missing persons and recovered bodies in estuaries and coastal waters of Canada. Report for Winston Churchill Travelling Fellowship. http://www.wcmt.org.uk/public/reports/90_1.pdf

Fenning, P.J. & Donnelly, L.J. 2004. Geophysical techniques for forensic investigation. In: Pye, K. & Croft, D.J. (eds) *Forensic Geoscience: Principles, Techniques and Applications*. Special Publication of the Geological Society of London, 232, 11–20.

Fildes, J. & Williams, C. 2001. Glacier girl flies again. *New Scientist*, **2322**, 5.

Flanigan, D.F. 1986. Detection of organic vapors with active and passive sensors: a comparison. *Applied Optics*, **25**, 4253–4260.

Fowler, M.J.F. 2004. Archaeology through the keyhole: the serendipity effect of aerial reconnaissance revisited. *Interdisciplinary Science Reviews*, **29**: 118–134.

France, D.L., Griffin, B.A., Swanburg, J.G., Lindemann, J.W., Davenport, G.C. et al. 1992. A multidisciplinary approach to the detection of clandestine graves. *Journal of Forensic Sciences*, 37, 1445–1458.

Freeland, R.S., Miller, M.L., Yodera, R.E. & Koppenjan, S.K. 2003. Forensic application of FM-CW and pulse radar. *Journal of Environmental and Engineering Geophysics*, 8(2), 97–103.

Frohlich, B. & Lancaster, W.J. 1986. Electromagnetic surveying in current middle eastern archaeology – Application and evaluation. *Geophysics*, 51, 1414–1425.

Fushinuki, Y., Hamasaki, S., Nagahama, K., Nanamegi, K. & Sakamoto, H. 2001. Discrimination of hot spring water from Southern Satsuma Peninsula, Japan. Abstracts of 50th Annual Meeting of the Japan Society for Analytical Chemistry, 2001.

Goggin, D.J., Thrasher, R.L. & Lake, L.W. 1988. A theoretical and experimental analysis of minipermeameter response including gas slippage and high velocity flow effects. *In Situ*, 12, 79–116.

Goldberg, P. & Macphail, R.I. 2006. *Practical and Theoretical Geoarchaeology*. Blackwell Science Ltd., Oxford.

Gomez-Lopez, A.M. & Patino-Umana, A. 2006. Who is missing? Problems in the application of forensic archaeology and anthropology in Colombia's conflict. In: Ferllini, R. (ed), *Forensic Archaeology and Human Rights*. Charles C. Thomas, Springfield, Illinois.

Goovaerts, P. 1997. *Geostatistics for Natural Resources Evaluation*. Oxford University Press, New York.

Graham, G.A., Kearsley, A.T., Drolshagan, G., McDonnell, J.A.M., Wright, I.P. & Grady, M.M. 2004. Mineralogy and microanalysis in the determination of cause of impact damage to spacecraft surfaces. In: Pye, K. & Croft, D.J. (eds) *Forensic Geoscience: Principles, Techniques and Applications*. Special Publication of the Geological Society of London, 232, 137–146.

Green, G. 2006. Call in the special branch. *Metro*, 14th November, 15.

Grip, W.M., Grip, R.W & Morrison, R. 2000. Application of aerial photography in environmental forensic investigations. *Environmental Forensics*, 1, 121–129.

Gross, H. 1891. *Handbuch fur System der Kriminalistik Untersuchun gsrichter als Kriminalistik (Criminal Investigation: A practical Textbook for Magistrates, Police Officers and Lawyers)*. Adam & Adam (1906) and Jackson (1962). Sweet & Maxwell, 1962.

Hammon, W.S., McMechan, G.A. & Zeng, X.X. 2000. Forensic GPR: finite-difference simulations of responses from buried human remains. *Journal of Applied Geophysics*, 45(3), 171–186.

Haneberg, W.C. 2004. Forensic geology in military or intelligence operations. In: *Encyclopedia of Espionage, Intelligence, and Security*. The Gale Group (Thomson Gale), Inc.

Hanson, I. 2003. The importance of stratigraphy in forensic investigations. In: Pye, K. & Croft, D.J. (eds), *Forensic Geoscience: Principles, Techniques and Applications*. Conference Abstracts, Geological Society of London, 3–4 March 2003.

Hanson, I. 2004. The importance of stratigraphy in forensic investigation, In: Pye, K, Croft, D.J. (eds), *Forensic Geoscience: Principles, Techniques and Applications*. Geological Society Special Publication 232, 214–224.

Haralambos, M, Holborn, M. & Heald, R. 1990. *Sociology: Themes and Perspectives*. Harper Collins, London.

Hardisty, J. 2003. Hydrodynamic modelling as investigative and evidential tools in murder enquiries: examples from the Humber and Thames. In: Pye, K. & Croft, D.J. (eds), *Forensic Geoscience: Principles, Techniques and Applications*. Conference Abstracts, Geological Society of London, 3–4 March 2003.

Harrison, M. 2006. Search Methodologies. Geoscientists at Crime Scenes Conference Abstract. Geological Society of London, 20th December 2006.

Heywood, I., Cornelius, S. & Carver, S. 2002. *An Introduction to Geographical Information Systems*, second edition. Prentice Hall, Harlow.

Hildebrand, J.A., Wiggins, S.M., Henkart, P.C. & Conyers, L.B. 2002. Comparison of seismic reflection and ground-penetrating radar imaging at the controlled archaeological test site, Champaign, Illinois. *Archaeological Prospection*, **9**, 9–21.

Hirschfield, A. & Bowers, K. 2001. *Mapping and Analysing Crime Data: Lessons from Research and Practice*. Taylor & Francis, New York.

Holzer, T.L., Fletcher, J.B., Fuis, G.S., Ryberg, T., Brocher, T.M. & Dietel, C.M. 1996. Seismograms offer insight into Oklahoma City Bombing. *Eos, Transactions American Geophysical Union*, **77**, 393–397.

Horrocks, M. & Walsh, K.A.J. 1998. Forensic palynology: assessing the value of the evidence. *Review of Palaeobotany and Palynology*, **103**, 69–74.

Horrocks, M., Coulson, S.A. & Walsh, K.A.J. 1999. Forensic palynology: Variation in the pollen content of soil on shoes and in shoeprints in soil. *Journal of Forensic Sciences*, **44**, 119–122

Houck, M. (ed.) 2001. *Mute Witnesses. Trace Evidence Analysis*. Academic Press, London. http://www.harcourt-international.com/e-books/pdf/85.pdf

Houde, G. 2006. Elgin forensic scientist aids Ramsey case. *Chicago Tribune*, **237**, 25th August 2006.

Hower, J.C., Schram, W.H. & Thomas, G.A. 2000. Forensic petrology and geochemistry: tracking the source of a coal slurry soil, Lee County, Virginia. *International Journal of Coal Geology*, **44**, 101–108.

Hunter, J.R. & Cox, M. 2005. *Forensic Archaeology*. Routledge, London.

Hunter, J.R., Karaska, J.B.T., Scott, E.A., Tetlow, E.A. & Reddick, A. 2005. 'The identification of mass graves in former Yugoslavia using geophysics and remote sensing'. Sarajevo: International Commission for Missing Persons.

Hutchinson, M.F. & Gessler, P.E. 1994. Splines – more than just a smooth interpolator. *Geoderma*, **62**, 45–67.

Imaizumi, M. 1974. Locating buried bodies. *FBI Law Enforcement Bulletin*, **43**(8), 2–5.

INTERPOL 2001. Forensic Science Symposium, Lyon, France, 16–19th October 2001. http://www.interpol.int/Public/Forensic/IFSS/meeting13/Reviews/Soil.pdf

Isphording, W.C. 2004. Statistics in court: the right and wrong ways. In: Pye, K. & Croft, D.J. (eds), *Forensic Geoscience: Principles, Techniques and Applications*. Geological Society, London, Special Publications 232, 289–299.

Jackson, R.L. 1962. *Criminal Investigation. A Practical Textbook for Magistrates, Police Officers and Lawyers*. From the adapted Adam, J. and Adam, J.C. (1906) *System Ker Kriminalistik* by Hans Gross. Sweet & Maxwell Ltd, London.

Jago, G. 2002. EMRG Meeting Reports Forensic Geology (Geoforensics). 17th October 2002. Report by Geoffrey Jago, with acknowledgement and thanks to Dr. Donnelly. http://www.geolsoc.org.uk/template.cfm?name=EMRG2002B

Jarvis, K.E., Wilson, H.E. & James, S.L., 2004. Assessing element variability in small soil samples taken during forensic investigation. In: Pye, K. & Croft, D.J. (eds), *Forensic Geoscience: Principles, Techniques and Applications*. Geological Society, London, Special Publications 232, 171–182.

Johnston, K., Ver Hoef, J.M., Krivoruchko, K. & Lucas, N. 2001. *Using ArcGIS™ Geostatistical Analyst*. Redlands, CA, ESRI, 2001.

Jones, D., Limburg, J. & Dienst, R.O.E.Z. 2002. Use of gamma-ray spectrometry for archaeological site investigations. Abstract 'Recent Work in Archaeological Geophysics', Environmental and Industrial Geophysics Group Meeting, Burlington House, London, 17th December 2002.

Journel, A.G. & Huijbregts, C.J., 1978. *Mining Geostatistics*. Academic Press, London.

Juggins, S. & Cameron, N. 1999. Diatoms and archaeology. In: *The Diatoms: Applications for the Environmental and Earth Sciences* (Stoermer, F. (ed.), Cambridge University Press, Cambridge, United Kingdom, 389–401 (illus. incl. sketch maps, 56).

Junger, E.P. 1996. Assessing the unique characteristics of close-proximity soil samples: Just how useful is soil evidence? *Journal of Forensic Sciences*, **41**, 27–34.

Kaplan, S.J. 1960. Climatic factors and crime. *The Professional Geographer*, **12**, 1–4.

Keaney, A., McKinley, J. & Ruffell, A. 2006. Different scales of spatial sampling in forensic investigations: case studies from the macro to micro scale. Geoscientists at Crime Scenes Conference Abstract. Geological Society of London, 20th December 2006.

Kearey, P. & Brooks, M., Hill, I. 2002. *An Introduction to Geophysical Exploration*. Blackwell Scientific Publishers, Oxford.

Keating, H.R.F. 1991. *Great Crimes*. Longmeadow Press. Stamford, Connecticut.

Keppner, G. 1991. Ludger Mintrop. *Geophysics: The Leading Edge*. **10**, 21–28.

Killam, E.W. 2004. *The Detection of Human Remains*. Charles C. Thomas, Springfield, Illinois.

Kind, S. 1982. Criminal identification. In: Deutch, Y. (ed). *Science Against Crime*. Marshall Cavendish House, London, 6–29.

Koch, G.S. & Link, R.F. 1970. *Statistical Analysis of Geological Data*. Dover Publications, New York, Volumes I and II.

Koper, K. 2003. http://www.geo.arizona.edu/researchers/kkoper_ftp/for_cases.html

Koper, K.D., Wallace, T.C., Taylor, S.R. & Hartse, H.E. 2001. Forensic seismology and the sinking of the Kursk. *Eos, Transactions, American Geophysical Union*, **82**, 37.

Krill, T. 2003. Hot on the suspect's trail. *On Site: the world of x-ray analysis*. **1**, 18–20.

Krinsley, D.H. & Doornkamp, J.C. 1973. *Atlas of Quartz Sand Surface Textures*. Cambridge University Press, Cambridge.

Krumbein, W.C. 1962. Open and closed number systems in stratigraphic mapping. *American Association of Petroleum Geologists Bulletin*, **46**, 2229–2245.

Kugler, W. 2003. X-ray diffraction analysis in the forensic science: the last resort in many criminal cases. *Advances in X-ray Analysis*, **46**, 1–16.

Lane, B. & Gregg, W. 1992. *The Encyclopedia of Serial Killers*. Berkley Publishing Group, New York.

Lee, B.D., Williamson, T.N. & Graham, R.C. 2002. Identification of stolen rare palm trees by soil morphological and mineralogical properties. *Journal of Forensic Sciences*, **47**, 310–315.

Lee, C.W. 2004. The nature of, and approaches to, teaching forensic geoscience on forensic and earth science courses. In: Pye, K. & Croft, D.J. (eds), *Forensic Geoscience: Principles, Techniques and Applications*. Special Publication of the Geological Society of London, 232, 301–333.

Lee, E.M., 2001. Geomorphological mapping. In: Griffiths, J.S. (ed.), *Land Surface Evaluation for Engineering Practice*. Geological Society, London, Engineering Geology Special Publications, 18, 53–56.

Lee, H.C. & Labriola, J. 2006. *Dr. Henry Lee's Forensic Files*. Prometheus Books, New York.

Lewis, P. 2002. Walter McCrone, debunker of legends, dies aged 82. http://membership.acs.org/C/Chicago/mccrone.html

Lillesand, T.M., Kieffer, R.W. & Chipman, J.W. 2004. *Remote Sensing & Interpretation*. John Wiley & Sons, Inc., Hoboken, New Jersey.

Locard, R. 1929. The analysis of dust traces. *Revue Internationale de Criminalistique*, **I**, 4–5. (English translation in *American Journal of Police Science*, **I** (1930), 416–418.

Lombardi, G. 1999. The contribution of forensic geology and other trace evidence analysis to the investigation of the killing of Italian Prime Minister Aldo Moro. *Journal of Forensic Sciences*, **44**, 634–642.

Longley, P.A., Goodchild, M.F., Maguire, D.J. & Rhind, D.W. 2006. *Geographic Information Systems and Science*. John Wiley & Sons, Ltd, Chichester.

Mack, L., Riley, L. & Smyth, F. 1986. *Unsolved Cases*. Orbis Book Publishing Corporation Ltd, London.

Magnuson, S. 2007. Daunting challenges face those waging subterranean warfare. *National Defense Magazine*, February 2007.

Martin, M. 2007. *Earth Evidence (Forensic Crime Solvers)*. Capstone Press.

Marumo, Y., Nagatsuka, S. & Oba, Y. 1986. Clay mineralogical analysis using the 0.05-mm fraction for forensic science investigation – Its application to volcanic ash soils and yellow-brown forest soils. *Journal of Forensic Sciences*, 31, 92–105.

Matza, D. 1964. *Delinquency and Drift*. John Wiley & Sons, Inc., New York.

McBratney, A.B. & Webster, R. 1986. Choosing functions for semi-variograms of soil properties and fitting them to sampling estimates. *Journal of Soil Science*, 37, 617–639.

McDowell, P.W. 1975. Detection of clay-filled sink holes in chalk by geophysical methods. *Quarterly Journal of Engineering Geology*, 8, 303–310.

McGrath, S.P. & Loveland, P.J. 1992. *The Soil Geochemical Atlas of England and Wales*, Blackie Academic and Professional, Glasgow.

McKinley, J. & Ruffell, A. 2007. Contemporaneous spatial sampling at a scene of crime: advantages and disadvantages. *Forensic Science International*, **172**, 196–202.

McPhee, J. 1996. The gravel page. *New Yorker Magazine*, New York, 44–69.

McVicar, M.J. & Graves, W.J. 1997. The forensic comparison of soils by automated scanning electron microscopy. *Canadian Society of Forensic Science Journal*, 30, 241–261.

Mikesh, R.C. 1990. *Japan's World War II Balloon Bomb Attacks on North America*. Farrar, Smithsonian Institution Press, Washington, DC.

Mildenhall, D., Wiltshire, P.E.J. & Bryant, V. 2006. Forensic palynology. *Forensic Science International*, **163**, 161–248.

Miller, P.S. 1996. Disturbances in the soil: Finding buried bodies and other evidence using ground penetrating radar. *Journal of Forensic Sciences*, 41(4), 648–652.

Minár, J. 1992. The principles of the elementary geomorphological regionalization. *Acta Facultatis Rerum Naturalium Universitatis Comenianae, Geographica*, 33, 185–198.

Minár, J. & Evans, I.S. 2007. Elementary forms for land surface segmentation: The theoretical basis of terrain analysis and geomorphological mapping. *Geomorphology*, doi:10.1016/j.geomorph.2007.06.003

Minár, J. & Mimian, L. 2002. Complex geomorphological characteristics of the Devînska Kobyla mountains. In: *Landscape Atlas of the Slovak Republic*, first edition. Bratislava: Ministry of Environment of the Slovak Republic; Banská Bystrica: Slovak Environmental Agency, 92–93.

Mitáš, L. & Mitášová, H.H. 1999. Spatial interpolation. In: Longley, P.A., Goodchild, M.F., Maguire, J.D & Rhind, D.W. (eds) *Geographical Information Systems. Volume I: Principles and Technical Issues*, second edition, John Wiley & Sons, Inc., New York, 481–492.

Monastersky, R. 2000. All mixed up over birds and dinosaurs – feathered dinosaur fossil found in China likely a fake. *Science News*, 15th January 2000.

Morgan, R.M. & Bull, P.A. 2006. Data interpretation in forensic sediment and soil geochemistry. *Environmental Forensics*, 7, 325–334.

Morgan, R.M. & Bull, P.A. 2007. The philosophy, nature and practice of forensic sediment analysis. *Progress in Physical Geography*, 31, 1–16.

Morgan, R.M., Wiltshire, P.E.J., Parker, A. & Bull, P. 2006. The role of forensic geoscience in wildlife crime detection. *Forensic Science International*, **162**, 152–162.

Morris, T. 1957. *The Criminal Area*. Routledge. Kegan & Paul, London.

Moussa, M. 2001. Gamma-ray spectrometry: a new tool for exploring archaeological sites: a case study from East Sinai, Egypt. *Journal of Applied Geophysics*, **48**, 137–142.

Murray, R. 2004. *Evidence from the Earth*. Mountain Press Publishing Co. Missoula, Montana.

Murray, R.C. 2005. Collecting crime evidence from the Earth. *Geotimes*, January 2005. http://www.geotimes.org/jan05/feature_evidence.html#links

Murray, R.C. & Solebello, L.P. 2002. Forensic examination of soil. In: *Forensic Science Handbook. Volume 1,* second edition, Saferstein, R. (ed.). Prentice Hall, New Jersey, USA.

Murray, R.C. & Tedrow, J.C.F. 1975 (republished 1986). *Forensic Geology: Earth Sciences and Criminal Investigation.* Rutgers University Press, New York.

Murray, R.C. & Tedrow, J.C.F. 1991. *Forensic Geology.* Englewood Cliffs, Prentice Hall.

Murray, R.C. & Tedrow, J.C.F. 1992. *Forensic Geology.* Englewood Cliffs: Prentice Hall.

National Research Council 2006. Successful Response Starts with a Map: Improving Geospatial Support for Disaster Management Committee on Planning for Catastrophe: A Blueprint for Improving Geospatial Data, Tools, and Infrastructure. The National Academies Press, Washington, DC. http://www.nap.edu/catalog/11793.html

Nobes, D.C. 2000. The search for 'Yvonne': A case example of the delineation of a grave using near-surface geophysical methods. *Journal of Forensic Sciences*, **45**(3), 715–721.

O'Connell, S. 2006. The pollen detectives. *Focus*, **164**, 69–74.

Olhoeft, G.R. 1985. Low frequency electrical properties. *Geophysics*, 50, 2492–2503.

Oliver, M.A., Webster, R. 1990. Kriging: a method of interpolation for geographical information systems. *International Journal of Geographical Information Systems*, 4, 313–332.

Omelyanyuk, G. & Alekseev, A. 2001. Application of the computer searching system 'SOILRAILWAY' in practice of the railway police. *Problems of Forensic Sciences (Zadnien Nauk Sadowych)*, 47, 17–19.

Owsley, D.W. 1995. Techniques for locating burials, with emphasis on the probe. *Journal of Forensic Sciences*, **40**(5), 735–740.

Palenik, S. 1982. Microscopic trace evidence – the overlooked clue. Part III Max Frei – Sherlock Holmes with microscope. *The Microscope*, 163–169.

Palenik, S.J. 2000. Microscopy. In: *Encyclopedia of Forensic Science*, Academic Press. London.

Pawlowsky-Glahn, V. & Egozcue, J.J. 2006. Compositional data and their analysis: an introduction. In: Buccianti, A., Mateu-Figueras, G. & Pawlowsky-Glahn, V. (eds), *Compositional Data Analysis in the Geosciences: From theory to practice.* Geological Society, London, Special Publications 264, 1–10.

Pearce, F. 1976. *Crimes of the Powerful: Marxism, Crime, and Deviance.* Pluto Press, London.

Pearson, K. 1897. Mathematical contributions to the theory of evolution. On a form of spurious correlation which may arise when indices are used in the measure of organs. *Proceedings of the Royal Society of London*, **60**, 489–502.

Pebesma, E.J. & Wesseling, C.G. 1998. Gstat, a program for geostatistical modelling, prediction and simulation. *Computers in Geoscience*, 24, 17–31.

Petraco, N. & Kubic, T. 2000. A density gradient technique for use in forensic soil analysis. *Forensic Science International*, **4**, 872–873.

Pinsker, L.M. 2002. Geology Adventures in Afghanistan. *Geotimes* Web Feature. February 2002. *http://www.agiweb.org/geotimes/feb02/Feature_Shroderside.html>(13 March 2003)*

Pirrie, D., Butcher, A.R., Gottlieb, P. & Power, M.R. 2003. QemSCAN – a fully automated quantitative SEM-based mineral analysis system. In: Pye, K. & Croft, D.J. (eds), *Forensic Geoscience: Principles, Techniques and Applications.* Conference Abstracts, Geological Society of London, 3–4 March 2003.

Pirrie, D., Butcher, A.R., Power, M.R., Gottlieb, P. & Miller, G.L. 2004. Rapid quantitative mineral and phase analysis using automated scanning electron microscopy (QemSCAN); potential applications in forensic geoscience. In: Pye, K. & Croft, D.J. (eds), *Forensic Geoscience: Principles, Techniques and Applications.* Geological Society, London, Special Publications 232, 123–136.

Popp, G. 1910. Bomben, erdspuren un instrumenten-untersuchung. *Zeitschrift für öffentliche Chemie,* **XXI**.

Popp, G. 1939. Botanische spuren und mikroorganismen im kriminalverfahren. *Archiv für Kriminologie*, XIV, 231–37.

Pye, K. 2004a. Forensic geology. In: Selley, R.C., Cocks, L.R.M. & Plimer, I.R. (eds), *Encyclopedia of Geology*. Elsevier, Amsterdam (5 volumes).

Pye, K. 2004b. Forensic examination of rocks, sediments, soils and dusts using scanning electron microscopy and x-ray chemical analysis. In: Pye, K. & Croft, D.J. (eds), *Forensic Geoscience: Principles, Techniques and Applications*. Special Publication of the Geological Society of London, 232, 103–122.

Pye, K. 2004c. Isotope and trace element analysis of human teeth and bones for forensic purposes. In: Pye, K. & Croft, D.J. (eds), *Forensic Geoscience: Principles, Techniques and Applications*. Special Publication of the Geological Society of London, 232, 215–236.

Pye, K. 2007. *Geological and Soil Evidence*. CRC Press/Taylor & Francis Group, Oxford.

Pye, K. & Blott, S.J. 2004. Comparison of soils and sediments using major and trace element data. In: Pye, K. & Croft, D.J. (eds), *Forensic Geoscience: Principles, Techniques and Applications*. Geological Society, London, Special Publications 232, 183–196.

Pye, K. & Croft, D.J. 2004. (eds), *Forensic Geoscience: Principles, Techniques and Applications*. Special Publication of the Geological Society of London, 232.

Rawlins, B.G. & Cave, M. 2004. Investigating multi-element soil geochemical signatures and their potential for use in forensic studies. In: Pye, K. & Croft, D.J. (eds), *Forensic Geoscience: Principles, Techniques and Applications*. Geological Society, London, Special Publications 232, 197–206.

Rawlins, B.G., Kemp, S.J., Hodgkinson, E.H., Riding, J.B., Vane, C.H. *et al.* 2006. Potential and pitfalls in establishing the provenance of earth-related samples in forensic investigations. *Forensic Science International*, 51, 832–845.

Redmayne, D.W. & Turbitt, T. 1990. Ground motion effects of the Lockerbie air crash impact. *Geophysical Journal International*, 101, 293.

Rees, L. 2005. *The Nazis: a Warning from History*. BBC Books, London.

Reyment, R.A. 2006. On stability of compositional canonical variate vector components. In: Buccianti, A., Mateu-Figueras, G. & Pawlowsky-Glahn, V. (eds), *Compositional Data Analysis in the Geosciences: From theory to practice*. Geological Society, London, Special Publication 264, 59–66.

Reynolds, J.M. 1997. *An Introduction to Applied and Environmental Geophysics*. John Wiley & Sons, Ltd, Chichester.

Rinehart, D. 2003. Excavations of skeletal remains from an anthropological point of view. http://www.crime-scene-investigator.net/excavation.html

Robertson, G. 2007 (ed). *Fairness and Evidence in War Crimes Trials*. Berkeley Electronic Press. http://www.bepress.com/ice/vol4/iss1/art1

Rodriguez, W.C. & Bass, W.M. 1985. Decomposition of buried bodies and methods that may aid in their detection. *Journal of Forensic Sciences*, 30, 836–852.

Rogers, D.A. 2007. http://web.umr.edu/~rogersda/forensic_geology/cdmg_helps_find_kidnapper.pdf

Rogers, D.J. & Koper, K.D. 2007. Some practical applications of forensic seismology. http://web.umr.edu/~rogersda/umrcourses/ge342/Forensic%20Seismology-revised.pdf

Rollinson, H.R. 1995. *Using Geochemical data: Evaluation, Presentation, Interpretation*. Longman Geochemistry series, Longman Group Ltd, Essex.

Rowe, T., Ketcham, R.A., Denison, C., Colbert, M., Xu, X. & Currie, P.J. 2001. Forensic palaeontology: the Archaeoraptor forgery. *Nature*, 410(6828), 539–40.

Ruffell, A. 2002. Remote detection and identification of organic remains: an assessment of Archaeological potential, *Archaeological Prospection*, 9, 115–122.

Ruffell, A. 2004. Burial location using cheap and reliable quantitative probe measurements. Diversity in forensic anthropology. Special Publication of *Forensic Science International*, **151**, 207–211.

Ruffell, A. 2005. Searching for the I.R.A. Disappeared: ground-penetrating radar investigation of a churchyard burial site, Northern Ireland. *Journal of Forensic Sciences*. **50**, 414–424.

Ruffell, A. 2006. Geoscientists at crime scenes: Conference abstract. Geological Society of London, 20th December 2006.

Ruffell, A. 2007. Freshwater under water scene of crime reconstruction using freshwater ground-penetrating radar (GPR) in the search for an amputated leg and jet-ski, Northern Ireland, *Science & Justice*, **46**(3), 133–145.

Ruffell, A. & McKinley, J. 2005. Forensic Geology & Geoscience. *Earth Science Reviews*. **69**, 235–247.

Ruffell, A. & Wilson, J. 1998. Shallow ground investigation using radiometrics and spectral gamma-ray data. *Archaeological Prospection*, **5**, 203–215.

Ruffell, A. & Wiltshire, P. 2004. Conjunctive use of quantitative and qualitative X-ray diffraction analysis of soils and rocks for forensic analysis. *Forensic Science International*, **145**, 13–23.

Saferstein, R.E. 2001. *Forensic Science Handbook*. Prentice Hall, New Jersey.

Savvides, A. 2006. Scale dependant correlations between cation exchange capacity and visible and near infrared spectroscopy in soils of Bedfordshire. Unpublished MSc Thesis, The University of Reading.

Saye, S.E. & Pye, K. 2004. Development of a coastal dune sediment database for England and Wales: In: Pye, K. & Croft, D.J. (eds), *Forensic Geoscience: Principles, Techniques and Applications*. Geological Society, London, Special Publication, 232, 75–96.

Scientific American. 1856. Science and art: Curious use of the microscope. *Scientific American*, **11**(30), 240. http://cdl.library.cornell.edu/cgi-bin/moa/pageviewer?coll+moa&root=%2Fmoa%2Fs

Scott, J. & Hunter, J.R. 2004. Environmental influences on resistivity mapping for the location of clandestine graves. In: Pye, K. & Croft, D.J. (eds), *Forensic Geoscience: Principles, Techniques and Applications*. Special Publication of the Geological Society of London, 232, 33–38.

Sever, M. 2005. Murder and mud in the Shenandoah. *Geotimes*, 26th September 2002. http://www.geotimes.org/jan05/feature_Shenandoah.html

Shaw, C.R. & McKay, H.D. 1942. *Juvenile Delinquency and Urban Areas*. University of Chicago Press, Chicago.

Sheridan, M. 2006. *Bloody Evidence*. Mentor Books, Dublin.

Silvestri, S., Omri, S. & Rosselli, R. 2005. The use of remote sensing to map illegal dumps in the Veneto Plain. Proceedings Sardinia 2005, *Tenth International Waste Management and Landfill Symposium*. Santa Margherita di Pula, Cagliari, Italy; 3–7th October 2005. CISA Environmental Sanitary Engineering Centre, Italy.

Smale, D. & Truedale, N.A. 1969. Heavy mineral studies as evidence in a murder case in outback Australia. *Journal of Forensic Science*, **9**, 3–4.

Small, D. 2006. New service for gemstone forensics. *Police Professional Magazine*, **56**, 38.

Small, I.F., Rowan, J.S., Franks, S.W.,Wyatt, A. & Duck, R.W. 2004. Bayesian sediment fingerprinting provides a robust tool for environmental forensic geoscience applications. In: Pye, K. & Croft, D.J. (eds), *Forensic Geoscience: Principles, Techniques and Applications*. Special Publication of the Geological Society of London, 232, 207–213.

Sounes, H. 1995. *Fred and Rose*. Time Warner Books, London.

Spindler, K. 1993. *The Man in the Ice*. Weidenfeld & Nicholson, London.

Stam, M. 2002. The dirt's on you. *News of the California Association of Criminalists*, **2** (Second Quarter), 8–11. http://cacnews.org/pdfs/2ndq02.pdf

Stam, M. 2004. Soil as significant evidence in a sexual assault/attempted homicide case. In: Pye, K. & Croft, D.J. (eds), *Forensic Geoscience: Principles, Techniques and Applications*. Special Publication of the Geological Society of London, 232, 295–299.

Stockdale, R. 1982. The forensic science laboratory. In: Deutch, Y. (ed.), *Science Against Crime*. Marshall Cavendish House, London, 30–43.

Strahler, A.N. & Strahler, A.H. 2002. *Elements of Physical Geography*. John Wiley & Sons, Ltd, Chichester.

Strongman, K.B., 1992. Forensic applications of ground penetrating radar. In: Pilon, J. (ed). *Ground Penetrating Radar*. Geological Survey of Canada, Paper 90-4, 203–211.

Suggs, J.A., Beam, E.W., Biggs, D.E., Collins, W., Dusenbury, M.R. *et al.* 2002. Guidelines and resources for conducting an environmental crime reconstruction in the United States. *Environmental Forensics*, 3, 91–113.

Suggs, J.A. & Yarborough, K.A. 2003. Environmental crime. In: Niamh Nic Daéid (ed), 14th International Forensic Science Symposium *Interpol (Lyon) Review Papers*, 359–392. http://www.interpol.int/Public/Forensic/IFSS/meeting14/ReviewPapers.pdf

Sugita, R. & Marumo, Y. 2004. 'Unique' particles in soil evidence. In: Pye, K. & Croft, D.J. (eds) *Forensic Geoscience: Principles, Techniques and Applications*. Special Publication of the Geological Society of London, 232, 97–102.

Sugita, R. & Suzuki, S. 2003. Forensic geology. A review: 2001 to 2004. In: Niamh Nic Daéid (ed), 14th International Forensic Science Symposium, *Interpol (Lyon) Review Papers*, 125–135. http://www.interpol.int/Public/Forensic/IFSS/meeting14/ReviewPapers.pdf

Swan, A.R.H. & Sandilands, M. 1995. *Introduction to Geological Data Analysis*. Blackwell Science, Oxford.

Szibor, R., Schubert, C., Schoning, R., Krause, D. & Wendt, U. 1998. Pollen analysis reveals murder season. *Nature*, 395, 449–50.

Talent, J.A. 1989, The case of the peripatetic fossils. *Nature* 338, 613–615.

Taroni, F., Biedermann, A., Garbolino, P. & Aitken, C.G.G. 2004. A general approach to Bayesian networks for the interpretation of evidence. *Forensic Science International*, 39, 5–16.

Thorwald, J. 1967. *Crime and Science: The New Frontier in Criminology*. Harcourt Brace Jovanovich.

Topham, J. & McCormick, D. 2000. A dendrochronological investigation of stringed instruments of the Cremonese School (1666–1757) including 'The Messiah' violin attributed to Antonio Stradivari. *Journal of Archaeological Science*, 27(3), 183–192.

Trueman, C.N. 2004. Forensic geology of bone material: geochemical tracers for post-mortem movement of bone remains. In: Pye, K. & Croft, D.J. (eds), *Forensic Geoscience: Principles, Techniques and Applications*. Special Publication of the Geological Society of London, 232, 249–256.

Tucker, M.E. 1988. *Techniques in Sedimentology*, Blackwell Scientific Publications, Oxford.

Tyler, A.N. 1999. Monitoring anthropogenic radioactivity in salt marsh environments through *in situ* gamma ray spectrometry. *Journal of Environmental Radioactivity* 45(3), 235–252.

Van Den Wyngaert, C. 2006. Disparities between evidentiary rules before international courts and tribunals. Can a clash be avoided? *Commentary on Evidence*, 4(1), 7. www.bepress.com/ice/vol4/iss1/art6

van Schoor, M. 2002. Detection of sinkholes using 2D electrical resistivity imaging. *Journal of Applied Geophysics* 50(4), 393–399.

Wackernagel, H. 2003. *Multivariate Geostatistics. An Introduction with Applications*, third edition, Springer, Berlin.

Wald, P. 2006. Fair trials for war criminals. *International Commentary on Evidence*, 4(1), 6. www.bepress.com/ice/vol4/iss1/art6

Waters, R.S. 1958. Morphological mapping. *Geography*, 43, 10–17.

Watters, M. & Hunter, J.R. 2004, Geophysics and burials: field experience and software development. In: Pye, K. & Croft, D.J. (eds) *Forensic Geoscience: Principles, Techniques and Applications*. Special Publication of the Geological Society of London, 232, 21–33.

Webster, R. & Oliver, M.A. 1990. *Statistical Methods in Soil and Land Resource Survey*. Oxford University Press, Oxford.

Webster, R. & Oliver, M.A. 2001. *Geostatistics for Environmental Scientists*. John Wiley & Sons, Ltd, Chichester.

Wienker, C.W., Wood, J.E. & Diggs, C.A. 1990. Independent instances of 'Souvenir' Asian skulls from the Tampa Bay area. *Journal of Forensic Sciences*, 35, 637–643.

Williams, G.M. & Aitkenhead, N. 1991. Lessons from Loscoe: the uncontrolled migration of landfill gas. *Quarterly Journal of Engineering Geology*, 24, 191–207.

Wiltshire, P.E.J. 2003a. Pollen and related botanical evidence – its recent contribution to forensic investigations. In: Pye, K. & Croft, D.J. (eds) *Forensic Geoscience: Principles, Techniques and Applications*. Conference Abstracts, Geological Society of London 3–4 March 2003.

Wiltshire, P.E.J. 2003b. Environmental profiling and forensic palynology. *http://www.bahid. org/docs/NCF_Env%20Prof.pdf*

Wiltshire, P.E.J. 2006. Consideration of some taphonomic variables of relevance to forensic palynological investigation in the United Kingdom. *Forensic Science International*, 163, 173–182.

Wiltshire, P.E.J. & Black, S. 2006. The cribriform approach to the retrieval of palynological evidence from the turbinates of murder victims. *Forensic Science International*, 163, 224–230.

Witten, A.J., Gillette, D.D., Sypniewski, J. & King, W.C. 1992. Geophysical diffraction tomography at a dinosaur site. *Geophysics*, 57, 187–195.

Witten, A.J. & King, W.C. 1990. Sounding out buried waste. *Civil Engineering*, 60, 62–64.

Zhang, J. & Goodchild, M. 2002. *Uncertainty in Geographical Information*. Taylor & Francis, London.

Zucca, J.J. 2003. *http://www.llnl.gov/str/Zucca.html*

Index

NB: numbers relate to pages, numbers with decimal places relate to figures and tables

aerial photography   92
animal cruelty   97, 102, 233, 234, 4.4
atmosphere   10, 13, 15, 16
atomic absorption spectroscopy (AAS)   206, 207

behavioural psychologists   312
bentonite   259
bin Laden, Osama   20
biosphere   13
Body Farm (Tennessee)   272, 273
Boxing Day Tsunami   287
burial locations   41, 291, 292, 2.15, 9.11

Camerena, Enrique   5
chemical analysis   206
Chicago School (deviancy)   127, 128
climatology   13, 15
conditional simulation   178
continuous (data)   162
correlation (statistics)   163

definition (of Geoforensics)   8, 9 10, 293
diatoms   16
Digital Elevation Model (DEM)   141
Digital Terrain Model (DTM)   141
disappeared (Colombia)   107
Disch, Eva   6
discrete (data)   162
domains (geographic)   30
Dragnet (software)   149

electrodes (geophysics)   67
elevation modelling   104
Exchange Principle (Locard)   6, 7
extra-terrestrial   236, 251

'feature-based search'   41, 145, 156
field notes   191, 192
footprint cast(s)   159
fossil fraud   225
fossils (macro, micro)   21, 221

gas monitoring   288, 289, 9.9, 9.10
Geoforensics (definition)   8, 9, 10, 293
Geographic Analysis Machine (GAM)   129
Geographic Explanation Machine (GEM)   129
geography   4
geography, physical   13
geological maps   51, 52
geomorphology   14, 18, 21, 23, 24, 26
geosphere   13, 15
'Glacier Girl'   287
gravity (surveying), gravimetrics   64
ground-penetrating radar (GPR; geophysics)   76, 203, 3.10, 281, 282
Gujarat (earthquake)   63, 287

heavy minerals   192
Heslin, Fr Patrick   5

hydrosphere   13
hyperspectral imaging   109

igneous rocks (definition)   185
incendiary balloons   16, 203, 204
induced polarity (geophysics)   69, 70
inductively-coupled plasma mass spectroscopy (ICP-MS)   206, 253
infrared (thermal imaging)   102, 106
International Organization for Standardization (ISO)   136
interpolation   169
Inverse Distance Weighting (IDW)   141, 169, 172
isotopes (& isotope ratios mass spectrometry)   254, 265

Jet Stream   16

kriging   169, 173, 178
Kursk (submarine sinking)   63

lakes   47, 48
landform mapping   18, 19, 20, 21, 22, 24, 26, 29, 30
landscape interpretation   18, 92, 93, 94, 275
Laser Raman Spectroscopy   253
laser scanning   117
LiDAR   114, 117, 139
Lockerbie (crash)   63

magnetic surveying   70
magnetometer   70, 3.5, 3.6
'Man in the Ice'   287
mass burials   83, 3.12, 108, 110, 229
mass murder(ers)   125
matches (safety)   264
metamorphic rocks (definition)   186
methane (probes)   288, 289, 9.9, 9.10
microscope (light)   4, 197
mineralogical determination   194
mining fraud   40, 96
Moors Murders   73
multi-proxy analysis   207, 223
multi-sensor platforms (geophysics)   89, 3.17

multispectral imaging   106
Munsell Colour Chart   184

Nairobi (bombing)   63
National Institute for Justice (NIJ)   151
nodes (geographic)   136
nuclear (tests, test-ban)   62

oil industry fraud   120
Open GIS Consortium (OGC)   136

palynology   2, 228, 229, 231, 232, 233
particle size   192, 204
photogrammetry   105
physical geography   13
pigs (burial)   271
pollution   97, 110
population (sample)   161
precipitation   14, 23, 267, 268
probe   35, 36

QemScan   251
quartz grains   244

radar interferometry (aerial)   105, 106
radioactivity   86, 87
radioactivity (and health)   88
Raman   253
relief (topographic)   23
resistivity (geophysics)   66, 67, 74
rivers   47

satellite (positioning)   104, 109
scanning electron microscoepe (SEM)   196, 249
'scenario-based search'   41, 145, 156
seas   48
sedimentary rocks (definition)   185
seismic (monitoring, surveying)   58, 61
self-potential (geophysics)   69
September 11 attacks   20, 64
serial killers   125
smuggling   187
Snow, John   127
soil   33, 34, 44, 50, 305, 306
soil analysis   188, 189, 7.2, 7.3
spree killer(s)   125
stratigraphy   32, 38
stratigraphy (macro)   274

stratigraphy (micro)   44, 45
synthetic aperture radar (SAR)   105, 106

Theissen Polygons   169
Thin Plate Spline with Tension (TPST)
   169, 177
'Tiger' kidnaps   284
Triangulated Irregular Network (TIN)
   141
tunnels   65, 79, 80

unique particles   259

variogram   6.2, 6.3
very-low frequency (geophysics)   73
Vietcong   65

Waco (Seige)   36
water (searching)   277
water (and landforms)   21, 23, 47
water (and sources)   53
water (trace evidence)   268
weather   14, 17, 126
World War I   53
World War II   16, 50, 95, 103, 229

X-ray (imaging)   117
X-ray diffraction (XRD)   194, 196, 253, 7.5, 7.7
X-ray fluorescence   123, 206

NB: Please note that *published case studies have the author's names in the section title: those from the author's personal files have no author names. The latter have no other source as many are subject to legal constraint, permission from either investigating authorities or victims or suspects. In these cases, non-scientific aspects of the descriptions have been intentionally obscured or mixed with other cases. The scientific principles of such cases have been maintained in order to demonstrate the Geoforensic application(s).*